J. Emil Morhardt

TID-26881

Vascular Plants of the Nevada Test Site and Central-Southern Nevada: ecologic and geographic distributions

Janice C. Beatley

Laboratory of Nuclear Medicine and Radiation Biology
University of California, Los Angeles

1976

Prepared for the Division of Biomedical and Environmental Research,
Energy Research and Development Administration

Published by Technical Information Center, Office of Technical Information,
Energy Research and Development Administration

Library of Congress Cataloging in Publication Data

Beatley, Janice Carson, 1919-
Vascular plants of the Nevada test site and central-southern Nevada.

"(TID-26881)"
Bibliography: p.
Includes index.
1. Botany—Nevada. 2. Desert flora—Nevada. I. United States. Energy Research and Development Administration. Division of Biomedical and Environmental Research. II. Title. III. Title: Nevada test site and central-southern Nevada.
QK173.B4 582'.09'79334 76-21839
ISBN 0-87079-033-1

Available as TID-26881 for $9.75 (foreign, $12.25) from

National Technical Information Service
U. S. Department of Commerce
Springfield, Virginia 22161

ERDA Distribution Category UC-11

July 1976

To

John N. Wolfe
ecologist, teacher, poet,

in whose wisdom description and explanation
were the inseparable pursuits of ecology,
and who knew the large pleasures
in achievement of both,

this work is gratefully dedicated.

Preface

The region described by this work is one of extraordinary biological interest because of its geographic location straddling the boundaries of two large deserts of the western United States. Characterized by conspicuous biological and environmental contrasts and their transitions, this region was essentially unknown biologically at the outset of the study. Because of the kind and manner of land use, many of its biological/environmental systems have been seriously modified, and access to much of the region has been restricted now for over a third of a century. The task of establishing what plant species are in the region and where they are was undertaken not only because it was necessary for the work of Nevada Test Site biologists, but more importantly—in consideration of all of these circumstances—because there was a professional obligation to the botanical community and the nation, and to what seemed the long-term interests of the Atomic Energy Commission (now part of the Energy Research and Development Administration) and the United States Air Force in whose custody most of the region resides.

In a botanically unknown region such as this, the expertise of plant group specialists was requisite, and the present work could not have been accomplished without the sustained help and support of plant taxonomists throughout the country. This volume therefore is the product of the expertise of the many people who have contributed to it through the years. Especially does it reflect the efforts of James L. Reveal, University of Maryland and the Smithsonian Institution, whose contributions began with the first intensive plant collecting on the Air Force Range and in Ash Meadows in the summer of 1968, and continued with his description of most of the new taxa for the literature, and his involvement, to the present, in the solution of the numerous taxonomic problems in the flora. The author assumes responsibility, however, for all errors. In a treatise of this kind it is inevitable there will be some errors, and it is certain that there will be new taxonomic interpretations and additional plant collections in the future which will make changes necessary in what is presented here.

Although the plant collecting, identification, and many related activities were superimposed on the author's principal work at the Test Site, they were nevertheless a highly satisfying and exciting accompaniment to the ecological studies. To have the opportunity to engage in pioneer botanical collecting in the latter half of the twentieth century in a region of rugged and great beauty and extraordinary biological and geological diversity—in some areas, where perhaps none other had penetrated since the aborigenes— has indeed been a privilege.

The whole region is today seriously threatened by the activities of man. The landscape of the Test Site and vicinity is one of the most intensively studied and best understood in the country, and its potential is unique for studies critical to our understanding of arid lands. Its continuing availability, therefore, as a field laboratory, in reasonably intact condition, seems the rightful heritage of yet other generations.

Janice C. Beatley

Department of Biological Sciences
University of Cincinnati
Cincinnati, Ohio

September 1975

Contents

Introduction

In 1940 a large area of southern Nevada was removed from the public domain for the Las Vegas Bombing and Gunnery Range (now Nellis Air Force Range); this area included most of southern Nye County, the third largest county in the United States, and adjacent parts of Clark and Lincoln Counties. In 1950 Frenchman Flat and Yucca Flat, on the south side of the Range, were set aside as the continental proving grounds for atmospheric testing of nuclear devices by the U. S. Atomic Energy Commission. Jackass Flats was added in 1958 for the Nuclear Rocket Development Station, and in 1964 Pahute Mesa was added to meet the needs of underground testing, thereby further increasing the size of the area. The entire 1350 sq. miles (3500 km^2) constitutes the Nevada Test Site—the scene of testing of hundreds of nuclear and other devices and of supportive geological, hydrological, biological, and many other kinds of investigations since the first nuclear detonation, the 1 kt Ranger Able event on Jan. 27, 1951, in Frenchman Flat. Temporary and permanent support facilities are maintained at several locations over the area, the largest of which is the base town of Mercury.

Before the Air Force Range was removed from the public domain, the area was botanically and zoologically almost unknown. With the coming of the AEC activities and the need to assess their effects on the native plant and animal species and assemblages came the requirement of establishing what species of plants and animals inhabited the area. It was thus that the inventories of the plants and animals on the Test Site were begun by AEC contractors in the late 1950s, as ancillary and requisite to the ecological studies undertaken concurrently.

Plant collecting was begun in 1957 by Lora M. Shields and William H. Rickard, New Mexico Highlands University, and the animal inventories were begun in 1959 by zoologists of Brigham Young University under the direction of Dorald M. Allred. Establishing what the species are and where and with what other species they occur continues to the present. Presented here is a concluding report for the vascular plants based on the plant collecting of the author and her associates during the period 1959 through 1975; this supercedes the several progress reports compiled at intervals during this period (Beatley, 1962; 1965; 1969a; 1971a; 1971b; 1973b).

The area covered by the plant collecting in central-southern Nevada is indicated in Fig. 1, and the locations of the physiographic, political, and cultural features are shown in Fig. 2. The region covers most of the area between the Spring Mountains in southeastern Nevada, including the northwest one-third of this 50-mile-long (80-km-long) mountain range, and the White Mountains of California and western Nevada. Flora of the remainder of the Spring Mountains (with some overlap) was covered by Clokey (1951), and that of the White Mountains, by Lloyd and Mitchell (1973). Most of the taxa are included in the California floras of Munz (1959; 1968; 1974) and in the four-volume flora for the Pacific states by Abrams (1923—1951) and Abrams and Ferris (1960); some are covered only in the Arizona flora of Kearney and Peebles (1960); and the northern half of the area, which lies within the Intermountain Region described by Cronquist et al. (1972), will be included in that flora as it becomes available. The whole region is included in the 50-year-old incomplete flora of Tidestrom (1923) for Utah and Nevada and in a series of 50 publications by various authors prepared as *Contributions Toward a Flora of Nevada* from 1940 through 1965.

The objectives of the plant collecting were to establish what taxa were present in this region, where each occurred with respect to mountain ranges and drainage basins, and with what kind of vegetation each was associated. The taxonomic and distribution record is probably nearly complete for the 1350 sq. mile (3500 km^2) area of the Test Site, where collecting was a continuous, but on-the-side, activity of the author during most of a 16-year period. In the remaining more than 8000 sq. miles (21,000 km^2), where nearly all the collecting was accomplished within four seasons (1968—1971), every mountain range and basin floor was sampled one or more times, but the record is patently incomplete.

The objective of this compilation is to make available the total collecting record for the region as a reference for those who need to know what taxa are known from central-southern Nevada, where each is known to occur, and in what kind of vegetation it occurs. The approach is environmental, rather than the taxonomic or evolutionary emphasis of most floristic compilations, and reflects the orientation of the author. Keys and descriptions for the taxa are not included, since the tools for identification are (or will be) available for nearly all the taxa in the various floras of adjacent areas. Other matters pertaining to the plant collecting and the records are discussed in Part 2.

In Part 1 subjects are selectively treated as seems appropriate for the objectives here. The literature for the various fields of knowledge, supportive to the needs of field biologists, is either "feast or famine" for the Test Site area—choosing what is relevant from many detailed studies or, in most of the disciplines, recognizing the paucity of information that is available. Both are related to the past operating emphases of the AEC at the Test Site on the engineering aspects of the testing program, the needs of which have determined the direction and kinds of studies and the order of priorities. Much of the information presented, therefore, is necessarily based on the field experience and published or unpublished studies of the author.* The pertinent U. S. Geological Survey literature was assembled by Frank M. Byers, Jr., U. S. Geological Survey, Denver, Colo., who also made numerous suggestions that improved the sections on physiography and geology and identified the geologic features of Figs. 16, 18, and 22.

---

*Data for each of 68 permanent study sites located in eight drainage basins of the Test Site were collected during the years 1962—1975. The sites were selected as representative of the major kinds of ecosystems in the region and were at elevations of 3000 to 7500 ft (900 to 2250m). The locations of the sites are shown in Fig. 4 in relation to the vegetation mosaic.

ENVIRONMENTAL DATA. Rainfall and maximum—minimum air temperature measurements were made from 1962 through 1972; those used are for the calendar years 1963—1972 (321 readings). Instruments were read weekly (from south to north over a 3-day reading period) through most of 1966 and thereafter usually biweekly, with approximate equal distribution through the solar seasons. For rainfall measurements a 20-cm-diameter funnel fed into a 2- or 4-liter bottle buried in the soil to the bottle neck; water was measured with a graduate cylinder to the nearest milliliter, and ice and snow were melted for water content. Air temperature was measured by partially exposed Taylor or Weksler Six thermometers mounted on narrow stakes; the thermometers faced north on the northeast side of a shrub-clump, and the sensors (bulbs) were 30 cm above the surface and shaded throughout the year. Precipitation measurements were for the absolute amounts of water. Air temperatures registered were the extremes for the time interval between readings; they are not comparable to Weather Bureau data based on daily readings. Mean temperatures are used here as indices to the relative daytime warmness and nighttime coldness of air (and soils) among the sites of the network. The ratio of mean precipitation to mean temperature (P/T) calculated for the 10-year period is an index to the pattern of climates in the Test Site drainage basins.

Soil moisture was measured, 1963—1971, by calibrated Bouyoucos gypsum blocks and thermocouples buried at 8-, 15-, and 25-cm depths; readings in ohms and °F were made at time of the rainfall and air-temperature readings. Field data were converted to atmospheres of moisture tension. Particle-size characterizations are for the upper 25 cm of soils by the method of Bouyoucos (1962).

PLANT DATA. Percent cover by perennial species was calculated from 335 m of modified line interception data (11 parallel lines, 30.5 m long by 2.5 cm wide, spaced at 3-m intervals); the height of each shrub intercepting the line was measured. Data for shrubs were collected in 1963 and again in 1975, and data for herbaceous perennials, in each of the spring seasons of 1963 through 1967 and 1975. Winter annual cover and density data are derived from cover class estimates (six classes) and from counts by species in fifty 0.1-m^2 quadrats at 0.6-m intervals along a 30.5-m line at the time of spring reproductive peaks in 1963 through 1968; standing-crop biomass was determined from sampling along a 30.5-m parallel line in the springs of 1964—1966 (Beatley, 1969a). Summer annual cover and density data were obtained in the same manner in September or October of the years 1963 through 1968.

The more than one million total field data items were computer-reduced and -analyzed; Frank G. Wood and Stanley D. Zellmer were the principal programmers. Except as noted, photographs were taken by the author in June and July 1975.

Elevations (in feet) are based on U. S. Geological Survey topographic maps (with altimeter measurements on the study sites). The English units of mile and square mile are used with metric equivalents. Temperatures are given in °F, the unit in which the measurements were made, and rainfall in inches, both conforming to Weather Bureau usage, with metric equivalents (°C, and mm rainfall) where feasible. All other units of measure are metric.

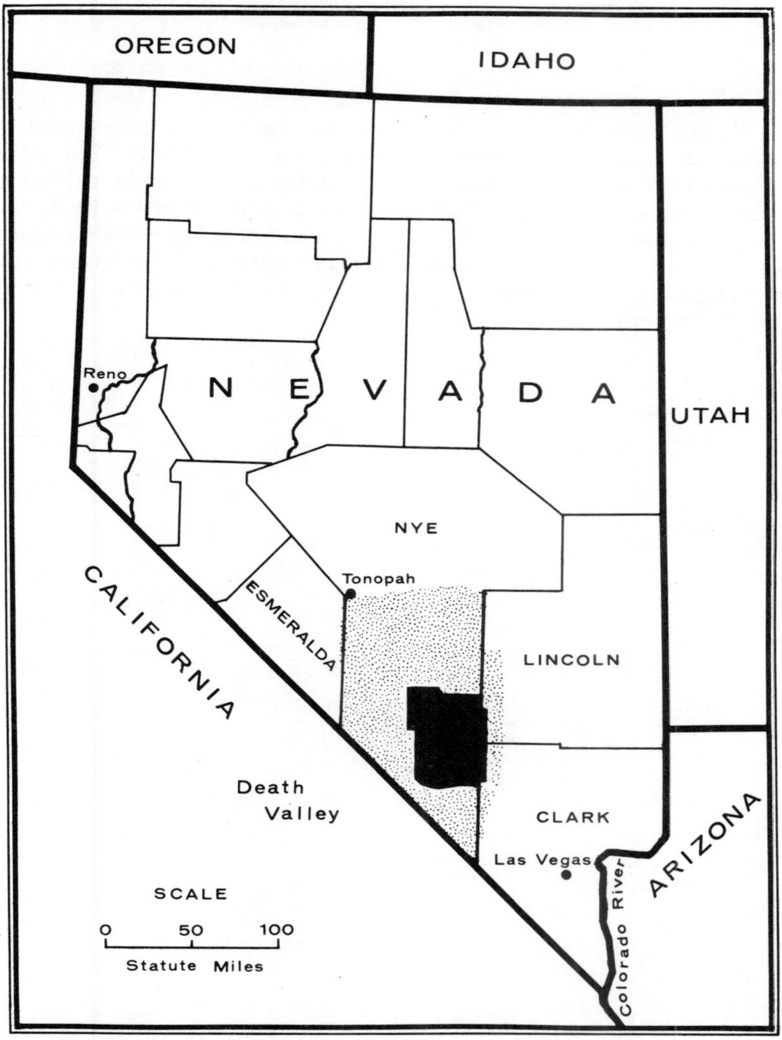

Fig. 1. State of Nevada, showing location of areas of plant collecting on the Nevada Test Site and near vicinity (indicated in black), and beyond the Test Site, in southern Nye County and adjacent parts of Clark, Lincoln, and Esmeralda Counties (indicated by stippling).

Fig. 2. Location of the major physiographic and political features of the region of plant collecting, southern Nye County, and adjacent parts of Clark, Lincoln, and Esmeralda Counties, Nevada. Plant collections are from areas with township boundaries indicated on map. Base maps include portions of the Death Valley, Las Vegas, Caliente, Goldfield, Tonopah, and Lund, Nevada—California—Utah, Army Map Service 1° by 2° topographic sheets. Most elevations (ft) are rounded. Positioning of mountain range names usually indicates direction of axis and length of the range.

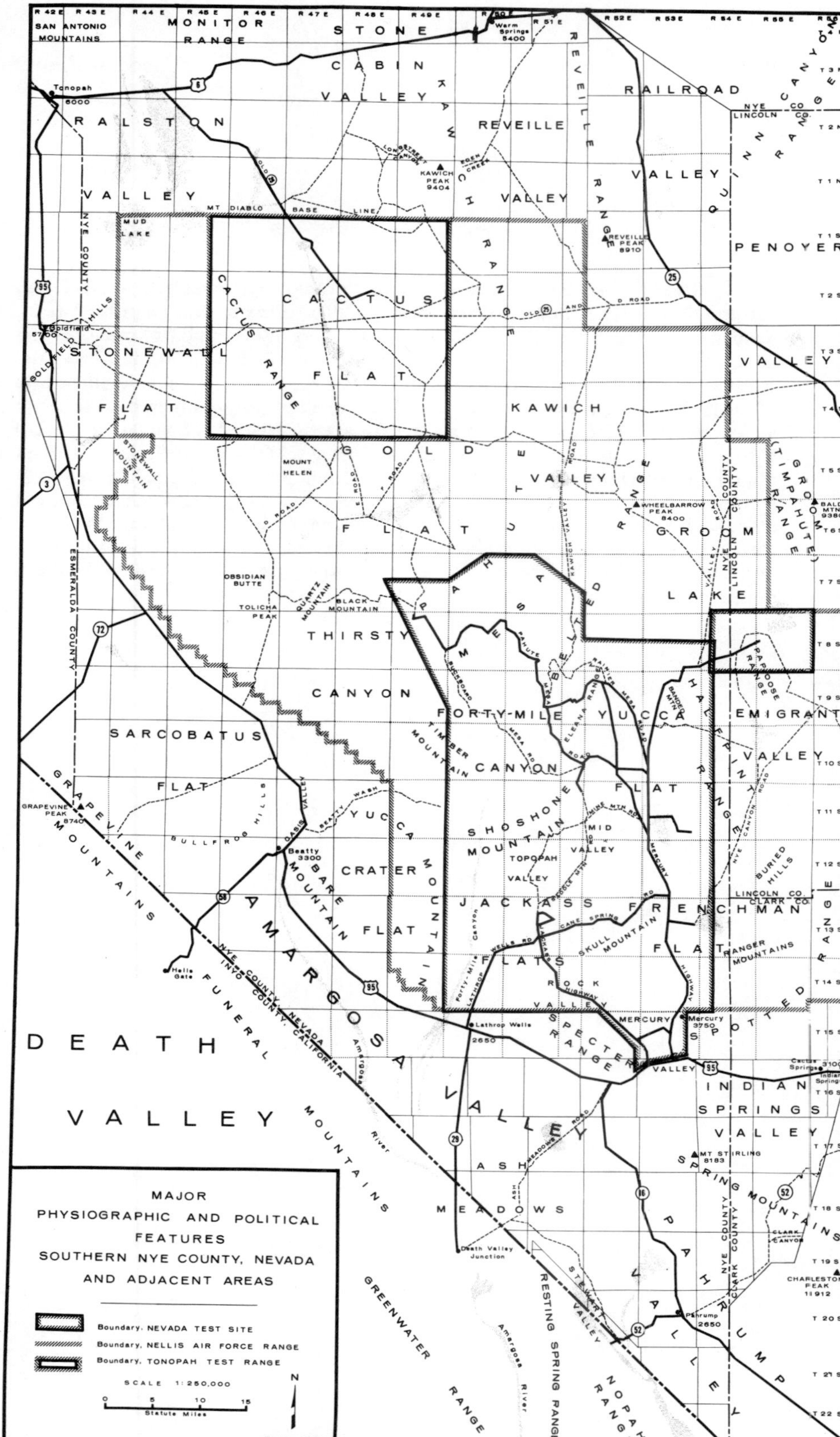
MAJOR PHYSIOGRAPHIC AND POLITICAL FEATURES SOUTHERN NYE COUNTY, NEVADA AND ADJACENT AREAS
Boundary, NEVADA TEST SITE
Boundary, NELLIS AIR FORCE RANGE
Boundary, TONOPAH TEST RANGE
SCALE 1:250,000
0 5 10 15
Statute Miles
N
Beatley, 1975
SAN ANTONIO MOUNTAINS
MONITOR RANGE
STONE CABIN VALLEY
Tonopah 6000
Warm Springs 5400
RALSTON VALLEY
KAWICH RANGE
KAWICH PEAK 9404
REVEILLE VALLEY
REVEILLE RANGE
REVEILLE PEAK 8910
RAILROAD VALLEY
QUINN CANYON RANGE
PENOYER VALLEY
MUD LAKE
MT DIABLO BASE LINE
CACTUS FLAT
CACTUS RANGE
Goldfield 5700
GOLDFIELD HILLS
STONEWALL FLAT
STONEWALL MOUNTAIN
KAWICH VALLEY
GOLD FLAT
MOUNT HELEN
BELTED RANGE
WHEELBARROW PEAK 8400
GROOM LAKE
GROOM RANGE
TIMPAHUTE RANGE
BALD MTN 9380
OBSIDIAN BUTTE
TOLICHA PEAK
QUARTZ MOUNTAIN
BLACK MOUNTAIN
THIRSTY CANYON
PAHUTE MESA
RAINIER MESA
PAPOOSE RANGE
HALFPINT RANGE
EMIGRANT VALLEY
FORTY-MILE CANYON
TIMBER MOUNTAIN
YUCCA FLAT
SARCOBATUS FLAT
GRAPEVINE PEAK 8740
GRAPEVINE MOUNTAINS
BULLFROG HILLS
OASIS VALLEY
YUCCA MOUNTAIN
Beatty 3300
BARE MOUNTAIN
CRATER FLAT
SHOSHONE MOUNTAIN
TOPOPAH VALLEY
MID VALLEY
BURIED HILLS
JACKASS FLATS
FRENCHMAN FLAT
SKULL MOUNTAIN
RANGER MOUNTAINS
ROCK VALLEY
SPOTTED RANGE
AMARGOSA VALLEY
Hells Gate
NYE COUNTY, NEVADA
INYO COUNTY, CALIFORNIA
FUNERAL MOUNTAINS
Lathrop Wells 2650
SPECTER RANGE
MERCURY VALLEY
Mercury 3750
Cactus Springs 3100
Indian Springs
INDIAN SPRINGS VALLEY
DEATH VALLEY
Amargosa River
ASH MEADOWS
MT STIRLING 8183
SPRING MOUNTAINS
CHARLESTON PEAK 11912
CLARK CANYON
Death Valley Junction
STEWART VALLEY
Pahrump 2650
PAHRUMP VALLEY
GREENWATER RANGE
RESTING SPRING RANGE
NOPAH RANGE
LINCOLN CO.
CLARK CO.
NYE COUNTY
ESMERALDA COUNTY

Fig. 3. Nevada Test Site, central-southern Nevada. Physiographic, political, and cultural features are those referred to in the plant distribution records (Part 2). Shaded, irregular boundary, surrounding the Test Site boundary, indicates the additional areas for which the plant distributions are included with the Test Site. Road system and all boundaries are approximate. Base maps include all or part of 16 U. S. Geological Survey Nevada 15-min quadrangles, numbered on the map as follows: (1) Big Dune (Nev.—Calif.); (2) Lathrop Wells; (3) Specter Range; (4) Mercury; (5) Frenchman Lake; (6) Cane Spring; (7) Topopah Spring; (8) Bare Mountain; (9) Thirsty Canyon; (10) Timber Mountain; (11) Tippipah Spring; (12) Papoose Lake; (13) Groom Mine; (14) Wheelbarrow Peak; (15) Silent Canyon; and (16) Black Mountain. Stippled areas are mountains and foothills. Elevations (ft) are for the high peaks of mountains or mountain ranges and are rounded to the nearest 100 ft, for the divides separating or otherwise defining the limits of the drainage basins.

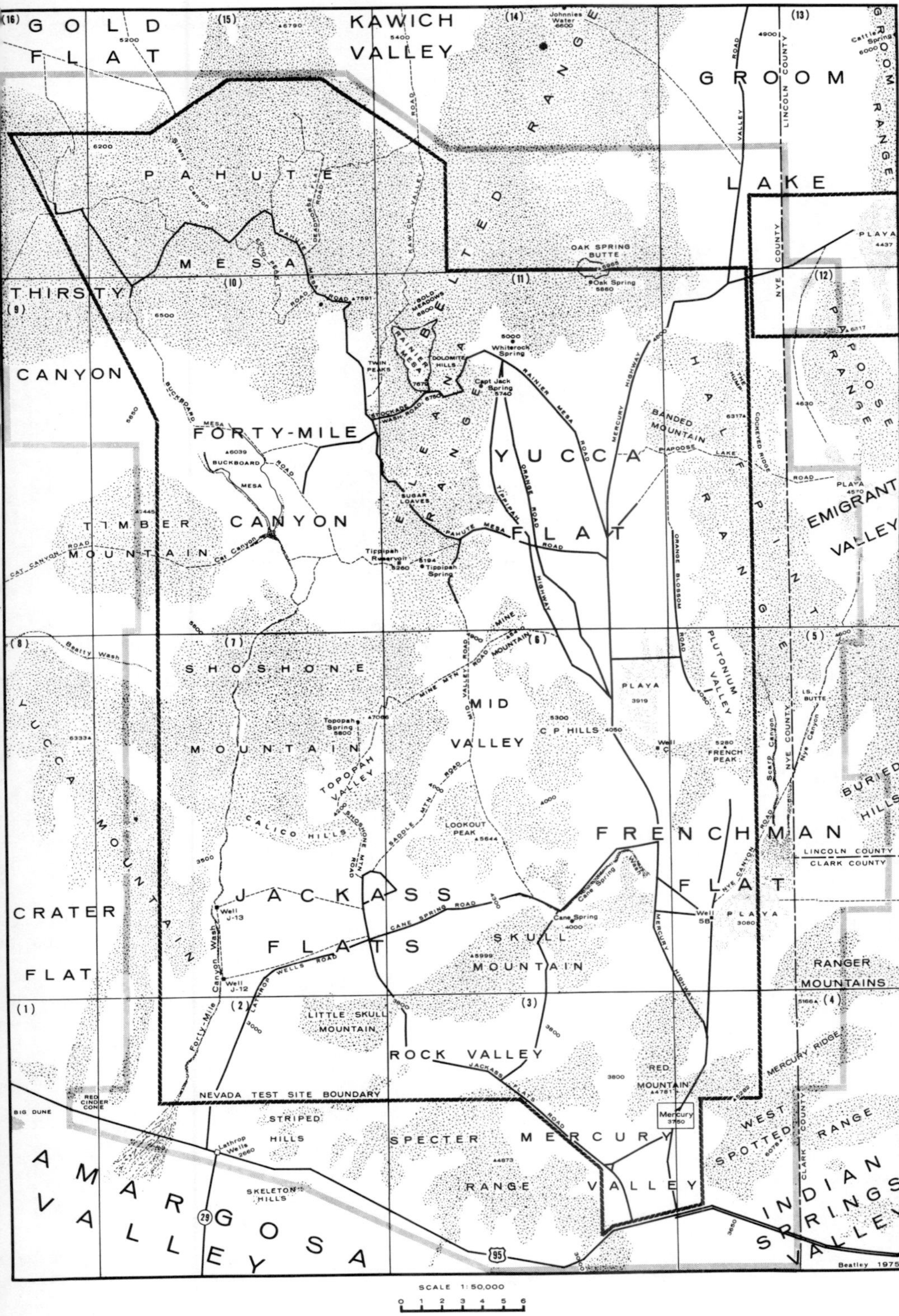
GOLD FLAT
KAWICH VALLEY
GROOM LAKE
GROOM RANGE
LINCOLN COUNTY
NYE COUNTY
PAHUTE MESA
BELTED RANGE
OAK SPRING BUTTE
THIRSTY CANYON
RAINIER MESA
DOLOMITE HILLS
TWIN PEAKS
Whiterock Spring
Capt Jack Spring
FORTY-MILE CANYON
BUCKBOARD MESA
YUCCA FLAT
BANDED MOUNTAIN
HALFPINT RANGE
PAPOOSE RANGE
PAPOOSE LAKE
EMIGRANT VALLEY
TIMBER MOUNTAIN
ELEANA RANGE
SUGAR LOAVES
Tippipah Reservoir
Tippipah Spring
PLUTONIUM VALLEY
SHOSHONE MOUNTAIN
Beatty Wash
YUCCA MOUNTAIN
MID VALLEY
CP HILLS
Topopah Spring
TOPOPAH VALLEY
FRENCH PEAK
BURIED HILLS
CALICO HILLS
LOOKOUT PEAK
FRENCHMAN FLAT
CLARK COUNTY
CRATER FLAT
JACKASS FLATS
Cane Spring
SKULL MOUNTAIN
RANGER MOUNTAINS
LITTLE SKULL MOUNTAIN
ROCK VALLEY
MERCURY RIDGE
RED MOUNTAIN
NEVADA TEST SITE BOUNDARY
RED CINDER CONE
BIG DUNE
STRIPED HILLS
Mercury
SPECTER RANGE
MERCURY VALLEY
WEST SPOTTED RANGE
Lathrop Wells
SKELETON HILLS
AMARGOSA VALLEY
INDIAN SPRINGS VALLEY
Beatley 1975
SCALE 1:50,000
0 1 2 3 4 5 6
Statute Miles

Fig. 4. Vegetation (plant associations), and location of study sites in the vegetation mosaic, Nevada Test Site (and vicinity), central-southern Nevada. Base maps as in Fig. 3. All boundaries are generalized and approximate.

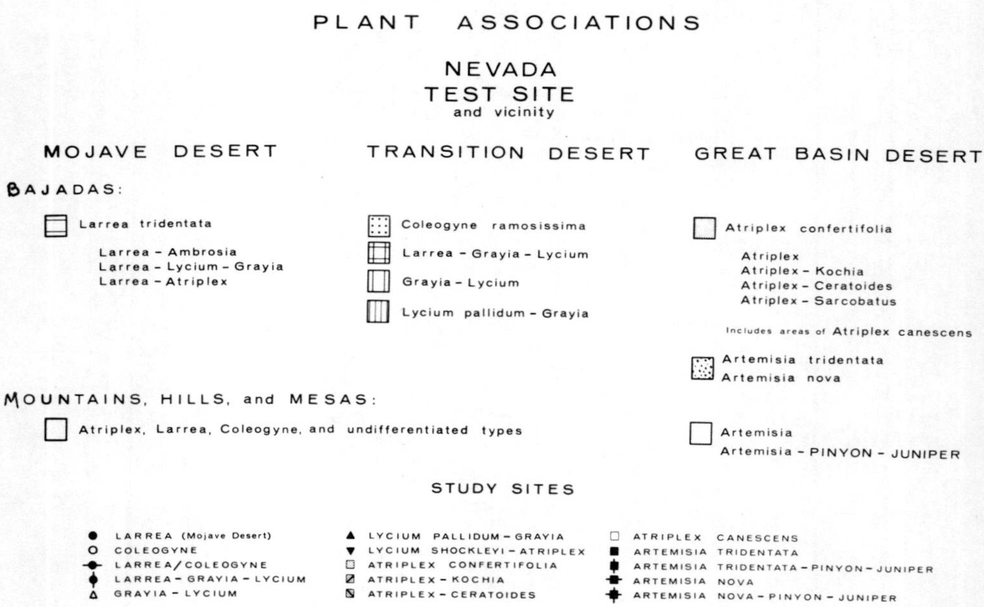

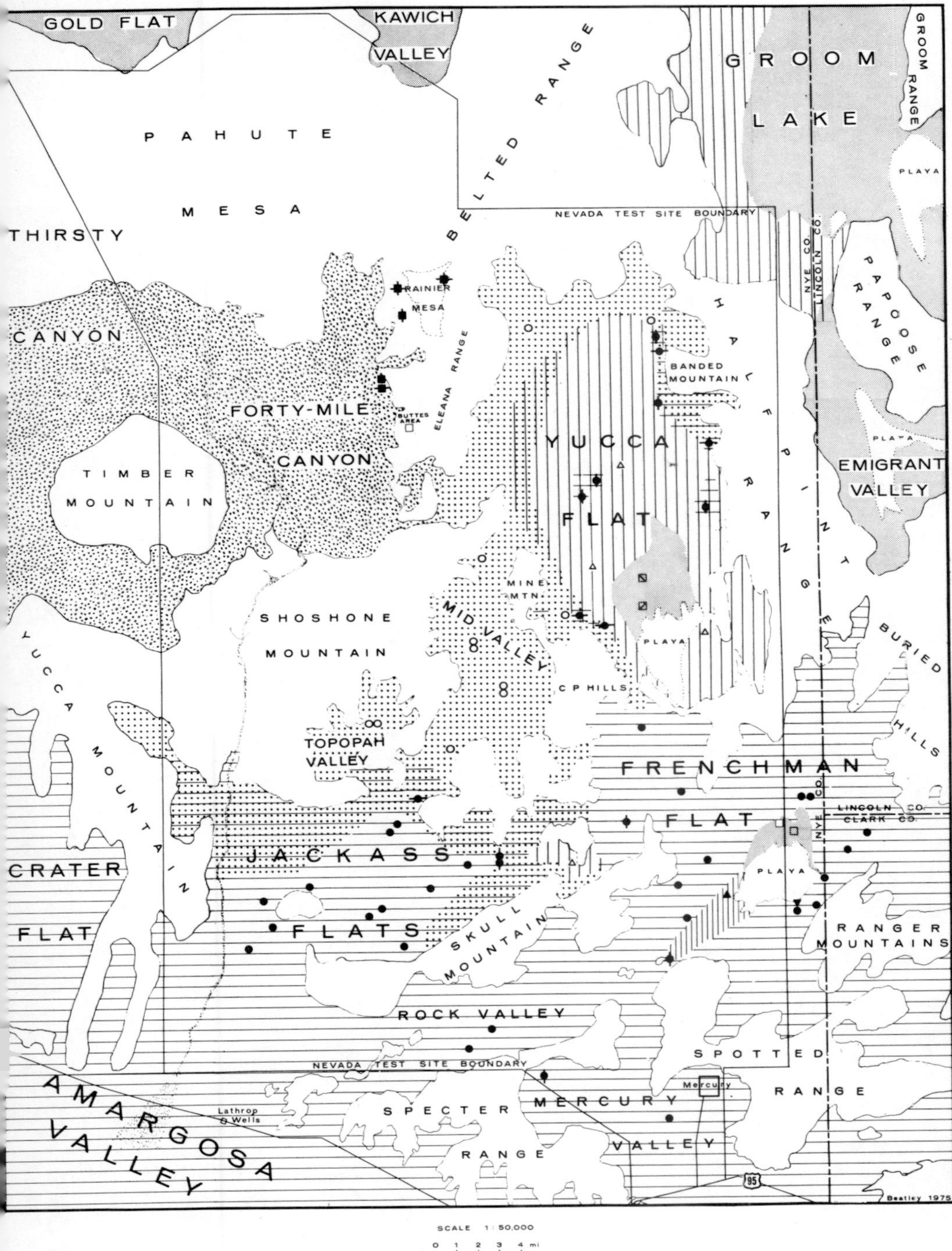
GOLD FLAT
KAWICH VALLEY
BELTED RANGE
GROOM LAKE
GROOM RANGE
PLAYA
PAHUTE MESA
THIRSTY CANYON
NEVADA TEST SITE BOUNDARY
NYE CO.
LINCOLN CO.
PAPOOSE RANGE
RAINIER MESA
ELEANA RANGE
HALFPINT RANGE
BANDED MOUNTAIN
FORTY-MILE CANYON
BUTTES AREA
YUCCA FLAT
EMIGRANT VALLEY
TIMBER MOUNTAIN
MINE MTN
SHOSHONE MOUNTAIN
MID VALLEY
PLAYA
BURIED HILLS
YUCCA MOUNTAIN
C P HILLS
TOPOPAH VALLEY
FRENCHMAN FLAT
LINCOLN CO.
CLARK CO.
JACKASS FLATS
CRATER FLAT
PLAYA
SKULL MOUNTAIN
RANGER MOUNTAINS
ROCK VALLEY
NEVADA TEST SITE BOUNDARY
SPOTTED RANGE
AMARGOSA VALLEY
Lathrop Wells
SPECTER RANGE
MERCURY VALLEY
Mercury
95
Beatley 1975
SCALE 1 : 50,000
0 1 2 3 4 mi
0 1 2 3 4 5 6 km

Part I The Desert Environment and Vegetation

1. The Physical Environment

PHYSIOGRAPHY

The area lies entirely within the Basin and Range physiographic province, between the Colorado Plateaus to the east and the Sierra–Cascade province to the west (Fenneman, 1931). It is peripheral to the Death Valley region to the south (Fig. 2). Throughout the province, mountain ranges are usually parallel with north-south axes and are separated by broad intermont valleys. The mountains rise above the coalesced alluvial fans (bajadas) and circumscribe the drainage basins. In the area of plant collecting (Fig. 2), there are around 25 mountain ranges or mountains, and as many drainage basins; in each basin the bajadas consist of alluvial materials derived usually from the mountains above them.

Where mountains or hills completely surround the basin (closed basin), surface waters (after rains) accumulate at the low point, and a playa is the conspicuous feature of the valley floor (Fig. 12); most basins of the region are of this kind. Playas are absent in basins where drainage waters have a surface outlet (open basins); most of these basins are in the southern part of the region, where stream discharge is ultimately into the Amargosa River of the large Amargosa Valley, and thence into Death Valley to the southwest. Waters from the east slopes of the mountain ranges of Lincoln and Clark counties drain to the Colorado River to the east.

In contrast with the nearly uniformly parallel north-south trending mountain ranges and broad intermontane valleys of central and northern Nevada, the mountain ranges of Fig. 2, and especially those of the Test Site, are somewhat less distant and the direction of their axes variable with the intervening valleys generally narrower and less uniform in shape and size. This reflects, in part, the deformation or movement of mountains along thrust faults and a major shear zone in southeastern Nevada (Ekren, 1968), which have modified the mountain–valley configurations. The basin floors, on the whole, occupy less than half the land area in this region.

Basin-floor levels increase by increments of 500 to 800 ft from south to north across the Test Site and the next tier of basins to the north (Fig. 3). Accompanying these altitudinal increases is the irregularly delineated northern boundary of the Mojave Desert and the southern boundary of the Great Basin Desert within a broad east—west corridor of transition. The desert transition area extends across southern Nevada, east into southwestern Utah and west into Owens Valley of southern California, and nearly everywhere is associated with the large incremental increases in basin floor levels that occur within 1° of latitude.

Lowest elevations of the lowlands in the region of Fig. 2 are around 2100 ft in Ash Meadows at the Nevada/California line. Bases of the mountains are commonly at 3000 to 4000 ft in the southern part, 4000 to 5000 ft in the central part, and 5000 to 6000 ft across the northern part. Highest elevation on the Test Site is 7679 ft on Rainier Mesa (Fig. 19), the southern terminus of the Belted Range. Beyond the Test Site, high peaks are Bald Mountain in the Groom Range, at 9380 ft (Figs. 14 and 28), and Kawich Peak of the Kawich Range, which reaches 9404 ft. Excluded from the region is Charleston Peak, high peak of the Spring Mountains (and of southern Nevada), which ascends to 11,912 ft, but included is a major canyon (Clark Canyon), with headwaters at 9300 ft, below the northwest side of the peak.

Springs occur in nearly all areas of the region. Most of these are at or near the bases of mountains. Many are perched springs that appear to be dependent on current rainfall and the water of percolation through bedrock of the nearby mountain mass, resulting in local seepage areas that are moist year-round (e.g., Capt. Jack Spring of northwest Yucca Flat). At springs from major aquifers, water issues from usually unknown depths, often along major fault lines, as at Cane Spring in west Frenchman Flat (Fig. 9). The area of greatest concentration of springs, and seepage sites in their vicinity, is in northern and eastern Ash Meadows, where many of the springs are thermal, and water origins are from the complex of groundwater basins linked to the large carbonate aquifer system that underlies southern Nevada and the whole of the Death Valley region at great depths below the surface (Winograd and Thordarson, 1975). Much of Oasis Valley has year-round moist soils, including swamps, from the underground movement of water originating on Pahute Mesa and elsewhere on or near the Test Site to the east (Blankennagel and Weir, 1973). Perennial streams occur at the base of the north slope of the Spring Mountains (Cold Creek and Willow Creek) and in canyons of the northern Kawich Range below Kawich Peak (Eden Creek and Longstreet canyons); elsewhere, so far as is known, surface waters accumulate in the drainage courses (arroyos, or washes) only intermittently, following rains.

GEOLOGY

Probably few, if any, areas in the United States are as well known geologically as the Nevada Test Site, where intensive and detailed structural, stratigraphic, and hydrologic studies have been required in connection with the underground detonation of nuclear devices. First published study was the reconnaissance geology of the original proving grounds (Frenchman and Yucca Flats) by Johnson and Hibbard (1957), which included a first geologic map of the area. Subsequent to this work, detailed geologic mapping by the U. S. Geological Survey, sponsored by the AEC, has been accomplished for all of the Test Site, and geologic maps (scale 1 : 24,000) published in recent years for nearly all of the area (many authors). Most parts of the Nellis Air Force Range included in Fig. 2 were mapped (scale 1 : 48,000) by Ekren et al. (1971) with accompanying detailed descriptions of the structure and stratigraphy. The Ash Meadows quadrangle (scale 1 : 62,500) was mapped, and the geology described in the accompanying Bulletin by Denny and Drewes (1965). The area of Fig. 2 is covered in its entirety by the more generalized geologic maps (scale 1 : 250,000) and their accompanying Bulletins of the Nevada Bureau of Mines and Geology: For southern Nye County, Cornwall (1972); Lincoln County, Tschanz and Pampeyan (1970); Clark County, Longwell et al. (1965); and Esmeralda County, Albers and Stewart (1972). The geological setting has been summarized by Ekren (1968) in a publication in which a number of other summary papers appear. Numerous other publications by many authors deal with various aspects of the geology of the region. The publications cited here are the information sources for the brief description that follows.

The region is readily divisible into two principal geologic areas, on the basis of age and composition of the mountain ranges. In the southern part the mountains are late Precambrian and Paleozoic (Cambrian through Pennsylvanian) sedimentary rocks, and in the northern part, the ranges are mostly Tertiary volcanics. Sedimentary rocks are usually limestones and dolomites, but include also some shales, quartzite, argillite, and conglomerate members; these in total have a thickness of more than 12,000 m (40,000 ft) and are above drainage in about one-third of the region. The volcanics are predominantly ash-flow tuffs of largely rhyolitic composition, from the 9 or 10 Tertiary volcanic centers of the region; in composite they are more than 9000 m (30,000 ft) thick and are exposed over about two-thirds of the region.

Folding, normal and thrust faulting, and intrusive lavas have resulted in many large-scale and local structural and stratigraphic complexities throughout. Foothills are often lithologically different from the mountain mass, and no mountain range or mountain consists of a completely uniform stratigraphic sequence. Several areas of subsidence (calderas) are known, with their centers marked by such mountains as Timber Mountain (Fig. 20), Black Mountain (Fig. 26), and Stonewall Mountain (Fig. 27). Cinder cones in the Crater Flat area and the basaltic cone and lava flow of Buckboard Mesa (Fig. 20) are distinctive volcanic features of Quaternary age.

The volcanic masses, in general, weather to sand-size particles and are the principal source of the sands that predominate in most of the region; where these accumulate, dunes result, the largest and best-formed of which is Big Dune in the Amargosa Valley. They are also the masses that give rise to mesas in this region (Figs. 19 and 21) and to the mountains with smooth and rounded contours along their high ridges and slopes. In the southern half of the region, these are the Grapevine Mountains, Bullfrog Hills, Yucca Mountain, mountains and buttes of the Thirsty Canyon drainage, Skull Mountain (Fig. 7), French Peak mountain and Raysonde Buttes (Fig. 11) as well as elsewhere in the Halfpint Range, Shoshone Mountain (Figs. 5, 15, and 18), Timber Mountain (Fig. 20), and the west slope of the Eleana Range (Figs. 16 and 17); in the northern half of the region, bedrock of all mountains, mountain ranges, and mesas, except for those in Lincoln County, is predominantly volcanic in origin. Within these ranges or their foothills, however, are many areas of outcropping Paleozoic sedimentary rocks, most of which are carbonate rocks (limestone and dolomite), and some (especially French Peak mountain) are inferred to be calcareous (or with an ecologically equivalent cation).

The predominantly limestone—dolomite mountains are restricted to the southern part of the region where the high ridges and spurs are commonly sharp (Fig. 10), there are often jagged peaks (Fig. 11), and the weathering products are usually fine sands, silts, and clays. These are the mountains of the Stewart Valley—Ash Meadows—Pahrump Valley area, including the Spring Mountains (Fig. 6); Bare Mountain; Striped Hills, where the southern bajada consists of deep sands of wind-blown origin; Specter Range; Ranger Mountains; both axes of the Spotted Range; and Buried Hills (Fig. 11), with its southern and western bajadas of deep sands of wind-blown origin from volcanic hills of the Halfpint Range to the west. The Papoose Range is almost uniformly Precambrian quartzite, but has local outcroppings of carbonate rocks.

In some of the ranges, both the Tertiary volcanics and underlying Paleozoic sedimentary rocks are well represented above drainage. On the Test Site, these ranges include the Halfpint Range, where Banded Mountain is a conspicuous limestone—dolomite hill on the northwest flank in northeastern Yucca Flat (Figs. 13 and 14); Eleana Range, where there are dolomite and limestone outcrops (and interlayered quartzite, argillite, chert, and shales) of the Eleana Formation on the east side, but the west side consists of a series of conspicuous volcanic buttes capped by the Timber Mountain and Paintbrush Tuffs (Figs. 17 and 18); Mine Mountain (Fig. 14) and the CP Hills, to the south of the Eleana Range, are mainly thrust-fault complexes of Paleozoic sedimentary rocks. Perhaps the most clear-cut example of a stratigraphically composite range is the volcanic east slope of the Groom Range, including the volcanic peak, Bald Mountain (Fig. 28), while on the west slope of this large east-tilted fault block, older sedimentary rocks (mostly late Precambrian quartzite) make up the exposed bedrock.

The Tertiary volcanic rocks, and late Mesozoic intrusives into the older Paleozoic sedimentary rocks, are in many places rich in valuable minerals and have been the basis of a mining industry with a long and colorful history in the past century. Within the region (or nearby) lie the famous Tonopah, Goldfield, and Rhyolite districts, and the mountains are still dotted nearly everywhere with mines, or prospect pits and shafts, most of them long abandoned but yet providing in many cases the only roads of access into the mountain ranges. Mining continues to the present, and today gold and silver are mined especially along fissures and veins in the volcanic tuffs of the Bullfrog Hills and along faults in sedimentary strata in the foothills of the northwest Spring Mountains. Fluorspar is mined in parts of the Yucca Mountain—Bare Mountain area, and bentonite is mined in Ash Meadows and elsewhere. Minerals yielding tungsten, molybdenum, mercury, and lead, as well as gold and silver, have been mined on or near the Test Site.

Many of the plant distributions—probably the great majority—are more or less closely correlated with the mountain masses from which the soil materials have been derived. In this region the chemical composition of the parent materials and the products of their weathering present an almost infinite array of possible relationships of plant distribution to soil chemistry. Some of these relationships have been investigated by Wallace and Romney (1972). The opportunities for this kind of study appear unlimited in this region, and are perhaps unsurpassed.

SOILS

In contrast to the many detailed studies and publications covering the geology of the region, little is known about the soils since they have not been of interest to the testing program of the AEC. Neither state nor federal agencies have conducted official soil surveys on the Test Site. The surficial deposits of Yucca Flat were mapped by the U. S. Geological Survey (Fernald et al., 1968), and limited soils studies were conducted by the U. S. Public Health Service and Environmental Protection Agency in areas of their interest (Leavitt, 1970, 1974; Leavitt and Mason, 1971). The principal field and laboratory characterizations are those of Romney et al. (1973), who have recorded 79 profile descriptions, with related physical and chemical data, for certain soils in the southern part of the Test Site; about one-third of the data sources were the present author's permanent study sites in Jackass and Frenchman Flats. The discussion that follows is based for the most part on the author's field work in the region and unpublished data for the network of 68 permanent study sites in eight drainage basins of the Test Site (see footnote in Introduction).

Chemical and physical characteristics of the soils vary according to the source of the parent materials, and these in turn are derived from the mountain masses—in the mountains as residual materials, and on the bajadas below, as transported alluvial materials whose thickness on the basin floors is often hundreds of feet, or more than 2000 ft (600 m) under the playa surface (Cornwall, 1972). Removal of particles from the mountains, and their deposition on the bajadas below, is a continuous process through time.

RESIDUAL SOILS. Soils from weathering products of the underlying rock (residual soils) are nearly always shallow; these occur on the more gentle slopes of the mountains and foothills, and on the mesas and other areas of undulating terrain. On the steeper mountain slopes, where the processes of weathering and erosion keep pace, bare rock outcrops are commonly the prevailing substrate; here plants are confined to depressions, fissures, crevices, and especially around boulders, or where there are other accumulations of weathered particles or channels in the rock penetrable by roots. Bare rock also is the surface of sometimes extensive areas of more level terrain, and these are the "flatrock" areas, on the Test Site especially well developed on Rainier Mesa, along the south rim of Pahute Mesa, and in the upper Forty-Mile Canyon drainage (Figs. 18 and 19); a number of rare plant species are associated with these environments.

ALLUVIAL SOILS. On the bajadas materials are, in general, sorted along a particle-size gradient at the time of deposition (from water), but the stratigraphic sequence of deposits may be so variable that surface and older subsurface layers may be contrastingly different. Boulders, cobbles, and gravels, however, characterize the materials of the upper bajadas, becoming less frequent in a gradient downslope. The smallest size particles (silts and clays) are those transported to the low end of the topographic gradient—in closed basins, commonly the playas, which are intermittently covered by water usually no more than a few centimeters deep. In most soils various proportions of sands, silts, and clays make up the matrix, and there are in addition variable numbers of larger rock fragments associated throughout; on the whole, soils have excellent internal drainage. Although the physical properties of the soil matrix are presumed to be of greatest significance to processes of water absorption by the roots, many of the plant distributions appear to be correlatable with the degree of stoniness, which in this region is evidently a variable of considerable ecological significance. Saline soils occur in Ash Meadows and locally elsewhere.

Where the fine materials have been removed at the surface by wind and water, the larger fragments make up a pavement, which reaches its best development on the bajadas below the limestone mountain ranges. Here also a carbonate hardpan (caliche) is usually well developed and often near the surface, although hardpans probably occur at some depth in most, if not all, alluvial soils (except in active washes) in this region of low rainfall. Soils are little-modified parent materials: Profile development is minimal, horizons (if present) are indistinct since organic matter accumulations are negligible (or moderate under shrubs) and eluviation is slight, color is like that of the parent materials, structure is weakly (if at all) developed, and most are extremely young by the criteria applied to soils of humid regions. The washes, where surface particles are in motion following each heavy rain and are without organization, are the extremes of substrate youthfulness; a number of plant species are restricted to these environments, and elsewhere to disturbed soils, which in some respects appear to be edaphically equivalent to washes.

Most soils are readily permeable to water, once it penetrates the surface, and most are brought to field capacity to a depth of at least

25 cm following a heavy rain of as little as 1 in. (25 mm). Those with relatively high proportions of silts and clays, and a well-developed pavement, retain the moisture significantly longer in the upper levels than do sandy soils without a pavement. Where recharge is to depths of 1 m or more, as may occur during prolonged periods of rainfall, the water reservoir may persist at these depths for many months. Contrary to popular belief, desert subsurface soils rarely, if ever, become air dry, at least in this region; they are at field capacity during the periods of greatest biological activity, and the remainder of the period of activity contain enough available moisture that plants do not wilt.

No data are available in this region for the percent of rainwater lost in runoff. Observation indicates that this is highly variable, depending in combination on the nature of the soil surface, slope gradient, and rainfall intensity. Where desert pavement is well developed and the rain is of high intensity, most of the water is lost as sheetwash (with rapid accumulations in the drainage courses)—an apparent anomaly in a region of low rainfall. However, most rains are not of high intensity in this region—and are rarely such during the growing season of plants—so that soils are recharged with water to variable, but adequate, depths during most periods of significant precipitation.

The character of the soils in relation to the distribution of plant communities is considered in the later discussion of the vegetation (Sec. 2).

CLIMATE

The pattern of climates over the Test Site, and its relation to vegetation pattern, has been described by Beatley (1974a; 1975) by way of summaries of 10 years of year-round rainfall and air temperature data for the eight drainage basins of the Test Site. Similar kinds of data have also been collected on a network of Weather Bureau stations, maintained since the late 1950s by the National Oceanic and Atmospheric Administration (and its predecessor) at various points on the Test Site, with various kinds of measurements made for differing durations; these have been in support of the needs of the testing program (include especially wind data), and the data are not recorded in the monthly Climatological Data for Nevada. Second-order stations, with varied histories of continuity of operation, have been maintained in some of the villages within the boundary limits of Fig. 2; these data are available in the monthly compilation for the state. The following characterization of the climate is based on the author's more than 10-year record for the Test Site, and for most of the major kinds of vegetation of the region (see footnote in Introduction).

RAINFALL

Mean annual rainfall for the 10-year period varied from 4.61 in. (117 mm), at the lowest elevation of 3000 ft in *Larrea—Ambrosia* vegetation of west Jackass Flats, to 11.25 in. (286 mm), at the highest elevation of 7500 ft in *Artemisia*—Pinyon—Juniper on Rainier Mesa. Extreme annual rainfalls were in 1964, the driest calendar year, from 1.61 in. (41 mm) in southern Yucca Flat, to 5.46 in. (139 mm) in west Frenchman Flat, and in 1969, the wettest year, from 7.69 in. (195 mm) in northeastern Frenchman Flat, to 25.94 in. (659 mm) on Rainier Mesa. These data illustrate the enormous fluctuations in rainfall which characterize the region from year to year, and the great range in variability at points of rainfall measurement. So variable is the amount of water delivered at the earth's surface from any given rain that it is not possible to extrapolate from one basin to another, or with accuracy within a drainage basin. Overall, there is a linear increase in annual precipitation with increase in elevation, but in individual storms the reverse may be true, and everywhere the amount of water reaching the earth's surface is subject to modifying influences of size and configuration of topographic features.

On the average, 60 to 65% of the rain comes between the middle of September and late March (autumn and winter). Rains during this period are associated with the cyclonic storms accompanying large air-mass movements, are usually of low to moderate intensity, and are regional. Spring is the season of low rainfall, with a predictability of only 10 to 15% of the annual total from late March until early June. Around 25% comes during the summer season, from early June through early September, often as high-intensity rain associated with convection phenomena, and summer rains are usually local and scattered over the region.

That deserts are regions of torrential rains, as they often have been characterized, is not supported by what is known of the nature of the rainfall and rainfall regimes in southern Nevada. Nor can it be said that rainfall is not predictable in desert regions, for it appears to be as predictable as elsewhere, once the bases of prediction are established. "Rainfall deficiency" and "drought," meaningful terms in regions of high average rainfall, are based on concepts scarcely applicable to regions characterized for many millenia by low total rainfall and prolonged periods between significant rains.

Winter precipitation is frequently snow—usually only snow flurries at lower elevations—which may persist in the lowlands of northern Yucca Flat (the persistence often beginning along a conspicuous east-west boundary here) and to the north. Higher mountains of the region are commonly snow-covered much of the winter into spring, and the high peaks are usually snow-capped until early summer. In about 1 of 3 years, Rainier Mesa and southern Pahute Mesa are under several inches or sometimes a foot or more of snow much or all of the late autumn and winter season. Snowfall is uncommon at the middle and lower elevations of the southern part of the region of Fig. 2 and seldom persists for more than a few hours on the basin floors.

On most days of all seasons, there is a partial cloud cover for at least a part of the day; clouds form over the mountains, especially over the Spring Mountains and the mountains in the northern part of the region. Many days are more or less overcast, but without rain, and a totally clear sky is uncommon for as long as 24 hr.

TEMPERATURE

The region is an intricate mosaic of air and soil temperature gradients at all seasons of the year; the patterns of some of these gradients have been sufficiently defined on the Test Site that certain generalizations can be made which are applicable elsewhere in the region. In open basins in the southern part, both daytime and nighttime air temperatures generally decrease upslope according to the environmental lapse rate. However, in the closed basins which predominate in the region, nocturnal accumulations of cold air in the lowlands, from its movement downslope from the surrounding mountains and foothills, result in large temperature depressions. Minimum temperatures in the lowlands are commonly lowered at least 10° F (6° C) within the cold air lake that forms apparently every night over the playas and nearby lower bajada slopes. Depth and area covered by the lake of cold or cool air, and the temperatures, vary with the season and location along the slope gradient of the basin floor; in basins such as Yucca Flat, where the bajadas are of low gradient most of the length of the basin, the largest part of the basin floor is under the influence of the cold air accumulations. In much of the year, but especially in the autumn, minimum temperatures are lower than on the mesas at 3000 to 4000 ft higher elevations. Lowest minimum temperature of the 10-year record was −18° F (−27.8° C) and occurred near the playa of Frenchman Flat, where even in July and August temperatures may go down into the 30's (° F) at night. Sharp vegetation boundaries in the lowlands of closed drainage basins coincide with levels of these cold air lakes (Figs. 10 and 11). The subject has been dealt with in detail in Beatley (1974a; 1975).

The entire region, as in other continental areas, is characterized by large daily ranges in air temperature. These are especially pronounced near the playas, where not only are the nighttime temperatures often the lowest in the region, but also the daytime temperatures are usually elevated. Just as the region is subject to periods of extreme low temperatures everywhere in the winter, associated with extremely cold air masses from polar regions to the north, it is also subject to periods of regional extreme high temperatures associated with tropical air masses from the south. Highest air temperature recorded is 121° F (49.4° C), on the same site near Frenchman Flat playa where the lowest minimum occurred, but air temperatures over 110° F (43.3° C), in the shade, are infrequent in the summer record; during the summer most maximums are in the 90's, or less frequently in the low 100's (° F). Because of the low vapor content of the air, and consequent high evaporation rates (at the lower elevations relative humidity is often below 10%), high temperatures of summer are not sensibly uncomfortable to most human beings. At the lower and middle elevations, summer is the season of apparent dormancy of most plants and reduced aboveground daytime activity of most animal species.

Subsurface soil temperatures, in the soil volume occupied by most roots of plants, follow the daily air temperature regimes with a time lag. Each 24-hr period soil temperatures fluctuate within a range that decreases with depth below the surface, but increases at all depths as soil moisture contents decrease. On the bajadas at the lower and middle elevations, subsoils apparently rarely freeze (at least long enough to freeze water in bottles buried to the bottle neck). They may, however, frequently freeze and thaw at the higher elevations (above 6000 ft), sometimes with obvious soil-creep and disruption of root systems where soils are nearly saturated.

Relationships between the climatic variables and plant distribution are discussed in Sec. 2, The Vegetation.

2. The Vegetation

The vegetation of the region is made up of communities in which shrubs, or shrubs and trees, are the dominant plants on the basis of size and percentage cover and stable occupancy of the land surface; only the playas are without vegetation of some kind.

At the lower elevations, shrubs usually occur in clumps in which the above-ground and below-ground parts of from one to several species live in close association. The shrub-clump is the fundamental unit of vegetation in the Mojave Desert areas of the region, each representing a more or less separate system in itself. The shrubs are increasingly less clumped with the increases in rainfall and decreases in temperature and the changes in species composition which accompany increases in elevation. On the whole, intricacy of branching decreases and size of leaves and the plants, themselves, increase along these same environmental gradients.

Every community consists also of herbaceous plant species, which are usually represented in greatest numbers near the shrubs, but on most sites are also represented in the inter-shrub areas, and in some communities, may be almost entirely on the soil surfaces between the shrubs. Herbaceous plants are perennials—many of them short-lived, but the life duration spanning more than one growing season—or winter annuals (and on some sites, summer annuals). Biennials are essentially restricted to the higher elevations, except for unusual circumstances elsewhere (see below).

As measured in undisturbed communities of the Test Site (see footnote in the Introduction), percentage cover of the soil surface by the shrub (or shrub and tree) component varies from 5 to 51% (mostly 15 to 30%), and the average heights in the all-shrub communities, from 0.2 to 0.9 m (mostly between 0.4 and 0.5 m). Percentage cover by herbaceous perennials is usually less than 3% (most commonly less than 1%). Winter annuals may be represented by as many as 30 species/1000 m^2, cover as high as 30%, densities to 975 plants/m^2, and a standing-crop biomass at time of the reproductive peak as much as 616 kg/ha (Beatley, 1969b). On the infrequent sites of their occurrence, the percentage cover by summer annuals may be as high as 8%, and their densities to 84 plants/m^2.

PHENOLOGY

For most plant species of the Mojave Desert of the lower elevations, and the transition desert communities of the middle elevations, the seasons of physiological activity are the autumn-winter-spring months, and summer is a period of apparent dormancy. The phenological events and timing of developmental phenomena in all plant components of the communities, in relation to their environmental triggers, have been described in detail, and a flow diagram presented for the relationships, in Beatley (1974b). In species of the higher elevations or those indigenous to spring areas of the region, where soils are more or less moist year-round, physiological activity is during the spring-summer-autumn seasons, and late autumn-winter-early spring is for most species a period of apparent dormancy.

Mojave and transition desert communities occupy nearly all the land surface below around 5000 ft elevation, which includes the valley floors and most of the mountains of the southern half of Fig. 2, and about two-thirds of the area within the Test Site boundaries (Fig. 3). It is in the communities of these parts of the region that the characteristics of the physical environments, and of the plants and animals, are those which distinguish warm deserts from other regions. It is also where the variables of the physical environment fluctuate over the greatest range, and their effects are the most clearly expressed in biological fluctuations.

Overall controls of Mojave communities are the timing and amounts of individual rains, and most (if not all) biological activity is synchronized with and restricted to periods following heavy rains whose seasonal timing has regional high predictability. Most consequential of these is the one (or more) rain of more than 1 in. (25 mm), predictable sometime between late September and early December, which brings soils to field capacity and triggers vegetative growth of perennial herbs and germination of winter annuals (under certain conditions this critical rain may be a winter rain). Heavy winter rains result in more profuse vegetative and reproductive growth in the spring, but they are usually not necessary for at least a moderate spring vegetative and reproductive success of shrubs, or spring reproductive success of herbaceous species.

Because the rains do not always come at the critical time in the life cycle and there are occasional seasons of infrequent or no rains, underground parts of perennial plants in these environments must have the potential to withstand prolonged periods of more or less low soil moisture levels, and entire years without a photosynthetic season above-ground; the populations must have the potential to be maintained through time with frequent years of reproductive failure.

In the higher mountains of the Mojave Desert, and in most of the northern half of the region of Fig. 2, which lies within the southern Great Basin Desert, most rains also come within the same predictable periods, but in greater amounts and under conditions of lower temperatures, so that the carryover of significant quantities of water in the environment from one period to the next is a fundamental distinguishing feature of these environments. Biological activity in these areas is synchronized, not with the periods of rainfall, but with the seasons of higher temperatures; the seasons of biological activity are essentially those of the more northerly and humid regions of the country. Underground parts of perennial plants in these environments apparently do not have a well developed potential for surviving prolonged periods of reduced soil moisture levels or growing seasons without a photosynthetic period in the aboveground parts; vegetative and reproductive success are predictable for most plants of all species, in most years.

The winter annuals, which are the widely fluctuating plant component of the Mojave and transition communities, are a distinctive feature of warm-desert vegetation. Although winter annuals are a part of the vegetation of humid regions, native species there are inconspicuous and usually sparingly represented. The requirements of the winter annual life cycle—which begins with germination following rainfall in the autumn, followed by slow vegetative growth of the seedlings (consisting of an unelongated stem, and leaves in rosettes or tufts) and overwintering in the low temperatures of winter when the seedling low-temperature requirements are fulfilled, and completion of the life cycle with flowering, fruiting, and death the following spring—all precisely fit the climatic regimes of the Mojave (and other) deserts in southwestern United States. The processes of speciation have resulted in many species, in many families, with the winter annual habit, all derived from the floras of adjacent regions.

The majority of winter annual seedlings, following mass germination in the autumn, probably do not survive to maturity. Survival is high through the winter months (the species are winter-hardy), and most of the losses occur in early spring at the time of rapid growth of all the vegetation components and the associated rapid depletion of moisture in the soils (Beatley, 1967).

This large group of plants (around 150 species in this region), with a growing season of five to eight months, has been erroneously called ephemerals. Use of this term appears to have had its origin with Shreve (1936; 1942; 1951), in the belief that the plants had, in fact, just germinated prior to the time of flowering in the spring. They had been identified, however, many years earlier as winter annuals by Spalding (1909), who also compared their rainfall and temperature controls with those of summer annuals. The group includes nearly all the winter annual species of the field and controlled-environment studies of Went and colleagues (Went, 1948; 1949; 1953; 1955; Went and Westergaard, 1949; Juhren et al., 1956).

Summer annuals, of which there are only a few species in the Mojave Desert, germinate following a heavy rain in usually late summer, and flower and fruit without delay, continuing until subfreezing temperatures occur a few weeks later. These species are, in fact, ephemeral, but they are rarely a prominent component of

Mojave and transition desert communities and are absent on most undisturbed sites of the region.

Annuals are usually a minor and inconspicuous component of the vegetation of the higher elevations (above about 6000 ft). The species here appear usually to germinate in early or mid-spring and complete the life cycle by late spring, or, in some species, through the summer months. A number of the vernal species are associated with the volcanic "flatrock" areas (Figs. 18 and 19), and the summer-flowering species are identified especially with the litter under conifer trees or the deeper soils elsewhere.

Certain of the herbaceous perennials of the lower and middle elevations are capable of behaving like biennials. This happens when mass germination of their seeds is triggered by heavy early spring rains (with low predictability at this season), and most of the plants mature and die in the spring of the following year (Beatley, 1970). In other cases, mass germination may occur in the autumn, and nearly all the plants, then, have a life cycle identical to that of the winter annuals.

VEGETATION TYPES AND ASSOCIATIONS

All undisturbed communities are climax in the sense that they are self-perpetuating and in equilibrium with the present climate and soils of the sites they occupy. Regional distributions of the species are under the major control of climate, but their local distributions are inferred usually to be under the control of edaphic variables.

The structure and aspect (physiognomy) of the shrub communities are similar in most of the shrub vegetation types of the region. The communities, however, are extremely diverse with respect to floristic composition; on no two sites is species composition the same (as documented, from 20 to 75 species in 1000 m^2). Population size in all the plant components fluctuates within a range, as expressions of climatic (especially rainfall) fluctuations, and there may be seasonal shifts in species representation, but the floristic composition of the community is a site characteristic and does not change from year to year in any of the components.

Shrubs that are the dominant species over large areas of the region (on the basis of numbers of plants or percentage cover) are species of large ecological (physiological) amplitudes, and grow on a wide variety of soils in a broad range of climatic regimes; these are *Larrea tridentata, Ambrosia dumosa, Grayia spinosa, Lycium andersonii, Atriplex confertifolia* and *A. canescens, Artemisia tridentata* and *A. nova, Pinus monophylla, and Juniperus osteosperma.* Others, dominant only under certain climatic or edaphic conditions, are species of lesser ecological amplitudes; these contribute to much of the community diversity apparent in the vegetation mosaic. Associated shrub species are usually those also of lesser ecological amplitudes and under population controls which preclude a status of numerical dominance. Many of the shrub species, particularly those of the mountains, are restricted to certain kinds of geologic substrates. Number and size of plants fluctuate within a usually narrow range, with new plants and stem growth of older plants equalled through time by death and abscission of older branches. The shrubs (and trees, where they occur) are the most stable component of the communities.

Herbaceous perennials also occur in relatively stable populations, but have a greater range in fluctuation of numbers with fluctuations in rainfall, as compared with shrubs, and a less controlled balance between death of established plants and seedling survival. In several species, mass germination and seedling survival give rise to periodic enormous, but short-lived, populations (with the biennial or winter annual habit), and the few individuals surviving give continuity to the presence of the species in the communities (Beatley, 1970). The areal distributions within the region are usually associated with the geologic origin of the soil materials.

The large fluctuations are in the winter annual component of the communities of the lower and middle elevations. When germination is a mass phenomenon, not only are the same species present year after year and wholly predictable on a given site, but they are present in essentially the same proportions (each within its own range of predictability), and each occurs on the same local microsites. A single species (*Nama pusillum*) is known with certainty to be present in occasional years only (germinating after a winter, rather than the autumn rain of other species). The geographic distributions of the species are everywhere more or less restricted and generally correlatable with specific site variables, especially the bedrock origin of the soils; the pattern of local site distribution is evidently under the control of the local patterns of expression of edaphic variables.

The classification and description of the vegetation which follows are based on the author's field observations over the region, including information recorded at the time of collection of herbarium specimens, and data from the permanent study sites on the Test Site (see footnote in the Introduction). The treatment of the vegetation follows first, a division of the region into Mojave, transition, and Great Basin desert vegetation, within which the plant associations (abstractions, of which the communities are concrete expressions) are distinguished according to the physiographic features with which they are associated. Only the major vegetation categories are considered, and the minor ones are omitted if they cannot be included with others.

Although most species occur across a spectrum of the regional communities, in the lists each of the taxa is assigned to the kind, or kinds, of vegetation with which it is most closely identified and in which it is most predictable; all the community types in which each is known to occur are cited in the compilation of Part II. Order of species in the lists is that of the families, genera, and species in Part II. In the species lists, shrubs include also suffrutescent perennials.

MOJAVE DESERT

The overall Mojave Desert vegetation pattern is related to the pattern of the physiographic features. Hence the vegetation categories are those of (1) bajadas, (2) mountains, (3) arroyos (washes), and (4) spring and seepage areas. Excluded from consideration are the *Atriplex confertifolia* (Shadscale) communities of the basin floors of the closed basins lying within the Mojave Desert and the higher elevation communities of the Grapevine and Spring Mountains; these are included later with the similar kinds of vegetation in the Great Basin Desert part of the region.

BAJADAS

Communities in which *Larrea tridentata* (Creosote-Bush) is a dominant are the prevailing shrub type of the bajadas of the region and define the area of the Mojave Desert (Fig. 4). These are usually below about 4000 ft elevation. *Larrea* occurs as high as 5200 ft on south exposures along its northern boundary across the region, where it is as a member of communities that are transitional in character and are here included with the other shrub types transitional between the Mojave and Great Basin deserts.

The bajadas are mosaics of communities in which *Larrea* occurs with various codominants and associated species. Its status as dominant is usually on the basis of its size—its canopy nearly always occupies the upper layer of two-layered shrub communities—although it may occur in low numbers and with low percentage cover in relation to numbers and cover by other shrub species. Codominants and associated shrubs are accorded their status on the basis of numerical representation. Total shrub cover, as measured on the Test Site, varies from 7.2 to 23.1%, and average height of all shrubs varies from 0.2 m, where *Larrea* is scarcely represented and all shrubs are of low stature, to 0.9 m, where most of the shrubs are large *Larrea* plants.

Herbaceous perennials are well represented with regard to numbers of species, and it is in *Larrea* vegetation where in occasional years mass germinations result in enormous biennial populations of *Astragalus lentiginosus* var. *fremontii* (at maturity, to 14% cover), and annual populations of the normally perennial *Erioneuron (Tridens) pulchellum* (to 4.1% cover). Perennial grasses are rather consistently present in significant numbers, but are absent on some sites. It is the winter annual component that distinguishes these communities as a group; most of the winter annual species belong to some kind of *Larrea* vegetation, where they fluctuate in numbers from year to year and site to site. The maximums of all winter annual parameters (cited earlier in this section) occur in *Larrea* vegetation; on all sites the winter annuals may be absent some years.

Mean rainfall in these communities, as measured on 24 sites during the period 1962–1972, was 4.7 to 6.2 in. (119 to 157 mm); annual rainfalls varied from 1.9 to 12.7 in. (49 to 324 mm). Mean maximum air temperatures for all seasons were 81.1 to 87.4°F (27.2 to 30.7°C), and mean minimums were from 29.1 to 39.8°F (−1.6 to 4.3°C). The extreme maximum temperature recorded for these communities during the 10-year period was 117°F (47.2°C), and the extreme minimum was −8°F (−22.2°C). Mean precipitation/mean temperature (P/T) ratios varied from 7.4 to 10.1.

Over most of the Mojave region, *Ambrosia dumosa* (Bur-Sage), *Lycium andersonii* (Desert-Thorn), *Grayia spinosa* (Hop-Sage), and *Atriplex confertifolia* (Shadscale) are the usual codominants, and their degree of representation in the communities appears to be under the major controls of climate and/or physical properties of the soils. These occur in various combinations, with variable representation of each, in many communities that are variations of the well defined combinations (associations) recognized below.

Larrea—Ambrosia

Where soils are deep, loose sands without a surface pavement, a plant association in which *Larrea* and *Ambrosia dumosa* are the dominants reaches its best development and most clear-cut expression. On the Test Site this is the prevailing vegetation of the long, low-gradient bajada extending from Shoshone Mountain in northwestern Jackass Flats (Fig. 5), south into the Amargosa Valley, and the deep sands in northeastern Frenchman Flat south and west of the Buried Hills; elsewhere, the sands below the Striped Hills and the Big Dune area of the northern Amargosa Valley are large areas that are edaphically similar.

Common shrub associates are *Psorothamnus (Dalea) fremontii*, *Krameria parvifolia*, *Ephedra nevadensis*, *Ceratoides (Eurotia) lanata*, and in some areas *Acamptopappus shockleyi* or *Menodora spinescens*. In extreme cases, as for example in the Lathrop Wells area and below the red cinder cone of the northern Amargosa Valley (Fig. 3), only *Larrea* is represented. Perennial grasses (especially *Oryzopsis hymenoides*) are characteristic of this association. A number of other herbaceous species characterize the communities as a whole, but most are of more or less restricted distribution and only some are predictable everywhere. Among these species are the following:

Cymopterus ripleyi
Anisocoma acaulis
Chaenactis fremontii
Chaetadelphia wheeleri
Eriophyllum pringlei
Geraea canescens
Glyptopleura setulosa
Malacothrix sonchoides
Monoptilon bellidiforme
Monoptilon bellioides
Stephanomeria exigua
Coldenia nuttallii
Coldenia plicata
Cryptantha micrantha
Streptanthella longirostris
Nemacladus rubescens
Euphorbia parishii
Stillingia spinulosa
Astragalus didymocarpus var. *dispermus*
Astragalus lentiginosus var. *micans*
Dalea mollissima
Lupinus shockleyi
Nama aretioides
Nama depressum
Eremalche exilis
Eremalche rotundifolia
Sphaeralcea emoryi var. *variabilis*
Abronia turbinata
Camissonia brevipes ssp. *brevipes*
Oenothera deltoides
Eriastrum eremicum
Gilia campanulata
Langloisia schottii
Linanthus arenicola
Polygala subspinosa var. *heterorhyncha*
Eriogonum insigne
Calyptridium monandrum
Penstemon albomarginatus
Penstemon thurberi var. *anestius*
Oryzopsis hymenoides
Sporobolus flexuosus

Larrea—Lycium—Grayia

Communities of the higher elevations (3500 to 4000 ft) and on soils having a more or less imperfectly developed surface pavement, generally without a hardpan near the surface, and with a sandy loam matrix with larger rock fragments scattered throughout, usually have as their dominants *Larrea*, *Lycium andersonii*, and *Grayia spinosa*, which exhibit all degrees of numerical codominance. The associated species are those of the *Larrea—Ambrosia* communities, including *Ambrosia*, but in addition, *Haplopappus cooperi* may be prominent, and a conspicuous associate in some areas is *Yucca brevifolia* (Joshua

Tree) which adds a third stratum to the shrub community when it is present.

On the Test Site this is the plant association of the lower north and east slopes of Jackass Flats and much of the western half of Frenchman Flat (Fig. 8), where soils are derived from volcanic mountains and hills. The association occurs elsewhere in the region where there is similar origin of soil materials and at elevations of usually less than 4000 ft.

It is in these communities that there are large biennial populations of *Astragalus lentiginosus*, associated with mass germinations in occasional years. Characteristic herbaceous species of the communities that make up the association are the most widely distributed species in the northern Mojave Desert and include especially the following:

Calycoseris wrightii
Chaenactis macrantha
Chaenactis stevioides
Glyptopleura marginata
Malacothrix glabrata
Rafinesquia neomexicana
Amsinckia tessellata
Cryptantha circumscissa
Cryptantha dumetorum
Cryptantha nevadensis
Cryptantha pterocarya
Cryptantha recurvata
Arabis glaucovalvula
Caulanthus cooperi
Caulanthus lasiophyllus
Descurainia pinnata
  ssp. *glabra*
Lepidium lasiocarpum
Astragalus acutirostris
Astragalus lentiginosus
  var. *fremontii*
Lupinus flavoculatus
Nama demissum
Phacelia fremontii
Phacelia vallis-mortae
Mentzelia obscura
Camissonia kernensis
  ssp. *gilmanii*
Eschscholzia glyptosperma
Gilia cana
  ssp. *speciformis*
  ssp. *triceps*
Gilia stellata
Ipomopsis polycladon
Linanthus demissus
Linanthus jonesii
Chorizanthe brevicornu
  var. *brevicornu*
Eriogonum deflexum
  var. *nevadense*
Eriogonum maculatum
Eriogonum nidularium
Eriogonum reniforme
Oxytheca perfoliata
Delphinium parishii
Oryzopsis hymenoides

Larrea—Atriplex

Communities in which *Larrea* and *Atriplex confertifolia* are consistent dominants, or among the dominants, cover the bajadas of probably the major part of the Mojave area. This is related to the calcareous soil materials originating in the limestone—dolomite mountain ranges and foothills, which predominate in the area of lower elevations in this region. Here the surface pavement is usually well developed, and the soil matrix is generally finer-textured—fine sandy or silt loams—associated nearly everywhere with gravels or larger fragments in the root zone; a more or less indurated calcareous hardpan (caliche) is commonly near the surface and restrictive to root depth.

The communities are regionally those with the greatest floristic diversity; species composition varies with the mountain range, or part of the range, the soil materials come from, and accordingly the plant distribution boundaries are often sharply defined. On many bajadas the communities occur in mosaic with *Larrea—Ambrosia*, and *Larrea—Ambrosia—Atriplex* is a well-defined variant of certain bajadas. In most communities, *Psorothamnus fremontii* and *Krameria parvifolia* are prominently represented. In certain ones, *Ephedra torreyana* or *E. funerea* is numerically prominent, both of which have highly restricted and non-overlapping distributions in the region; in others, the ecological equivalent is the widely distributed *E. nevadensis*. Either *Lycium shockleyi* or *L. pallidum* may be codominants in this association, although neither is exclusively identified with it. *Yucca schidigera* (Mojave Yucca) occurs in certain areas, where it may be associated with either *Larrea—Atriplex* or *Larrea—Ambrosia* vegetation and often has the status of a community dominant (Fig. 6). *Opuntia ramosissima* is known only from this kind of vegetation; *Opuntia basilaris*, perhaps the most widely distributed cactus of the region, is frequent in these communities. Suffrutescent perennials essentially restricted to these communities are *Sphaeralcea ambigua* ssp. *ambigua* and *Stephanomeria pauciflora*.

Herbaceous species most characteristic of this association, or restricted to it, are the following:

Tidestromia oblongifolia
Atrichoseris platyphylla
Chaenactis carphoclinia
Cryptantha angustifolia
Cryptantha decipiens
Cryptantha recurvata
Cryptantha virginensis
Pectocarya heterocarpa
Pectocarya platycarpa
Pectocarya recurvata
Lepidium flavum
Lepidium fremontii
Nemacladus glanduliferus var. *orientalis*
Nemacladus sigmoideus
Astragalus tidestromii
Nama pusillum
Phacelia crenulata var. *crenulata*
Allionia incarnata
Selinocarpus diffusus
Camissonia boothii ssp. *condensata*
Camissonia brevipes ssp. *pallidula*
Camissonia munzii
Oenothera primiveris
Orobanche cooperi
Plantago insularis var. *fastigiata*
Gilia clokeyi
Gilia filiformis
Gilia transmontana
Langloisia setosissima
Chorizanthe rigida
Eriogonum bifurcatum
Eriogonum deflexum var. *deflexum*

Eriogonum inflatum
Eriogonum rixfordii
Eriogonum trichopes
Antirrhinum filipes
Mimulus parryi
Androstephium breviflorum
Bouteloua barbata
Erioneuron pulchellum
Hilaria rigida
Vulpia octoflora

MOUNTAINS

Since over most of the Mojave area the mountains are limestones and dolomites, the mountain flora is that of highly calcareous substrates. In the lower mountain ranges (excluding the Spring Mountains), plants everywhere grow along rock crevices, rock ledges, on cliffs, in depressions, or on loose talus. Most perennial species apparently have roots extending down or back into fissures or crevices in the bedrock. Over most of these mountains, there is little organization of plants to suggest communities, beyond small aggregations on microsites—most plants appear to be living largely independently of others. Here, however, the diversity of environments enables a very large number of species to live. Many of these are species also of the Death Valley region of Inyo County, Calif., some are rare (or have been only rarely collected), and the species differ somewhat from one mountain range to the next.

A large total number of shrub species is known from these mountains, but the one that occurs to some extent in most parts of all the mountains is *Atriplex confertifolia.* The vegetation is therefore here referred to as *Atriplex* because this is the unifying, if often inconspicuous, species of the limestone–dolomite ranges and calcareous hills or outcrops elsewhere, and not because it is necessarily the most common shrub present locally. In some mountain areas or particular canyons, *Atriplex hymenelytra* (Desert Holly) replaces *A. confertifolia.*

Most species listed previously for the *Larrea–Atriplex* communities of the bajadas below these ranges also occur in the mountains above. In addition to these, most of those listed below are restricted to the mountains or foothills (or sometimes extending down on to the uppermost bajadas) whose ridges and peaks are below 6000 ft; included are those of special interest to systematists, geographers, or others concerned with the evolution or distribution of Mojave Desert species.

Shrubs and Cacti

Sarcostemma hirtellum
Amphipappus fremontii
Artemisia bigelovii
Gutierrezia microcephala
Encelia farinosa
Perityle intricata
Pleurocoronis pluriseta
Coldenia canescens
Coryphantha vivipara var. *deserti*

Echinocactus polycephalus
Echinocereus engelmannii var. *engelmannii*
Mammillaria tetrancistra
Opuntia erinacea var. *ursina*
Scopulophila rixfordii
Mortonia utahensis
Buddleja utahensis
Fraxinus anomala
Leptodactylon pungens ssp. *hallii*
Eriogonum heermannii var. *heermannii*
Cercocarpus intricatus
Galium magnifolium
Galium stellatum
Fendlerella utahensis
Penstemon petiolatus
Agave utahensis var. *eborispina*

Herbs

Cymopterus gilmanii
Lomatium scabrum
Dyssodia pentachaeta var. *belenidium*
Enceliopsis nudicaulis var. *nudicaulis*
Machaeranthera tortifolia var. *imberbis*
Senecio douglasii
Cryptantha confertiflora
Cryptantha maritima
Cryptantha racemosa
Plagiobothrys jonesii
Arabis pulchra var. *munciensis*
Draba cuneifolia var. *integrifolia*
Thysanocarpus laciniatus var. *hitchcockii*
Arenaria macradenia var. *parishiorum*
Astragalus funereus
Astragalus mohavensis
Eucrypta micrantha
Hedeoma nanum ssp. *californicum*
Mentzelia oreophila
Mirabilis bigelovii
Camissonia chamaenerioides
Camissonia walkeri ssp. *tortilis*
Arctomecon merriamii
Gilia latifolia
Gilia ripleyi
Gilia scopulorum
Eriogonum glandulosum
Eriogonum thomasii
Anemone tuberosa
Mimulus montioides
Mohavea breviflora
Nicotiana trigonophylla
Aristida adscensionis
Aristida glauca
Aristida purpurea
Bouteloua trifida
Erioneuron pilosum
Poa scabrella
Stipa arida
Tridens muticus

Species of crevices or ledges of limestone cliff faces, or at cliff bases, are:

Shrubs

Brickellia arguta
Brickellia desertorum
Brickellia longifolia
Brickellia watsonii
Haplopappus brickellioides
Haplopappus gooddingii
Hecastocleis shockleyi
Peucephyllum schottii
Crossosoma bigelovii
Buddleja utahensis
Eriogonum heermannii var. *sulcatum*
Petrophytum caespitosum

Herbs

Cheilanthes feei
Notholaena jonesii
Notholaena parryi
Cymopterus aboriginum
Cryptantha barbigera
Arabis perennans

Phacelia rotundifolia
Eucnide urens
Galium proliferum
Physalis crassifolia

Species identified with *Atriplex hymenelytra* in certain mountain areas or locally on bajadas are:

Ephedra fasciculata
Kochia californica
Phacelia beatleyae
Salvia funerea
Eremalche rotundifolia
Camissonia megalantha
Eriogonum contiguum

ARROYOS

Plants of the arroyos (washes) are generally those of the nearby bajadas or mountain slopes, but there is also a rather well-defined part of the flora which is restricted to these environments, and the latter are often the conspicuous occupants of disturbed soils elsewhere. The species are somewhat different in washes of the limestone and volcanic areas. Some, such as *Chrysothamnus paniculatus* and *Atriplex polycarpa*, may be the numerically dominant species along the floodplains of major canyons, and *Ambrosia eriocentra* or *Eriogonum fasciculatum* var. *polifolium* the prevailing species in the shallow upland drainage courses. Species restricted, or essentially so, to the arroyos (and disturbed sites), and at elevations usually below 5000 ft, are the following:

Shrubs

Cynanchum utahense
Ambrosia eriocentra
Bebbia juncea
Brickellia incana
Brickellia multiflora
Chrysothamnus paniculatus
Encelia virginensis ssp. *virginensis*
Haplopappus linearifolius
Hymenoclea salsola
Viguiera reticulata
Atriplex lentiformis
Atriplex polycarpa
Salazaria mexicana
Salvia dorrii ssp. *gilmanii*
Salvia mohavensis
Petalonyx nitidus
Eriogonum heermannii var. *humilius*
Eriogonum fasciculatum var. *polifolium*
Fallugia paradoxa
Prunus fasciculata
Thamnosma montana

Herbs

Amsonia tomentosa
Asclepias erosa
Baileya multiradiata
Cirsium neomexicanum
Dyssodia cooperi
Eriophyllum ambiguum
Porophyllum gracile
Cryptantha utahensis
Sibara deserti
Cleome lutea
Lotus rigidus
Lupinus concinnus var. *orcuttii*
Phacelia pedicellata

Salvia columbariae
Mentzelia reflexa
Sphaeralcea grossulariaefolia var. *pedata*
Camissonia refracta
Gaura coccinea
Argemone corymbosa
Eschscholzia covillei
Eschscholzia minutiflora
Eriogonum brachypodum
Eriogonum pusillum
Mimulus bigelovii
Mimulus spissus
Penstemon fruticiformis ssp. *amargosae*
Datura meteloides
Physalis crassifolia
Verbena gooddingii
Muhlenbergia porteri

Spring and Seepage Areas

In the spring and seepage areas, where soils are more or less moist year-round, or seasonally saturated, the native flora is distinct from that in adjacent areas. Many of the regional introduced species are confined to these sites. For most native and introduced species, the growing season is within the spring—summer—autumn period, unlike that elsewhere at the lower elevations of the region.

The large spring-fed area in the region is the lowlands known as Ash Meadows (Fig. 2), in the northern and eastern parts of which there are (or were) around 25 springs, most of them issuing from deep underground sources in lime-encrusted, sometimes circular pools at the surface, or in pools at the base of the nearby limestone mountains (Devils Hole). The spring-fed Oasis Valley, with headwaters of the Amargosa River, is the second-largest such area in the region, and much of its flora is similar to that of Ash Meadows. Elsewhere, springs are infrequent below 5000 ft, but, where they occur, the plants of the moist soils are usually those of the Ash Meadows communities where the species are usually better represented.

Ash Meadows is an extensive lowland plains area of the Amargosa Valley (around 75 sq. miles), which is the more or less dissected remnants of a large Pleistocene playa that extended well into California. Hence the soils are extremely fine textured (silts and clays) and have slow internal drainage, and the water table is near the surface in much of the area (Denny and Drewes, 1965). The soils are light colored and have a high salt content; many are heavily salt encrusted at the surface. Most of the springs occur in the region where these are the soils of the drainage courses and adjacent uplands. These are also the soils where a large agricultural operation was established in the mid 1960s, over about half the area of Ash Meadows, pumping the water from the springs for large-scale irrigation. Both the unusual nature of the plants endemic to these soils and the failure of past and present attempts to use them for pasture or crop purposes reflect the soil physical and chemical properties which are the textural and salt-content extremes of the region.

In the western and southern parts, there are extensive sand deposits, also more or less salt encrusted at the surface, and low sand dunes, all consisting of materials deposited surficially on the old underlying playa deposits.

As the large oasis of southern Nevada, the area has been of exceptional biological interest since the time of the Death Valley Expedition of 1891 (Coville, 1893). The vegetation in the vicinity of all the lower elevation springs of the region is related to the plant communities of Ash Meadows. Its spring waters are the home of the now greatly endangered Desert Pupfish (genus *Cyprinodon*) (Miller, 1948; Minckley and Deacon, 1968; Brown, 1971), as well as equally endangered rare or endemic plant species. Many of the plant species are narrowly endemic to the Death Valley region (Inyo County, Calif.). Species restricted, or essentially so, to Ash Meadows, are the following:

Enceliopsis nudicaulis var. *corrugata*
Grindelia fraxino-pratensis
Ivesia eremica
Machaeranthera ammophila
Nitrophila mohavensis
Centaurium namophilum
Astragalus phoenix
Mentzelia leucophylla

Those of uncertain taxonomic status are the Ash Meadows populations of *Castilleja linariaefolia* and *Elymus cinereus*. Populations of *Chrysothamnus nauseosus* appr. ssp. *mohavensis* are hybrids, known only from Ash Meadows, the other inferred parental taxon being the mountain ssp. *hololeucus*.

Ash—Screwbean—*Baccharis*

In northern and eastern Ash Meadows, where water comes to the surface as springs, the areas are marked by small streams with running water year-round, meadows where soils are saturated or continuously moist, and groves of small trees, which are the conspicuous vegetation feature of the area. The name Ash Meadows is descriptive of the scattered Ash groves and nearby level terrain of the northern and eastern parts. The dominant tree species are Velvet Ash (*Fraxinus velutina* var. *coriacea*) and Screwbean (*Prosopis pubescens*), neither of which is known elsewhere in the region. Nearly always the large shrub *Baccharis emoryi* is the dominant species of the shrub stratum. Many other species are a part of this plant association, most of which are also unknown elsewhere in the region. Included are aquatic, riparian, swamp, and lowland *Distichlis* meadow species, all of which are, in the broad interpretation used here, a part of the tree communities.

The species listed below constitute the native flora of the spring areas, as it is presently known. Species occurring also in Oasis Valley are indicated with an asterisk (*); those which are known also at or near springs elsewhere in the region are indicated by a dagger (†). Species in parentheses are known from springs (or water impoundments) of the lower or middle elevations elsewhere but have not been collected in Ash Meadows.

Trees, Shrubs, and Woody Vines

Pluchea sericea
†**Sarcobatus vermiculatus*
†*(*Suaeda torreyana* var. *torreyana*)
Prosopis pubescens
Lythrum californicum
Fraxinus velutina var. *coriacea*
†*Populus fremontii*
†*Salix exigua* var. *stenophylla*
†**Salix gooddingii*
†*Vitis arizonica*

Herbs

†**Amaranthus blitoides*
†**Berula erecta*
Hydrocotyle verticillata
Apocynum cannabinum var. *glaberrimum*
Asclepias fascicularis
Asclepias speciosa
Aster exilis
†(*Aster frondosus*)
**Aster pauciflorus*
†*Conyza canadensis*
†*Conyza coulteri*
Crepis runcinata ssp. *hallii*
Grindelia fraxino-pratensis
Helianthus nuttallii
†*(*Iva axillaris* var. *robustior*)
Pluchea purpurascens
†**Solidago spectabilis*
†**Xanthium strumarium* var. *canadense*
†**Heliotropium curassavicum* var. *oculatum*
Plagiobothrys stipitatus
†*Hutchinsia procumbens*
†*Rorippa nasturtium-aquaticum*
Cleomella brevipes
†**Atriplex phyllostegia*
Nitrophila mohavensis
†**Nitrophila occidentalis*
Glycyrrhiza lepidota
†*(*Trifolium wormskioldii*)
Centaurium namophilum
†(*Forestiera neomexicana*)
Gaura parviflora
†*Oenothera hookeri* ssp. *angustifolia*
†(*Rumex salicifolius*)
Dodecatheon pulchellum
Samolus parviflorus
†*(*Ranunculus cymbalaria* var. *saximontanus*)
†**Anemopsis californica*
†*Castilleja linariaefolia*
†**Mimulus guttatus*
†*Veronica americana*
†(*Veronica peregrina* ssp. *xalapensis*)
†(*Lippia incisa*)
†*Verbena bracteata*
†**Carex praegracilis*
Cladium californicum
†**Eleocharis parishii*
Eleocharis rostellata
**Fimbristylis thermalis*
Schoenus nigricans
†**Scirpus olneyi*
†**Scirpus robustus*
†*Sisyrinchium demissum*
†**Juncus balticus*

Juncus cooperi
†**Juncus nodosus*
**Triglochin concinnum* var. *debile*
Najas marina
Spiranthes romanzoffiana
†(*Deschampsia danthonioides*)
†**Distichlis spicata* var. *stricta*
Muhlenbergia utilis
Panicum virgatum
†**Phragmites australis*
Spartina gracilis
†(*Vulpia myuros*)
†(*Potamogeton pectinatus*)
Ruppia maritima
†**Typha domingensis*
†(*Typha latifolia*)

Atriplex and *Atriplex—Haplopappus*

On the uplands, where soils are largely silts and clays, or where they are sand deposits on the old playa surface, communities are usually numerically dominated by *Atriplex confertifolia*, or, especially on the sandy soils, the association is more often *Atriplex—Haplopappus acradenius.* Included here also are the upland *Distichlis* meadows, where soils appear to be water-saturated seasonally rather than year-round. Associated shrubs are especially other *Atriplex* species and *Suaeda.*

Following is the native flora of the Ash Meadows uplands, as known to date. Taxa either endemic strictly to Ash Meadows, or restricted to Ash Meadows in this region, are indicated by an asterisk (*); those known also from Oasis Valley, Pahrump Valley, or rarely elsewhere in the region (excluding roadside plants), are marked with a dagger (†). Taxa of similar sites in Oasis Valley, but not known from Ash Meadows, are in parentheses.

Shrubs

Ephedra funerea
†*Chrysothamnus albidus*
**Chrysothamnus nauseosus* ssp. *mohavensis* × ssp. *hololeucus*
**Encelia frutescens*
Haplopappus acradenius
* ssp. *acradenius*
* ssp. *eremophilus*
**Iva acerosa*
†*Stanleya pinnata* var. *inyoensis*
Atriplex canescens
Atriplex confertifolia
Atriplex hymenelytra
†*Atriplex parryi*
†*Atriplex torreyi*
**Suaeda intermedia*
†*Petalonyx thurberi*
**Polygala acanthoclada*
Lycium shockleyi

Herbs

†*Aster intricatus*
**Cirsium mohavense*
**Enceliopsis nudicaulis* var. *corrugata*
Haplopappus racemosus
(ssp. *glomeratus*)
* ssp. *sessiliflorus*
**Machaeranthera ammophila*
Cryptantha confertiflora
**Lepidium montanum* ssp. *cinereum*
**Thelypodium integrifolium* ssp. *affine*
†*Cleomella obtusifolia*
†*Oxystylis lutea*
**Suaeda occidentalis*
**Cressa truxillensis* var. *minima*
**Astragalus phoenix*
Astragalus preussii
**Phacelia calthifolia*

Mentzelia leucophylla
†*Sida hederacea*
**Camissonia claviformis* ssp. *funerea*
Camissonia heterochroma
Gilia hutchinsifolia
Langloisia schottii
Langloisia setosissima
†*Eriogonum contiguum*
Eriogonum reniforme
**Ivesia eremica*
**Cordylanthus tecopensis*
**Muhlenbergia asperifolia*
†*Sporobolus airoides*

Mesquite

In areas of low sand dunes or strongly undulating sandy terrain, which occur especially in the southwestern part of Ash Meadows, Mesquite (*Prosopis glandulosa* var. *torreyana*) often forms a more or less dense cover over the sand surface. Mesquite thickets occur also on dunes at Cactus Springs in Indian Springs Valley, and the small trees are especially well developed in northern Pahrump Valley. Only a few taxa appear to be identifiable with Mesquite:

Solidago spectabilis
Cleome sparsifolia
Suaeda torreyana var. *ramosissima*
Oligomeris linifolia
Elymus cinereus

The vegetation and soils of Ash Meadows are related to, but not the same as, those described in Death Valley by Hunt (1966), at over 2000 ft lower elevation.

TRANSITION DESERT

Plant communities that characterize the transition from the lower elevation Mojave Desert, in the southern part of the region of Fig. 2, to the higher elevation Great Basin Desert, in the northern part, belong to one of two major categories: (1) those which occur on the bajadas above the *Larrea* communities, or over the basin floors of open drainage basins at the middle elevations, and are transitional because of their intermediate topographic position (*Coleogyne ramosissima*) or (2) those which characterize the valley floors of closed drainage basins at the middle and lower elevations (*Grayia*—*Lycium andersonii*, *Lycium pallidum*—*Grayia*, and communities in which *Lycium shockleyi* is a dominant). In addition, the communities in which *Larrea* occurs discontinuously near its northern limits (at 4000 to 4500 ft), which occur on the upper bajadas and basin floor of Yucca Flat and elsewhere at drainage divides or locally at elevations below 4000 ft, are a part of the vegetation mosaic of the transition region. All contribute to the distinctive character of the vegetation across the desert transition.

Although the vegetation of the categories is dissimilar, because of dissimilarity in the environmental controls on the upper bajadas and the basin floors, they have in common exposures to the environmental extremes of both the Mojave and Great Basin deserts and to the range of climatic fluctuations (especially rainfall) through time. Their floristic compositions reflect their intermediate position between, or genetic relationship to, both deserts. Several species appear to belong uniquely to the transition across southern Nevada; these endemics are unknown in central and northern Nevada or in nearby California or Arizona.

Upper Bajadas

Coleogyne

In its best expression, *Coleogyne ramosissima* vegetation is the simplest and most nearly uniform upland vegetation of the region. So nearly complete is the dominance of this shrub species that in areas that are not ecotonal there are only a few associated shrub species, and these occur usually as scattered plants in an otherwise pure stand of *Coleogyne.* On the Test Site, *Coleogyne* communities occupy the upper bajadas of the north and east slopes of Jackass Flats, the east slope of Frenchman Flat, much of western and northern Yucca Flat, and the entire basin floors of Topopah Valley and Mid Valley (Figs. 4 and 15). They are also well developed on the upper bajadas below the Spring Mountains. They (or their ecotones) are the communities of the bajadas from around 4000 to 5000 ft elevation, and are identified with soils at least in part derived from calcareous rocks (reaching 6000 ft locally in the dolomite hills of the Eleana Range in northwestern Yucca Flat).

Everywhere on the upper bajadas of the southern part of the region, they form broad ecotones at their lower limits with *Larrea—Atriplex* or *Larrea—Lycium—Grayia*, and with *Grayia—Lycium* communities of the basin floors (or *Larrea—Grayia—Lycium* of certain areas) of Yucca Flat and southern Groom Lake; at their upper altitudinal limits they form usually narrow ecotones, or in foothill areas occur in mosaic, with *Artemisia* communities. At the known northern limits of *Coleogyne*—below the central Groom Range, where soils are derived from sedimentary strata outcropping on the west slope—it occurs with *Grayia spinosa*, as it does locally elsewhere in the ecotones to the south (Fig. 12). The *Coleogyne* association occurs in the same topographic position in relation to *Larrea* and *Artemisia* vegetation through the Mojave Desert south to the mountains of southern California.

Soils of the upper bajadas (and lower foothills) where these communities and their ecotones occur are usually shallow, contain many gravels, boulders of various size, or rock outcrops, and have a predominantly sand matrix. On the floors of the middle-elevation open drainage basins, where the vegetation may cover many square miles, alluvial deposits are deeper and there are fewer large rock fragments in the surface and subsurface soils. Nearly everywhere they occur, however, these communities are associated with soils that would be characterized as stony or rocky soils.

The *Coleogyne* environment is also distinguished by its climate. On the Test Site, the year-round mean [33.0 to 35.4°F (0.8 to 1.9°C)] and extreme [−7°F (−21.7°C)] minimum air temperatures and the mean [81.8 to 83.7°F (27.7 to 28.7°C)] and extreme [118°F (47.7°C)] maximum temperatures lie usually well within the range of these same parameters in *Larrea* communities (see comparable data for *Larrea*). Therefore, these temperature parameters do not distinguish the *Coleogyne* environment from that of the Mojave Desert *Larrea* communities.

The rainfall regimes, however, do distinguish the environments of these two kinds of vegetation. *Coleogyne* enters the *Larrea* bajada communities, upslope along the topographic and rainfall gradient, at around 6.4 in. (160 mm) mean annual rainfall, and, where *Coleogyne* vegetation is best developed (in Topopah and Mid Valleys and on upper bajadas of northwestern Yucca Flat below the Eleana Range), the mean annual rainfall, as measured for 10 years, was 8.9 to 9.5 in. (225 to 240 mm); on these sites, however, annual rainfalls varied from 3.7 to 19.9 in. (94 to 505 mm). The P/T ratios are in the range 14.7 to 17.0. *Larrea*, and many of the Mojave species associated with it, do not occur where the mean rainfall exceeds 7.2 in. (180 mm) (Beatley, 1974a), and the *Larrea/Coleogyne* ecotones occur between mean rainfalls of 6.4 to 7.2 in. (160 to 180 mm); in the *Larrea/Coleogyne* ecotones annual rainfalls varied from 2.0 to 17.6 in. (51 to 447 mm).

Coleogyne vegetation is therefore interpreted as basically belonging to the Mojave Desert, but almost exclusively occupying sites where rainfall exceeds the tolerances of probably many Mojave species. *Coleogyne* is restricted to the upper bajadas in all the closed drainage basins where it occurs and is absent from the lowlands under the influence of cold air accumulations (see Climate in Sec. 1). Lower daytime or nighttime temperatures, or both, are probably limiting at its upper limits on the bajadas and at its northern limits in Groom Lake, where elevations are over 5000 ft.

Percentage shrub cover reaches its highest value in the region in the nearly pure stands of *Coleogyne* (as measured, 45 to 51% cover), in large part reflecting the high rainfall. The populations are exceedingly uniform in height (average around 37 cm) nearly everywhere they occur in the region. The plants are closely spaced and are scarcely or not at all in clumps. The single predictable associated shrub is the sparingly represented *Ephedra nevadensis;* around rock outcrops, which may punctuate the communities, *Chrysothamnus teretifolius* is also predictable. Toward its lower altitudinal limits, especially in the ecotones with *Larrea*, *Yucca brevifolia* (Joshua Tree) is often prominent (Fig. 12), and towards its upper altitudinal limits, especially in the ecotones with *Artemisia*, *Yucca baccata* is often represented. Otherwise, occasional plants of *Opuntia echinocarpa* (Golden Cholla) are commonly all that breaks the uniformity of this exceedingly well-defined plant association.

In the broad ecotones, other species are associated with *Coleogyne;* these are principally *Larrea*, *Lycium andersonii*, and *Grayia spinosa*. These, *Ephedra nevadensis*, and occasional other shrub species occur in discrete clumps within the matrix of closely spaced *Coleogyne* plants. Where *Atriplex confertifolia* shares dominance of a site with *Coleogyne*, as is commonly the case on the upper bajadas below limestone hills or mountain ranges, or site dominance is shared with *Menodora spinescens*, as occurs across the whole of the upper north slope of Jackass Flats, there is also a marked tendency for *Coleogyne* and the second shrub species to be somewhat segregated in their local distributions. The ecotonal communities are therefore mosaics, more or less, of two plant associations, each with its own soil environment and herbaceous flora.

Although many herbaceous species occur in the *Coleogyne* communities, they are seldom present in large numbers, except in the ecotones. As measured during the years 1963–1968, maximum values for native winter annuals were 7.8% cover and 97.6 plants/m^2 (density) (compare with *Larrea* communities); maximum cover value for herbaceous perennials was 0.4%, and on most sites was $<0.1\%$, with the perennial grasses uniformly scarcely represented, except in the ecotones with other kinds of vegetation. A number of herbaceous species, however, are predictable, or have their greatest predictability, in this association, even if present in small numbers; these include the following:

Cheilanthes covillei
Amaranthus fimbriatus
Cymopterus globosus
Lomatium nevadense
Balsamorhiza hookeri var. *neglecta*
Calycoseris parryi
Microseris linearifolia
Stephanomeria parryi
Stylocline micropoides
Syntrichopappus fremontii
Pectocarya setosa

Plagiobothrys arizonicus
Arabis pulchra
var. *munciensis*
Thysanocarpus curvipes
var. *eradiatus*
Euphorbia micromera
Euphorbia serpyllifolia
Euphorbia setiloba
Astragalus acutirostris
Astragalus casei
Astragalus layneae
Lotus humistratus
Lupinus brevicaulis
Emmenanthe penduliflora
Tricardia watsonii
Mentzelia veatchiana
Oenothera primiveris
Plantago purshii
var. *oblonga*
Gilia aliquanta
Gilia malior
Linanthus dichotomus
Chorizanthe thurberi
Chorizanthe watsonii
Eriogonum howellianum
Delphinium parishii
Castilleja chromosa
Penstemon utahensis
Dicholostemma pulchellum
Calochortus flexuosus
Sitanion hystrix
Sitanion jubatum
Stipa speciosa

The introduced winter annual, *Bromus rubens*, has become thoroughly integrated into the *Coleogyne* vegetation of certain areas during the past 50 years (introduced in Nevada apparently in the early 1920s). Some years it may occur on undisturbed sites in densities as high as 1500 plants/m^2, and with cover as much as 30%. *Coleogyne* vegetation is that of the region most susceptible to ground fire and total destruction over large areas, because of the tinder-like nature and close-spacing of the shrubs (Beatley, 1966); the presence of dense *Bromus* populations further contributes to fire-susceptibility. Fire scars today—all covered with *Bromus* in high density, with dead plants persisting for up to 3 years—include much of the basin floors of Topopah and Mid Valleys on the Test Site, from fires in 1958 and 1959, and extend (as *B. tectorum* populations) up the slopes of the south and east faces of Shoshone Mountain (Fig. 15).

Larrea—Grayia—Lycium

On sites between 4000 and 4500 ft, near the northern or upper altitudinal limits of *Larrea*, *Larrea—Grayia—Lycium* communities occur discontinuously or sporadically. These are on the upper bajadas, commonly below *Larrea/Coleogyne* ecotones, in central-eastern and northeastern Yucca Flat below especially Banded Mountain, and conspicuously as an island on the basin floor in the central-western part of the basin. They also occur near drainage divides (as at the saddle between west Frenchman Flat and east Jackass Flats), or in local areas below 4000 ft (as in northwest Mercury Valley).

All occur within a narrow range of mean rainfall and mean and extreme minimum temperatures. On the seven sites of measurement, mean annual rainfall for 10 years varied from 6.5 to 7.0 in. (166 to 179 mm), and annual rainfalls varied from 1.6 to 14.2 in. (41 to 361 mm). Year-round mean minimum temperatures were in the range of 32.5 to 36.0°F (0.8 to 2.2°C), and the extreme low minimums of the period were from −5 to 1°F (−20.5 to −17.2°C). Mean maximums were from 80.6 to 83.9°F (27.0 to 28.8°C), and the extreme maximums were from 109 to 118°F (42.7 to 47.7°C). The P/T ratio varied within the narrow range of 11.1 to 12.4. Thus these communities all have in common (1) rainfall at the upper end of the range for *Larrea* and *Larrea* communities and (2) mean and extreme minimum temperatures above those where there are cold air accumulations (see same parameters for the communities of the basin floors). The island of *Larrea—Grayia—Lycium* on the basin floor of western Yucca Flat is immediately below the major break in the Eleana Range—Mine Mountain axis to the west; here the higher minimum temperatures are inferred to reflect nocturnal air turbulence along the major channel of nighttime airflow from the hills to the west.

The communities are closely related to the *Larrea—Lycium—Grayia* described earlier for the Mojave Desert. However, *Grayia* is here usually the numerically dominant shrub, and *Ambrosia dumosa*, *Psorothamnus fremontii*, and *Krameria parvifolia*, all of which are at their northern limits in Yucca Flat, are present only locally. The communities are commonly enriched by *Ceratoides lanata*, *Tetradymia axillaris*, *T. glabrata*, or *Artemisia spinescens*, all of which are better represented in other transition associations or Great Basin Desert communities. It is in these *Larrea* communities (and the ecotonal *Larrea/Coleogyne*) where the *Larrea* plants are conspicuously large (averaging 1.0 to 1.8 m tall) and occur in low densities (Beatley, 1974a). Total shrub cover is from 25 to 33%, correlated with the relatively high rainfall. The herbaceous flora is essentially that of the *Larrea—Lycium—Grayia* communities, with little obvious difference in the species represented or their population size.

Lower Bajadas

It is the communities of the floors of closed drainage basins, under the influence of the nocturnal cold air lakes (Climate, Sec. 1), which distinguish the desert transition region. Soils on the lower bajadas are derived from usually deep alluvial deposits and are silt loams or sandy loams; they markedly differ on the Test Site with regard to geologic origin of the parent materials. Three distinctively different associations occur in the lowlands of Frenchman and Yucca Flats.

Grayia–Lycium

The plant association that most clearly characterizes the desert transition environmentally and vegetationally is *Grayia–Lycium andersonii* (Fig. 13), which covers large areas of the floors of closed drainage basins across the transition; it also occurs locally on upper bajadas or near drainage divides (as in the upper Cane Spring Wash area of the Test Site) and lower mountain slopes within the transition region. On the long low-gradient basin floor of Yucca Flat (and that of southern Groom Lake to the north), the communities are well developed over the many square miles where every night the lowlands are under the influence of cool or cold air accumulations. Here, most Mojave Desert species do not live. The association is confined to elevations between 4000 and 5000 ft.

Mean rainfall is regionally intermediate, on the sites of measurement being in the range of 6.5 to 7.1 in. (164 to 179 mm); annual rainfalls varied from 1.7 to 16.4 in. (43 to 417 mm). Minimum temperatures are comparable to those in the Great Basin Desert of central and northern Nevada: Year-round mean minimums were from 25.6 to 28.3°F (−3.6 to −2.1°C), and the extreme minimums of the 10-year record were −14 to −12°F (−25.5 to −24.4°C). Maximum temperatures are similar to those of the Mojave Desert communities to the south: Mean maximum temperatures were from 82.1 to 83.6°F (27.8 to 28.6°C), and the extreme maximums recorded were from 111 to 116°F (43.8 to 46.6°C). The P/T ratios were 13.1 to 13.5.

Total shrub cover in these communities, as measured, is from 32 to 37%; from 40 to 60% of this is attributable to the two dominant species, *Grayia spinosa* and *Lycium andersonii*, which grow in close association in usually well-developed clumps. Average shrub height is 0.4 to 0.5 m. In many of the communities, especially those with silty soils, *Ceratoides lanata* is prominently represented (locally may be the numerical dominant); in some areas, *Tetradymia glabrata*, *Artemisia spinescens*, *Atriplex canescens*, or *Acamptopappus shockleyi* are numerically important constituents. Other associated woody species are especially *Chrysothamnus viscidiflorus* ssp. *stenophyllus*, *Ephedra nevadensis*, *Tetradymia axillaris*, *Yucca brevifolia*, and *Opuntia echinocarpa; Opuntia stanlyi* var. *parishii* is known only from this kind of vegetation.

Among the herbaceous perennials, *Mirabilis pudica* (endemic to the desert transition region of southern Nevada), *Sphaeralcea ambigua* ssp. *monticola*, and *Stephanomeria parryi* are more or less consistently present in significant numbers in the undisturbed communities. *Oryzopsis hymenoides* or *Sitanion hystrix*, or both, are predictable constituents, but in most seasons have cover values of less than 0.2%. The communities are characterized by their small populations of winter annuals, because in most years soil temperatures at time of the critical autumn rain are limiting to germination; if mass germination occurs, however, there may be densities to 200 plants/m^2, cover to 19%, and as many as 20 species/1000 m^2. The winter annual species, in general, are those mostly widely distributed in the region.

Herbaceous species generally predictable in these communities are the following:

Amaranthus fimbriatus
Calycoseris wrightii
Chaenactis macrantha
Chaenactis stevioides
Chaenactis xantiana
Eriophyllum pringlei
Glyptopleura marginata
Machaeranthera tortifolia var. *tortifolia*
Malacothrix glabrata
Pectis papposa
Prenanthella exigua
Rafinesquia neomexicana
Stephanomeria parryi
Amsinckia tessellata
Cryptantha circumscissa
Cryptantha nevadensis
Cryptantha pterocarya
Cryptantha recurvata
Arabis glaucovalvula
Caulanthus cooperi
Caulanthus lasiophyllus
Descurainia pinnata ssp. *glabra*
Stanleya pinnata ssp. *pinnata*
Streptanthella longirostris
Chenopodium incanum
Euphorbia albomarginata
Euphorbia micromera
Astragalus acutirostris
Astragalus lentiginosus var. *fremontii*
Lupinus flavoculatus
Phacelia fremontii
Phacelia vallis-mortae
Mentzelia albicaulis
Mentzelia obscura
Sphaeralcea ambigua ssp. *monticola*
Mirabilis pudica
Camissonia boothii ssp. *condensata*
Camissonia claviformis ssp. *integrior*
Eriastrum eremicum
Gilia cana ssp. *triceps*
Ipomopsis polycladon
Langloisia schottii
Polygala subspinosa var. *heterorhyncha*
Eriogonum deflexum var. *nevadense*
Eriogonum maculatum
Eriogonum nidularium
Oxytheca perfoliata
Delphinium parishii
Oryzopsis hymenoides
Sitanion hystrix

Lycium pallidum—*Grayia*

Extending southwest from the playa in Frenchman Flat, in an area about 4½ miles long and 2 miles wide (7 by 3 km), at elevations of 3100 to 3250 ft, is one of the most conspicuous vegetation features of the Test Site (Fig. 10). Sharply defined by the *Larrea* communities surrounding it on three sides is a well-defined association in which shrub cover is about 75% *Lycium pallidum* var.

oligospermum and 20% *Grayia spinosa.* The boundaries are defined by the contrasting evergreen *Larrea* plants and the all-deciduous *Lycium pallidum—Grayia* or *Atriplex canescens* vegetation that occurs in mosaic with the *Lycium pallidum* in these lowlands.

Soils in southwestern Frenchman Flat, derived from materials from the north face of the limestone Spotted Range (Mercury Ridge and Red Mountain, Fig. 10) and the east end of the volcanic Skull Mountain, are everywhere sands, which do not distinguish the *Larrea* boundaries but do characterize the soils of the zone of *Lycium pallidium—Grayia.** Both the sharp *Larrea* boundaries and the *Lycium pallidum—Grayia* area, however, are correlated with the low nocturnal temperatures of the cold air lake in the Frenchman Flat lowlands. Mean minimum temperature for 10 years was 27.0°F (−2.8°C), and the extreme minimum, −11°F (−23.9°C). Mean maximum temperature was 86.0°F (30.0°C), and the extreme maximum was 119°F (48.3°C). Although daytime temperatures are somewhat higher than in the *Grayia—Lycium* communities of the Yucca Flat lowlands (at 1000-ft higher elevation), the similarity of the temperature regimes in the two plant associations is apparent. Mean annual rainfall, however, is much lower in the Frenchman Flat association—4.8 in. (121 mm), with the annual rainfall varying from 2.6 to 10.0 in. (66 to 254 mm)—and the P/T ratio of 8.9 places it climatically in the Mojave Desert.

---

*On a *Larrea* site, a few hundred feet beyond the sharp boundary at the southwestern limits of the *Lycium pallidum—Grayia* association, and at an elevation of 3260 ft, the *Larrea* community (in 1975) had a total shrub cover of 27.1%. Of this value, 10.3% was for *Larrea;* 4.9%, *Grayia spinosa;* 4.6%, *Ceratoides lanata;* 2.7%, *Acamptopappus shockleyi;* 2.6%, *Ambrosia dumosa;* 1.8%, *Lycium andersonii;* and 0.2%, *Atriplex canescens. Tetradymia axillaris* also occurs sparingly on the study site. The community is interpreted as a variation in the *Larrea—Grayia—Lycium* association, in part related to low nocturnal temperatures at an upper level of the cold air lake; the site is 145 ft higher than the *Lycium pallidum—Grayia* site (3115 ft) 3 miles (5 km) to the northeast (toward the playa).

Soils of both sites are conspicuously sandy and without a surface pavement. The particle-size distributions of the upper 25 cm of soils on the two sites, *Lycium pallidum—Grayia* and *Larrea—Grayia—Lycium,* were, respectively, total clays, 5.34 and 7.41%; total silts, 1.45 and 2.15%; total sands, 83.24 and 81.22%; total >2 mm, 9.97 and 9.22% (cf. Romney et al., 1973). Of the sand-size particles, the distributions were, respectively, 0.02 to 0.10 mm, 20.29 and 16.78%; 0.10 to 0.25 mm, 31.84 and 32.36%; 0.25 to 0.50 mm, 16.76 and 21.43%; 0.50 to 1.00 mm, 9.16 and 8.14%; and 1.00 to 2.00 mm, 5.19 and 2.51%. The particle-size distributions in these samples, from 3 miles (5 km) distant, are at least as similar as would ordinarily be the case with any two random samples from an area the size of the study plots (929 m^2).

Lycium pallidium, a Mojave Desert species of restricted distribution, is here at its northern limits, and here the plants are the largest known in the region (averaging over 0.7 m tall). Both *Lycium pallidium* and *Grayia* are prominent constituents of the surrounding *Larrea* vegetation of southwestern Frenchman Flat; *Lycium pallidium* continues in the *Larrea* vegetation to the southwest, on the sandy soils below the south face of the volcanic Skull Mountain (Fig. 7) and south across the crest of the limestone Specter Range, where soils have physical properties in notable contrast. Occurrence of the species in this region is inferred to be correlated with soil chemical properties. In the *Lycium pallidium—Grayia* association of Frenchman Flat, *Lycium pallidium* is the ecological equivalent of *Lycium andersonii* in the *Grayia—Lycium* of Yucca Flat, and the two associations are in fundamental ways ecological equivalents in the two drainage basins.

Associated with *Lycium pallidum* and *Grayia* is *Atriplex canescens*, which becomes the dominant shrub in local areas of the lowlands, and *Psorothamnus (Dalea) polydenius;* both are species of sandy soils, especially at the middle elevations of the region (see *Atriplex canescens* communities). As in the *Grayia—Lycium* of Yucca Flat, the environment is marginal for herbaceous species of the Mojave Desert because of the low nighttime temperatures. Most species of the *Grayia—Lycium* communities, especially those of the more sandy soils, also occur here to some extent. Of interest in this association is the lowland occurrence of *Machaeranthera canescens*, *Gilia hutchinsifolia*, and *G. sinuata*, all species of the higher elevations in this region; the narrowly restricted endemics, *Mirabilis pudica* and *Polygala subspinosa* var. *heterorhyncha;* the occurrence of *Abronia turbinata*, a species of restricted distribution and associated elsewhere in the region with *Larrea—Ambrosia* communities; and *Peteria thompsonae*, known from the region elsewhere only from the Goldfield Hills.

Lycium shockleyi—Atriplex

Perhaps the most noteworthy of the southern Nevada endemics is the shrub *Lycium shockleyi.* Described by Asa Gray in 1887 from a collection by William H. Shockley in Mineral County, Nev., it was omitted from the Flora of Utah and Nevada (Tidestrom, 1925) and, like the other southern Nevada endemics, is not included in floras for adjacent states. The plant is a distinctive member of the genus, with its 4-parted, extremely viscid corollas, enlarged calyx in fruit, and its fairy-ring growth pattern.

The species is restricted to the broad desert transition of southern Nevada, where it nearly always occurs with *Atriplex confertifolia*, in lowlands or in uplands. Its occurrence, like that of *Lycium pallidum*, is sporadic and is associated with certain parts of certain mountain ranges or hills, both limestone and volcanic, which in this region of great geologic complexity suggests that its local distribution is under the control of one or more soil chemical variables. The soils appear to span the range of physical characteristics. Elevations of its known occurrence are from 2600 ft below the Skeleton Hills of the northern Amargosa Valley (Fig. 2), to 3100—4000 ft near the southern boundary of the Test Site, to 5200 ft in Ralston Valley to the north.

The distribution of *Lycium shockleyi* on the Test Site is defined by the area of bajada soils originating from the northwest face of the Ranger Mountains in southeastern Frenchman Flat, where it occurs from the playa margin to the base of the mountains, mostly in otherwise *Larrea—Atriplex* vegetation. Here, it is best represented (cover values of 7.8 to 10.1%) in the lowlands near the playa, where it occurs with *Atriplex* along the south margin in a narrow zone below the *Larrea—Lycium shockleyi—Atriplex* of the bajadas (Beatley, 1975). It occurs in lowlands elsewhere, especially in the Mud Lake area to the north, but it is equally identifiable with middle and upper bajadas. The largest known area over which it predominates is in Stonewall Flat, on the south slope below the Goldfield Hills and west slope of the Cactus Range (Fig. 2), where it is associated with *Sarcobatus vermiculatus*, as well as *Atriplex confertifolia*.

Since *Lycium shockleyi* is restricted to the desert transition region in southern Nevada, climatic controls of its overall distribution seem likely. On the Test Site, the rainfall regimes are strictly those of the Mojave Desert, with mean annual rainfall, as measured on four sites in southeastern Frenchman Flat, of 4.9 to 5.4 in. (124 to 137 mm), and annual rainfalls of 2.3 to 9.9 in. (58 to 251 mm); rainfall, however, is certain to be greater where it occurs to the north in Nye, Mineral, and Esmeralda counties. Its best development on the Test Site, at the edge of the playa, is where the year-round mean minimums were from 26.2 to 29.1°F (—3.2 to —1.6°C), and the extreme minimums were —11 to —8°F (—23.9 to —22.2°C), as in the other communities lying near the bottom of the cold air lakes; however, on the upper bajada site near the base of the Ranger Mountains, at about 200 ft higher elevation and above the cold air lake, the mean minimum was 37.3°F (3.0°C), and the extreme minimum was 0°F (—17.8°C). The maximums are those of the Mojave Desert: Mean maximums were from 83.2 to 85.5°F (28.4 to 29.7°C), and the extremes were from 112 to 116°F (44.4 to 46.6°C).

No herbaceous species are known to distinguish the communities in which *Lycium shockleyi* is a dominant. Since the species is nearly always a codominant with *Atriplex confertifolia* in both upland and lowland communities, and with *Larrea* or *Sarcobatus*, as well, the herbaceous component of the communities is that of the basic community types.

GREAT BASIN DESERT

Drainage basins and their surrounding mountains, in most of the northern two-thirds of Fig. 2, belong environmentally and vegetationally to the Great Basin Desert (in the sense of Shreve, 1942). All belong to the Tonopah Section of the floristic areas of Cronquist et al. (1972). On the Test Site, this includes Shoshone and Timber mountains, the large Forty-Mile Canyon drainage, uplands of the northern Thirsty Canyon drainage, Pahute Mesa, and the southern Belted Range. To the west, across the southwestern Gold Flat drainage to central and northern Stonewall Flat (including Stonewall Mountain), and to the northeast, northern Groom Lake (but including the whole of the Groom Range), are the additional southern limits of the Great Basin Desert region, as here defined; these are in essential agreement with the southern boundary of the Intermountain Region as delimited by Cronquist et al. (1972, p. 82). Most of the region is above 5000 ft elevation; except for the Forty-Mile Canyon and Thirsty Canyon drainages, the basins are closed basins; and, except for the Papoose Range and much of the Groom Range, bedrock of the mountains is of volcanic origin.

The region is one of large geologic and floristic diversity. The pattern of its vegetation, however, is relatively simple and clearly defined. Basin floors are covered with *Atriplex confertifolia* communities—*Atriplex*–*Ceratoides*, *Atriplex*–*Kochia*, *Atriplex*–*Sarcobatus*, or *Atriplex* in pure stands. Over areas of deep loose sands at the middle elevations, *Atriplex canescens* is the dominant shrub. On the middle or upper bajadas, *Artemisia* communities—*A. tridentata* and *A. nova* in mosaic—are the prevailing vegetation. Commonly beginning at around 6000 ft, Pinyon Pine (*Pinus monophylla*) and Juniper (*Juniperus osteosperma*) enter the *Artemisia* communities, and continue as usually *Artemisia*–Pinyon–Juniper to around 8000 ft. These are replaced at the highest elevations of the region by forests of White Fir (*Abies concolor*) and certain other high-elevation conifers, or treeless mountain slopes where the vegetation is *Artemisia*–*Cercocarpus*. On the whole, the vegetation pattern is that of central Nevada. Included here for descriptive purposes are the lowland *Atriplex* communities of the

Mojave and transition desert basins and the Grapevine Mountains and large Spring Mountain range, which lie wholly within the Mojave Desert but whose vegetation is that also of the mountain ranges in the Great Basin Desert to the north.

In general, all the plant associations are characterized by increased representation of herbaceous perennials and decreased representation of annual species, as compared with the Mojave and transition deserts; these characteristics are associated with the higher rainfall and lower temperatures of the higher elevations.

Atriplex confertifolia

In the closed drainage basins of the upper half of Fig. 2, most of the basin floors are occupied by *Atriplex confertifolia* (Shadscale) communities, of which *Atriplex—Ceratoides* is the most widely represented and is here considered to be the association from which the other *Atriplex* types are derived in this region. This association occurs over many square miles of the basin floors of Gold Flat, Kawich Valley, Groom Lake, and other basins to the north. Lowland *Atriplex* communities in the corridor of desert transition, or in the Mojave Desert, are usually restricted (in this region) to discrete areas in the vicinity of playas, from which the Mojave *Larrea* communities are inferred to be excluded by the low temperatures of the lowlands (Beatley, 1974a; 1975). The regional ecological status of *Atriplex confertifolia* is discussed in Beatley (1975).

In Frenchman Flat of the Test Site, the association is represented by a nearly pure stand of *Atriplex* in a discrete, roughly triangular-shaped area north of the playa (Fig. 11); the *Lycium pallidum—Grayia* (and *Atriplex canescens*) communities extend to the south-west, and the *Lycium shockleyi* communities are along the south playa margin. Mean minimum temperature here, on the north side of the playa, was 26.6° F (—3.0° C), and the extreme for all winters was —8° F (—22.2° C). Maximum temperatures in these lowlands are among the highest in the region: Mean maximum was 85.9° F (29.9° C), and the highest summer temperature recorded was 120° F (48.8°C). Mean rainfall was 5.1 in. (129 mm), annual rainfalls ranged from 3.2 to 7.9 in. (81 to 201 mm), and the P/T ratio was 9.4. Shrub height is uniform, and averages only 0.3 m. Shrub cover is the lowest in the regional vegetation mosaic (5.5%), and likewise the number of herbaceous species represented. Number of individuals of the few species is variable: As measured, maximum cover by herbaceous perennials was 1.1% (consisting of two species, *Sphaeralcea parvifolia* and *Oryzopsis hymenoides*); for the winter annuals, maximum density was 2.6 plants/m^2, and 0.9% cover (five species), which is the lowest maximum occurrence (over a 6-year period) of winter annuals in any kind of vegetation in the region.

In Yucca Flat, at around 800 ft higher elevation, the communities, occurring in a somewhat less discrete but triangular-shaped area north of the playa, are *Atriplex—Kochia americana* adjacent to the playa, passing northward along a topographic and minimum temperature gradient into well-defined *Atriplex—Ceratoides lanata;* along the continuing gradient, the *Atriplex—Ceratoides* passes into the *Grayia—Lycium* communities covering most of the basin floor of Yucca Flat. In the *Atriplex—Kochia* are the heaviest textured soils in the basin, and the lowest nighttime temperatures: Mean minimum here was 24.7°F (−4°C)—still lower than in the lowlands of Frenchman Flat—and the extreme minimum recorded was −15°F (−26.1°C). Mean maximum temperature was 81.7°F (27.6°C), and the extreme summer temperature was only 109°F (42.7°C). Both the minimums and maximums and the mean annual rainfall of 7.0 in. (178 mm)—varying from 1.7 to 17.1 in. (43 to 434 mm) during the 10 years—reflect the overall lowering of temperature and increase in rainfall, with increase in basin elevation; the P/T ratio was 15.0. Cover by the shrub component, consisting entirely of the two codominants, was 14.9% (average height <0.3 m), and the maximum cover by herbaceous perennials, $<0.1\%$; winter annuals, also scarce here, had a maximum density of 23.6 plants/m^2 (three species) and cover of 2.4%. *Cymopterus globosus*, known from only one other site in the region (*Coleogyne* in northwestern Yucca Flat), in about one of four years occurs locally as a large population.

In the slightly higher elevation *Atriplex—Ceratoides* communities of southern Yucca Flat, other shrubs are associated species, especially *Artemisia spinescens*, *Grayia spinosa*, *Lycium andersonii*, and *Tetradymia glabrata; Chrysothamnus greenei* ssp. *filifolius* and *C. viscidiflorus* ssp. *stenophyllus* are prominent associates or codominants over large disturbed areas in drainage basins to the north. As measured, shrub cover varied from 17.5% in 1963 to 25.6% in 1975 on the same site, in part reflecting the effects of 16.7 in. (425 mm) of rainfall in 1969; percentage of the shrubs that were *Atriplex confertifolia* and *Ceratoides lanata* (74%), however, remained unchanged. As in *Atriplex—Kochia*, mean annual rainfall for the 10-year period was 7.0 in. (178 mm), varying annually from 1.9 to 16.7 in. (48 to 424 mm). Mean minimum temperature was slightly higher, 26.0°F (−3.3°C), and the extreme minimum was −10°F (−23.3°C). Mean maximum temperature was 82.0°F (27.8°C), and the extreme summer temperature, 111°F (43.8°C). The P/T ratio was 14.6. There was an increase in number of herbaceous species and individuals present, as compared with *Atriplex—Kochia:* Maximum cover by herbaceous perennials was 2.4% (consisting of *Astragalus lentiginosus* var. *fremontii*, *Mirabilis pudica*, and *Oryzopsis hymenoides*); maximum winter annual density was 76.4 plants/m^2 (nine species), and 8.3% cover.

In some areas, especially Sarcobatus and Stonewall Flats, *Sarcobatus vermiculatus* is a prominent associate of *Atriplex confertifolia*, or locally the dominant species. The species is not known to occur within the boundaries of the Test Site, but is conspicuous, if only local, in the playa areas of nearby southern Groom Lake. In certain areas it is codominant with *Atriplex parryi*, as at Twin Springs below the north end of the Reveille Range. *Suaeda intermedia* and *Chrysothamnus albidus*, species known in the region primarily from Ash Meadows, are known also from the *Atriplex—Sarcobatus* communities of the more northerly drainage basins. *Sarcobatus* sometimes occurs on residual soils, as in the east Goldfield Hills, with which one of the two known occurrences of *Peteria thompsonae* and the single occurrence of *Phacelia gymnoclada* are identified. It appears to be nearly restricted to areas of higher rainfall, on sites with predictably lower air and soil temperatures, and where the soils have relatively high salt contents and moisture levels year-round.

In the lowland *Atriplex* communities of the region, the paucity of species and individuals which characterize all plant components on most sites is inferred to be under perhaps locally inseparable controls of both climatic and edaphic variables (Beatley, 1975). The communities occur at the low end of the low-temperature gradients in closed drainage basins, and at the high end of the soil salt-content and fine-particle gradients. The combination of temperature and soil-variable extremes appears to exclude most species.

Although these plant associations are variously characterized by low numbers of species and individuals, there is also a large number of herbaceous species identified with the lowland *Atriplex* vegetation of the region. As elsewhere, species composition appears to be under the major control of the geologic origin of the soil materials, and the local or sporadic distributions of most of the species are attributable to local variations in edaphic variables.

Herbaceous species (and cacti) that either characterize the *Atriplex* communities as a whole (especially the widespread *Atriplex—Ceratoides*) or are of special interest because they are known in the region only in *Atriplex* vegetation are the following:

Iva nevadensis
Machaeranthera canescens
Psathyrotes annua
Caulanthus pilosus
Lepidium montanum ssp. *canescens*
Opuntia pulchella
Sclerocactus polyancistrus

Cleome serrulata
Chenopodium incanum
Chenopodium nevadense
Euphorbia albomarginata
Astragalus lentiginosus var. *fremontii*
Psoralea lanceolata ssp. *scabra*
Mentzelia torreyi
Sphaeralcea ambigua ssp. *monticola*
Hermidium alipes
Camissonia boothii ssp. *intermedia*
Camissonia claviformis ssp. *integrior*
Oenothera caespitosa ssp. *crinita*
Ipomopsis depressa
Langloisia punctata
Eriogonum deflexum var. *nevadense*
Penstemon arenarius
Penstemon venosus
Blepharidachne kingii
Hilaria jamesii
Oryzopsis hymenoides
Scleropogon brevifolius
Sporobolus cryptandrus

These plant associations are those which have sustained probably the heaviest grazing pressures in the region, since they cover large areas of the nearly level terrain of basin floors. Where both *Atriplex* and *Atriplex—Ceratoides* vegetation make up the vegetation mosaic, as in Kawich Valley to the north of the Test Site, large populations of the introduced annual *Halogeton glomeratus* are established on the *Atriplex* soils, but are scarce or absent in the disturbed *Atriplex—Ceratoides* vegetation, where *Salsola* species are the common noxious weeds.

Atriplex canescens

On deep loose sands, which occur over large and localized areas at the middle elevations of the region (4500 to 5500 ft), the unifying and most predictable shrub species is *Atriplex canescens* (Four-Winged Saltbush). The plant assemblages are the middle-elevation equivalents of the *Larrea—Ambrosia* of edaphically similar sites at the lower elevations, and usually include locally one or more species of unusual interest.

On the Test Site, soils of the southern and eastern Forty-Mile Canyon drainage are sands derived from the volcanic tuffs of the north slope of Shoshone Mountain and the distinctive light-colored tuffs of the buttes area on the west side of the Eleana Range (Figs. 16 and 17). Three endemic taxa have been described in recent years from these sands: *Gilia nyensis*, *Eriogonum concinnum*, and *Eriogonum umbellatum* var. *vernum* (the latter identified with *Artemisia tridentata* vegetation). In addition, the known ranges of some species are greatly extended by their occurrence on the sands of this area, including *Lesquerella ludoviciana* and *Oenothera pallida*.

Elsewhere, sands of southern Groom Lake (where the dominant shrub may be *Atriplex confertifolia*) and southwestern Penoyer

Valley are large sandy areas of distinctive floristic character, where *Penstemon venosus* and *Lupinus pusillus* var. *intermontanus* are locally common species. At certain springs of intermediate elevations, *Atriplex canescens* is the dominant shrub, e.g., Cane Spring of west Frenchman Flat (4000 ft), where the endemic *Camissonia megalantha* occurs. (Fig. 9).

At the middle elevations, mean annual rainfall, as measured on the southeastern Forty-Mile Canyon uplands west of the Sugar Loaves, was 7.2 in. (183 mm), and varied from 3.3 to 14.5 in. (84 to 368 mm). Mean minimum temperature was 26.5°F (−3.0°C), with −9°F (−22.8°C) the lowest temperature recorded in 11 winter seasons. Year-round mean maximum was 77.0°F (25.0°C), and the highest summer temperature was 108°F (42.2°C). The P/T ratio was 16.8.

Atriplex canescens communities also occur in mosaic with *Lycium pallidum—Grayia* and *Atriplex confertifolia* in the Frenchman Flat lowlands, at elevations around 2000 ft lower. In an *Atriplex canescens* area, on the north side of the playa, mean rainfall was 5.0 in. (127 mm) and varied among the years from 3.2 to 7.7 in. (81 to 196 mm). Mean minimum temperature was 25.7°F (−3.5°C), and the all-time low temperature reading for the Test Site, −18°F (−27.8°C), was set on this site in the winter of 1970. Mean maximum temperature was 86.8°F (30.4°C), and the all-time high-temperature reading for the Test Site, 121°F (49.4°C), was also set on this site (1969). The P/T ratio was 9.3.

In the middle-elevation *Atriplex canescens* community, shrub cover was around 15%, about two-thirds of which was the suffrutescent *Eriogonum kearneyi* var. *kearneyi*; associated shrub species were *Chrysothamnus viscidiflorus* ssp. *viscidiflorus*, *Psorothamnus polydenius*, and *Ephedra viridis*. Cover by herbaceous perennials fluctuated between 6 and 23% during the years 1963 to 1975, reflecting large fluctuations in rainfall during this period; most of the cover was due to *Oryzopsis hymenoides*, but also to *Dalea (Petalostemum) searlsiae* and *Cymopterus ripleyi*, both widely distributed species of sandy soils in this region, and especially to *Oenothera pallida* which some years contributes as much as 5% to the total cover. Only in certain years are winter annuals present; in most years of the period of observation, low temperatures apparently were limiting at the time of the regional critical autumn rain; maximum winter annual density, in the six years of measurement, was 15 plants/m^2 (seven species), and maximum cover was 3.5%.

The lower elevation *Atriplex canescens* communities are characterized by around 5.4% shrub cover, consisting almost entirely of the single species. Cover by herbaceous perennials, including perennial grasses, was comparable to that elsewhere in lowland communities of the Mojave Desert: Maximum was 2.1% (six species, all generally present in the Frenchman Flat lowlands). Maximum cover by winter annuals was 1.6%, with a maximum density of 20 plants/m^2 (two species). The middle and lower elevation communities, therefore, have in common low shrub cover and small winter annual populations, but because of the large difference in mean rainfall are dissimilar with respect to the representation of herbaceous perennials.

The following are the characteristic herbaceous species of the middle-elevation *Atriplex canescens* communities; those which do not occur also in either the lower elevation *Atriplex canescens* or *Larrea—Ambrosia* are indicated with an asterisk.

Cymopterus ripleyi
Ambrosia acanthicarpa
Anisocoma acaulis
Baileya pleniradiata
Chaetadelphia wheeleri
Iva nevadensis
Machaeranthera canescens
Malacothrix sonchoides
Prenanthella exigua
Stephanomeria exigua
Coldenia nuttallii
Coldenia plicata
Cryptantha micrantha
**Lesquerella ludoviciana*
Chenopodium desiccatum var. *leptophylloides*
**Dalea searlsiae*
**Lupinus pusillus* var. *intermontanus*
Lupinus shockleyi
Nama aretioides
Phacelia bicolor
Mentzelia nitens
Abronia turbinata
Camissonia heterochroma
**Camissonia megalantha*
**Oenothera pallida*
**Orobanche fasciculata*
Eriastrum eremicum
Gilia campanulata
Gilia leptomeria
**Gilia nyensis*
Ipomopsis depressa
Langloisia schottii
**Eriogonum brachyanthum*
**Eriogonum concinnum*
**Eriogonum hookeri*
**Penstemon venosus*
Oryzopsis hymenoides

Artemisia tridentata

Beginning at around 5000 ft elevation, *Artemisia tridentata* (Big Sagebrush) becomes the dominant species on the deeper (usually sandy) soils throughout the region, irrespective of geologic origin of the soil materials. On the Test Site, large areas of Pahute Mesa (and small areas on Rainier Mesa) are covered by these communities in which the *Artemisia* is often in nearly pure stand (Figs. 17, 24, and 26). In the Forty-Mile Canyon drainage, at around 1000 to 2000 ft lower elevation, the communities are more often the ecotonal *Artemisia—Grayia.* Elsewhere, in the northern half of Fig. 2, *Artemisia tridentata* occurs on the upper bajadas of the closed drainage basins, usually forming a broad band of uniform vegetation between

Atriplex communities of the basin floors, and *Artemisia*—Pinyon—Juniper above around 6000 ft. At its lower altitudinal limits, the species is present in narrow ecotones with *Coleogyne* and *Atriplex—Ceratoides*. In most areas the communities are in mosaic with *Artemisia nova* communities of the shallow soils; on some sites the two species are codominant, especially on the mesas, but more often they are separated by a sharp boundary (along which they appear to hybridize in some areas). Soils may be either alluvial or residual.

As measured in a sandy valley bottom of the eastern Forty-Mile Canyon drainage, west of Rainier Mesa, at 5740 ft, mean rainfall was 8.0 in. (203 mm), and annually varied from 5.1 to 17.0 in. (130 to 432 mm). Mean minimum temperature was 25.9°F (−3.4°C), and the extreme minimum recorded was −10°F (−23.3°C); mean maximum temperature on this topographic site was 80.7°F (27.0°C), and the extreme maximum was 110°F (43.3°C). The P/T ratio was 17.3. The data are adequate to indicate that *Artemisia* communities in this region are characterized by high rainfall, low year-round nighttime temperatures, and at least on some sites moderately high daytime temperatures.

Shrub cover was 32.6%, 95% of which was due to *Artemisia tridentata*; average height was 0.6 m, but the shrubs are commonly as much as 0.9 m tall. Associated shrub species, usually sparingly represented in these communities, may be *Ceratoides lanata*, *Chrysothamnus viscidiflorus* spp. *viscidiflorus* and ssp. *stenophyllus*, *C. nauseosus* ssp. *leiospermus*, *Ephedra viridis*, *Tetradymia glabrata*, *Eriogonum microthecum* var. *foliosum*, *Opuntia echinocarpa*, and *Echinocereus engelmannii* var. *armatus*; in some areas, as on Pahute Mesa, *Cowania mexicana* var. *stansburiana* is prominently associated. Especially along washes or around low rock outcrops, *Amelanchier utahensis*, *Purshia glandulosa*, *Symphoricarpos longiflorus*, and *Artemisia ludoviciana* ssp. *incompta* are predictable; of limited distributions along washes of volcanic areas are *Prunus andersonii* and *Eriogonum umbellatum* var. *vernum*, and along the major washes in limestone areas *Lepidospartum latisquamum* and *Fallugia paradoxa* are often common; *Peraphyllum ramosissimum*, rare and local on the Test Site, around calcareous rock outcrops in the Eleana and Halfpint Ranges, is common in washes (and on slopes) of the limestone Spring Mountains.

Herbaceous perennials are represented in these communities by a large number of species, since there is great diversity in the origin of the soil materials over the region as a whole. However, except for the grasses, they are seldom present as more than occasional plants on any given site. As measured, their maximum cover was 6.6%, nearly all of which was due to the perennial grasses *Stipa comata* and *Sitanion hystrix*. Winter annuals are not consistently associated with this kind of vegetation [as measured, maximum density of 34.4 plants/m^2 (nine species), and cover of 7.7%], since low temperatures are limiting to germination most years. In total, a number of winter annual species, however, occur in *Artemisia tridentata* vegetation, and these, together with the herbaceous perennials that are more or less frequent in this shrub type (or of particular interest because they are rare or uncommon in this region, indicated by an asterisk), are the following (including the washes):

Cymopterus ripleyi
Anisocoma acaulis
Baileya pleniradiata
Brickellia oblongifolia var. *linifolia*
Chaenactis xantiana
**Erigeron aphanactis*
**Erigeron breweri* var. *porphyreticus*
Leucelene ericoides
Malacothrix sonchoides
Stephanomeria exigua
Plagiobothrys kingii
Caulanthus crassicaulis
Caulanthus glaber
Caulanthus pilosus
Descurainia pinnata ssp. *halictorum*
Stanleya elata
Stanleya pinnata var. *pinnata*
Chenopodium incanum
Chenopodium leptophyllum
Euphorbia albomarginata
Astragalus beckwithii var. *purpureus*
Astragalus lentiginosus var. *fremontii*
Astragalus minthorniae var. *villosus*
Astragalus newberryi
Astragalus purshii var. *tinctus*
Dalea searlsiae
**Lathyrus hitchcockianus*
**Lupinus aridus*
Lupinus caudatus
**Lupinus holmgrenanus*
**Nama densum*
Phacelia bicolor
Phacelia crenulata var. *funerea*
Linum lewisii
Mentzelia albicaulis
Mentzelia veatchiana
Sphaeralcea ambigua ssp. *monticola*
Sphaeralcea emoryi var. *variabilis*
Abronia elliptica
Oxybaphus comata
Camissonia boothii ssp. *intermedia*
Camissonia claviformis ssp. *integrior*
Camissonia heterochroma
**Camissonia parvula*
Gayophytum ramosissimum
Oenothera avita
Orobanche corymbosa
Orobanche fasciculata
Eriastrum sparsiflorum
Eriastrum wilcoxii
Gilia brecciarum
Gilia hutchinsifolia
Gilia leptomeria
Gilia modocensis
Gilia sinuata
Phlox stansburyi
Eriogonum brachyanthum
Eriogonum cernuum var. *viminale*
**Eriogonum concinnum*
Eriogonum deflexum var. *baratum* var. *nevadense*
Eriogonum hookeri
**Eriogonum nutans*
Eriogonum ovalifolium var. *ovalifolium*

Eriogonum saxatile	*Oryzopsis hymenoides*
**Oxytheca dendroidea*	*Sitanion hystrix*
Antirrhinum kingii	*Sporobolus cryptandrus*
Penstemon floridus	*Stipa comata*
Hilaria jamesii	

Artemisia nova

On the shallow soils of the mountains, mesas, foothills, or rolling uplands elsewhere, *Artemisia nova* (Black Sagebrush) is the dominant shrub, beginning at around 5000 ft and occurring in mosaic with *Artemisia tridentata.* Like those of *A. tridentata, A. nova* communities occur in the altitudinal zone between *Coleogyne* (or *Atriplex confertifolia*) and *Artemisia*—Pinyon—Juniper, in both volcanic and limestone mountain areas. The soils, unlike those of the usual *A. tridentata* environment, are residual, stony, and usually a thin mantle of various size rock particles over the underlying rock.

Large areas of *A. nova* communities occur on especially northwestern Pahute Mesa of the Test Site, and *A. nova* vegetation is exceptionally well developed as a vegetation type over the extensive area of rolling hills in the eastern Forty-Mile Canyon drainage, where its aspect is much like that of *Coleogyne* vegetation: Plants are dark-stemmed, uniformly of low height (average 0.3 m), closely spaced, and make up about 90% of the total shrub cover (37.3%, as measured). As in *Coleogyne, Ephedra nevadensis* and *Yucca baccata* are rather consistently, if sparingly, present in these communities below 6000 ft. The most predictable shrub associate is *Chrysothamnus viscidiflorus* ssp. *puberulus.* Several cacti are characteristic of this plant association: *Coryphantha vivipara* var. *rosea, Echinocereus engelmannii* var. *chrysocentrus,* and *E. triglochidiatus* vars. *melanacanthus* and *mojavensis.*

As measured on the uplands, 20 ft higher and overlooking the *A. tridentata* valley-bottom site referred to above, mean annual rainfall was 8.0 in. (204 mm), varying annually from 5.3 to 17.9 in. (135 to 455 mm). Mean minimum temperature was 29.8°F (−1.2°C), and the extreme minimum was −3°F (−19.4°C); mean maximum was 78.5°F (25.8°C), and the extreme high temperature was 108°F (42.2°C). The P/T ratio was 16.6. On the *A. nova* site, nighttime temperatures are much higher, and daytime temperatures are lower than in the nearby lowland *A. tridentata*; moreover, the nights appear to be considerably warmer than in the lowlands of Frenchman and Yucca Flats, at 2000 to 3000 ft lower elevation.

Maximum cover by herbaceous perennials was 3.8%, most of which was due to the grass *Sitanion hystrix* and lesser numbers of *Stipa speciosa* and *Oryzopsis hymenoides.* Winter annuals, represented by 10 species, had a maximum density of 73.8 plants/m^2 and cover of 5.3%.

As in *A. tridentata* communities, there is a large number of herbaceous species associated with the *A. nova* communities of the region as a whole. Those species characteristic of *A. nova* vegetation (including the washes), or of interest because of their rarity or uncommon occurrence in this association somewhere in the region (indicated by an asterisk), are the following:

Lomatium nevadense
Balsamorhiza hookeri var. *neglecta*
**Erigeron aphanactis*
Erigeron pumilus ssp. *concinnoides*
Hymenoxys cooperi
Microseris linearifolia
Senecio multilobatus
Stylocline micropoides
Syntrichopappus fremontii
Viguiera multiflora var. *nevadensis*
**Cryptantha humilis* var. *ovina*
Pectocarya setosa
**Arabis dispar*
Arabis holboellii var. *pinetorum*
Arabis pulchra var. *gracilis*
**Arabis shockleyi*
**Phoenicaulis cheiranthoides*
Cleomella hillmanii
Euphorbia fendleri
**Astragalus beatleyae*
Astragalus calycosus
Astragalus casei
Astragalus purshii var. *tinctus*
Lupinus brevicaulis
**Lupinus uncialis*
Tricardia watsonii
Sphaeralcea ambigua ssp. *monticola*
Mirabilis bigelovii
Camissonia pterosperma
Gayophytum racemosum
Orobanche corymbosa
Orobanche fasciculata
Plantago purshii var. *oblonga*
Eriastrum sparsiflorum
**Gilia aliquanta*
**Gilia malior*
Gilia ophthalmoides
Microsteris gracilis ssp. *humilis*
**Phlox lanata*
Polygala subspinosa var. *subspinosa*
**Eriogonum beatleyae*
**Eriogonum ovalifolium* var. *multiscapum*
**Eriogonum rupinum*
**Eriogonum shockleyi*
Delphinium andersonii
Allium nevadense
Dichelostemma pulchellum
Calochortus flexuosus
Aristida fendleriana
Hilaria jamesii
Munroa squarrosa
Poa sandbergii
Sitanion hystrix

Artemisia—Pinyon—Juniper

At elevations of around 6000 ft Pinyon Pine (*Pinus monophylla*) and Utah Juniper (*Juniperus osteosperma*) enter the *Artemisia* communities and form a more or less open shrub—woodland association, which continues to elevations of around 8000 ft. This is the most extensive plant association of the region, since it covers the slopes and ridges of most of the region's volcanic mountain ranges and mesas, and much of the limestone Spring Mountains in the southeastern part of the region. Juniper enters at a slightly lower elevation than Pinyon Pine, often occurring first as scattered small trees on the *Artemisia* slopes, followed by Pinyon Pine within 200 to 300 ft increase in elevation; at the upper limits of the association, Pinyon Pine continues upslope in the White Fir communities usually a few hundred feet after Juniper has disappeared.

Over much of the high country occupied by this association, especially on Pahute Mesa and the upper Thirsty Canyon uplands on the Test Site, the trees are localized and discontinuous—around low cliffs or rock outcrops, or along shallow drainage courses, in what is otherwise *Artemisia* vegetation (Figs. 24 and 27). In local but discrete areas, Juniper may occur almost to the exclusion of Pinyon Pine, as on the "Juniper Flats" of northwest Pahute Mesa (Fig. 23), and, in other areas, the Pine may be in essentially pure stand. On mountain slopes, the community may be Pinyon–Juniper, without either of the *Artemisia* species in the shrub layer. While recognizing that there are these discrete variations—inherent to all plant associations—the communities in their totality seem best designated as belonging to the *Artemisia*–Pinyon–Juniper association.

At elevations of 6000 to 8000 ft, a significant part of the precipitation comes as snow, which may persist through much of the winter season. Mean annual precipitation on sites from 6100 to 7500 ft elevation was from 10.3 to 11.3 in. (261 to 286 mm); lowest and highest annual rainfalls on the sites were 5.1 and 26.1 in. (130 and 663 mm), both recorded on the high-elevation site (Rainier Mesa). Mean minimum temperatures were from 24.6 to 29.3° F (–4.1 to –1.5° C), and the extreme low minimum was –17° F (–27.2° C). The extremes of the minimums were all recorded on the site at 7500 ft, and the extreme low temperature was 1° F higher than the all-time low temperature of –18° F (–27.8° C) recorded at 4400 ft lower elevation in the Frenchman Flat lowlands. Mean maximum temperature varied from 71.1 to 79.5° F (21.7 to 26.4° C), and the high maximum of the record was 108° F (42.2° C). The P/T ratios were from 22.2 to 32.5.

Since the trees are variously represented in this open shrub–woodland vegetation, with canopies usually not closed even where trees occur in highest density, cover and density of trees are highly variable parameters: As measured, total tree and shrub cover varied from 34.8% (where much of the surface was outcropping bedrock) to 43.9%. Maximum herbaceous perennial cover varied among the sites from 7.0 to 31.1% (to 19 species/1000 m^2), with from 69 to 99% of the cover due to perennial grasses, especially species of *Stipa (S.*

comata, S. pinetorum, or *S. thurberiana)*. Winter annuals are represented usually only near the lower limits of the association, where on one site, one year, they were present in a density of 35.6 plants/m^2 (10 species), and cover of 15.7% reflecting the large size of plants in areas of high rainfall.

Around half the plant species in the whole region are known to occur somewhere in *Artemisia*—Pinyon—Juniper communities. The great diversity of geologic substrates and erosional and depositional features permits species separation on a number of bases. The lists below combine the objectives of recognizing the widely distributed or characteristic species in this region, and the species of the apparent environmental extremes of moisture availability; they also distinguish those which appear restricted either to the single limestone mountain range on which *Artemisia*—Pinyon—Juniper occurs (Spring Mountains) or to this range and sedimentary rocks elsewhere in the region (indicated by an asterisk). Since the association consists fundamentally of *Artemisia* communities, most of the herbaceous species listed earlier for these communities occur at least to some extent in the *Artemisia tridentata*—Pinyon—Juniper and *Artemisia nova*—Pinyon—Juniper, and these are not repeated in the lists below.

Species of *Artemisia*—Pinyon—Juniper (excluding those restricted to flatrock areas, cliffs, springs, or valley bottoms with perennial streams) are the following:

Trees, Shrubs, and Cacti

Juniperus osteosperma
Ephedra viridis
Pinus monophylla
Rhus trilobata
  var. *anisophylla*
Artemisia tridentata
Artemisia nova
Chrysothamnus nauseosus
  ssp. *hololeucus*
Chrysothamnus parryi
  ssp. *nevadensis*
Chrysothamnus viscidiflorus
  ssp. *puberulus*
  ssp. *viscidiflorus*
Gutierrezia sarothrae
Perityle megalocephala
  var. *megalocephala*
Petradoria discoidea
Petradoria pumila
Stephanomeria spinosa
Tetradymia canescens
Viguiera multiflora
  var. *nevadensis*
Berberis fremontii
Coryphantha vivipara
  var. *rosea*
Echinocereus engelmannii
  var. *chrysocentrus*
Echinocereus triglochidiatus
  var. *melanacanthus*
  var. *mojavensis*
Opuntia chlorotica
Opuntia erinacea
  var. *erinacea*
Opuntia phaeacantha
  var. *major*
Opuntia polyacantha
  var. *rufispina*
Sclerocactus polyancistrus
Symphoricarpos longiflorus
Symphoricarpos parishii
Forsellesia nevadensis
Arctostaphylos pungens

Lupinus excubitus
Quercus gambelii
**Garrya flavescens*
**Eriodictyon angustifolium*
Salvia dorrii
ssp. *argentea*
Leptodactylon pungens
ssp. *pulchriflorum*
Linanthus nuttallii
Polygala subspinosa
var. *subspinosa*
Eriogonum caespitosum
Eriogonum microthecum
var. *foliosum*
var. *lapidicola*
var. *laxiflorum*
Eriogonum umbellatum
var. *dichrocephalum*
var. *subaridum*
var. *versicolor*
**Ceanothus cordulatus*
**Ceanothus greggii*
var. *vestitus*
Amelanchier utahensis
Cercocarpus ledifolius
Cowania mexicana
var. *stansburiana*
Holodiscus microphyllus
**Peraphyllum ramosissimum*
Purshia tridentata
Galium hypotrichium
ssp. *nevadense*
**Galium multiflorum*
**Galium parishii*
Ribes cereum
Ribes niveum
Ribes velutinum
Keckiella rothrockii

Herbs

Pellaea mucronata
Pellaea truncata
Pityrogramma triangularis
Lomatium parryi
Achillea millefolium
var. *lanulosa*
Agoseris glauca
var. *laciniata*
Antennaria dimorpha
Antennaria rosea
Artemisia dracunculus
Brickellia californica
Brickellia microphylla
var. *scabra*
Chaenactis douglasii
Crepis intermedia
Crepis occidentalis
Haplopappus acaulis
Hulsea vestita
ssp. *inyoensis*
Hymenopappus filifolius
var. *megacephalus*
Lygodesmia dianthopsis
Machaeranthera canescens
Cryptantha ambigua
Cryptantha confertiflora
Cryptantha flavoculata
Cryptantha gracilis
Cryptantha jamesii
var. *abortiva*
Cryptantha scoparia
Cryptantha torreyana
Cryptantha virginensis
Cryptantha watsonii
Lappula occidentalis
Lithospermum ruderale
Arabis fendleri
Arabis hirsuta
var. *pycnocarpa*
Arabis holboellii
var. *retrofracta*
Arabis microphylla
Arabis pendulina
Arabis pulchra
var. *gracilis*
**Erysimum asperum*
var. *purshii*
Lesquerella kingii
ssp. *kingii*
ssp. *latifolia*
Physaria chambersii
Streptanthus cordatus
Polanisia trachysperma
Arenaria congesta
var. *congesta*
Silene verecunda
ssp. *andersonii*
Chenopodium atrovirens
Chenopodium berlandieri
var. *zschackei*
Chenopodium fremontii
Chenopodium gigantospermum
Chenopodium leptophyllum
Chenopodium missouriense

Monolepis spathulata
**Astragalus aequalis*
Astragalus inyoensis
**Astragalus nuttallianus*
var. *imperfectus*
Astragalus oophorus
*var. *clokeyanus*
var. *oophorus*
Astragalus purshii
var. *purshii*
**Lupinus alpestris*
Lupinus argenteus
var. *tenellus*
Lupinus palmeri
Frasera albomarginata
Frasera pahutensis
Phacelia affinis
Phacelia curvipes
Phacelia hastata
Phacelia lemmonii
Phacelia tetramera
Mentzelia congesta
Mentzelia laevicaulis
Mentzelia montana
Mirabilis froebelii
Gayophytum decipiens
Gayophytum ramosissimum
Oenothera caespitosa
ssp. *marginata*
Argemone munita
**Ipomopsis aggregata*
Ipomopsis congesta
Linanthus septentrionalis
Chorizanthe brevicornu
var. *spathulata*
Eriogonum baileyi
Eriogonum cernuum
var. *cernuum*
Eriogonum esmeraldense
Eriogonum palmerianum
Eriogonum panamintense
Eriogonum racemosum
Calyptridium parryi
var. *nevadense*
Ranunculus andersonii
**Thalictrum fendleri*
Ivesia sabulosa
Galium bifolium
Galium hilendiae
ssp. *hilendiae*
ssp. *kingstonense*
**Comandra umbellata*
ssp. *californica*
Castilleja linariaefolia
Castilleja martinii
var. *clokeyi*
Collinsia callosa
Mimulus densus
Mimulus pilosus
Mimulus rubellus
Pedicularis semibarbata
Penstemon bridgesii
Penstemon confusus
Penstemon eatonii
Penstemon humilis
Penstemon kingii
Penstemon pahutensis
Penstemon palmeri
**Penstemon thompsoniae*
Scrophularia desertorum
Nicotiana attenuata
**Physalis hederaefolia*
var. *cordifolia*
Allium atrorubens
Carex occidentalis
Calochortus bruneaunis
Fritillaria atropurpurea
Fritillaria pinetorum
Agropyron spicatum
Agropyron vaseyi
Aristida longiseta
var. *robusta*
Bouteloua gracilis
Bromus anomalus
Bromus carinatus
Elymus cinereus
Koeleria cristata
Melica stricta
Muhlenbergia richardsonis
Oryzopsis micrantha
Poa fendleriana
Stipa arida
Stipa coronata
var. *depauperata*
Stipa pinetorum
Stipa thurberiana
Vulpia microstachya
var. *pauciflora*

Species that characterize the volcanic flatrock areas of especially the Rainier Mesa and southern Pahute Mesa areas, including Gold

Meadows and the northern Forty-Mile Canyon drainage (Figs. 18 and 19), but also similar areas of different rock substrates elsewhere in the region, e.g., near the base of the northern Belted Range, are the following:

Lomatium foeniculaceum
ssp. *fimbriatum*
Haplopappus nanus
Hymenoxys cooperi
Townsendia scapigera
Cryptantha humilis
var. *ovina*
Arabis holboellii
var. *pinetorum*
Draba cuneifolia
var. *cuneifolia*
Lesquerella kingii
ssp. *kingii*
Trifolium andersonii
ssp. *beatleyae*
Phacelia saxicola
Camissonia pterosperma
Camissonia pusilla
Gayophytum diffusum
ssp. *parviflorum*
Plantago purshii
var. *oblonga*
Gilia aliquanta
Gilia brecciarum
Microsteris gracilis
ssp. *humilis*
Navarretia breweri
Chorizanthe watsonii
Eriogonum caespitosum
Eriogonum microthecum
var. *lapidicola*
Polygonum douglasii
var. *johnstonii*
Lewisia rediviva
Delphinium andersonii
Ivesia sabulosa
Lithophragma tenellum
Collinsia parviflora
Mimulus densus
Mimulus suksdorfii
Carex douglasii
Zigadenus paniculatus
Poa sandbergii
Stipa pinetorum

Species of cliff ledges and crevices (Fig. 22):

Pityrogramma triangularis
Haplopappus cuneatus
Haplopappus watsonii
Heterotheca villosa
var. *hispida*
Halimolobos diffusa
var. *jaegeri*
Thelypodium laxiflorum
Phacelia mustelina
Phacelia peirsoniana
Eriogonum heermannii
var. *argense*
Montia perfoliata
**Petrophytum caespitosum*
Purpusia saxosa

Species of springs and seepage areas, from around 6000 to 8000 ft, and mountain meadows or lowlands along perennial streams of the Spring Mountains and/or Kawich Range:

Trees, Shrubs, and Vines

Chrysothamnus nauseosus
ssp. *albicaulis*
Clematis ligusticifolia
Prunus virginiana
var. *melanocarpa*
Rosa woodsii
Populus fremontii
Populus trichocarpa
Salix exigua
var. *stenophylla*
Salix lasiolepis
Salix rigida
var. *watsonii*
Ribes aureum

Herbs

Equisetum hyemale
Equisetum laevigatum
Amaranthus blitoides
**Angelica kingii*
Osmorhiza occidentalis
Apocynum sibericum var. *salignum*
Arnica parryi ssp. *sonnei*
Artemisia biennis
Aster adscendens
Conyza canadensis
Erigeron divergens
Senecio integerrimus var. *exaltatus*
Plagiobothrys scouleri var. *penicillatus*
Descurainia californica
Descurainia obtusa
Hutchinsia procumbens
Cleome lutea
Sagina saginoides
Monolepis nuttalliana
Trifolium andersonii ssp. *beatleyae*
Trifolium monanthum
Trifolium wormskioldii
Epilobium ciliatum
Epilobium glandulosum var. *tenue*
Oenothera hookeri ssp. *angustifolia*
Gilia tenerrima
Leptodactylon caespitosum
Rumex salicifolius
Montia chamissoi
Aquilegia shockleyi
Ranunculus cymbalaria var. *saximontanus*
Geum macrophyllum
Potentilla biennis
Potentilla gracilis ssp. *nuttallii*
Mimulus guttatus
Veronica americana
Solanum triflorum
Urtica holosericea
Verbena bracteata
Viola nephrophylla
Allium bisceptrum
Carex alma
Carex aurea
Carex douglasii
Carex lanuginosa
Carex microptera
Carex nebrascensis
Carex scopulorum
Eleocharis macrostachya
Eleocharis parishii
Scirpus microcarpus
Iris missouriensis
Sisyrinchium halophilum
Juncus balticus
Juncus bufonius
Juncus ensifolius
Juncus longistylis
Juncus nodosus
Lemna minuta
Smilacina stellata
Agrostis exarata var. *monolepis*
Agrostis scabra
Alopecurus aequalis
Elymus cinereus
Elymus triticoides
Hordeum brachyantherum
Hordeum jubatum
Poa nevadensis
Zannichellia palustris

White Fir

White Fir (*Abies concolor*), largely replacing *Juniperus osteosperma*, begins at around 8000 ft near the base of Bald Mountain, high peak of the Groom Range, and at around 7000 ft in the major canyon of the northwest slope of the Spring Mountains, below Charleston Peak. It is not known from high peaks or ridges above 8000 ft elsewhere in the region, e.g., Wheelbarrow Peak (8400 ft) in the central Belted Range and the Kawich Peak area of the northern Kawich Range (9400 ft).

On Bald Mountain, a volcanic peak (Fig. 28), *Pinus monophylla* may continue for a few hundred feet up the mountain slopes, with *Abies concolor*; locally Limber Pine (*Pinus flexilis*) is an associated tree species. The Fir or Fir—Pinyon Pine open forests are discontinuous along the mountain contours, their continuity broken by large areas of loose, coarse talus alternating with irregular ribbons of forest. Peak of the mountain (9380 ft) is treeless and covered by a low stand of *Ceratoides lanata*.

In Clark Canyon of the limestone Spring Mountains, the Fir communities occur in a wholly different physiographic setting—a mesic canyon, near the head of which is a large perennial spring. *Abies concolor*, together with Western Yellow Pine (*Pinus ponderosa*), make up a closed-canopy forest on the floor of the canyon from around 7000 to 9300 ft at the head of the canyon. *Pinus monophylla* is a canopy species to around 8000 ft, and Rocky Mountain Juniper (*Juniperus scopulorum*) is an associated tree species, especially in the upper canyon. The tree species are variously represented locally along the canyon as Fir—Pinyon, Yellow Pine—Fir, or other variations of the dominants. *Cercocarpus ledifolius* (Mountain Mahogany) is a prominent associated small tree throughout.

Many of the species in the Fir communities are altitudinal extensions from *Artemisia*—Pinyon—Juniper. There is, however, a flora which distinguishes or otherwise characterizes these high-elevation communities. In the lists that follow, those known in the region of Fig. 2 only from the Spring Mountains are indicated by an asterisk.

Trees and Shrubs

**Juniperus scopulorum*
Abies concolor
Pinus flexilis
**Pinus ponderosa*
Acer glabrum
  var. *diffusum*
**Chrysothamnus parryi*
  ssp. *asper*
Petradoria discoidea
Petradoria pumila
Berberis fremontii
**Berberis repens*
Sambucus coerulea
Symphoricarpos vaccinioides
Forsellesia nevadensis
**Arctostaphylos pungens*
**Garrya flavescens*
**Eriodictyon angustifolium*
Leptodactylon pungens
  ssp. *pulchriflorum*
Eriogonum microthecum
  var. *laxiflorum*
Eriogonum umbellatum
  var. *subaridum*
  var. *versicolor*
Ceanothus greggii
  var. *vestitus*
Ceanothus martinii
Amelanchier utahensis
Cercocarpus ledifolius
Holodiscus microphyllus
**Jamesia americana*
**Philadelphus microphyllus*
  ssp. *stramineus*
Ribes cereum
Ribes velutinum

Herbs

Lomatium foeniculaceum
ssp. *macdougalii*
**Apocynum androsaemifolium*
Antennaria rosea
Artemisia dracunculus
Chaenactis douglasii
**Cirsium nidulum*
Crepis intermedia
**Hymenopappus filifolius*
var. *eriopodus*
**Hymenoxys lemmonii*
Senecio multilobatus
Senecio spartioides
**Solidago sparsiflora*
**Stephanomeria tenuifolia*
Cryptantha jamesii
var. *abortiva*
Lappula occidentalis
Arabis fendleri
Arabis fernaldiana
var. *stylosa*
Arabis holboellii
var. *pinetorum*
Arabis inyoensis
Arabis microphylla
Arabis pendulina
Descurainia pinnata
ssp. *halictorum*
**Erysimum asperum*
var. *purshii*
Lesquerella kingii
ssp. *latifolia*
Physaria chambersii
Paronychia jamesii
**Euphorbia robusta*
Corydalis aurea
var. *aurea*
Phacelia hastata
Hedeoma nanum
ssp. *nanum*
Monardella odoratissima
**Calyophus lavandulifolius*
**Oenothera brachycarpa*
**Ipomopsis aggregata*
Linanthus nuttallii
Eriogonum cernuum
var. *cernuum*
Eriogonum microthecum
var. *laxiflorum*
Eriogonum panamintense
Eriogonum umbellatum
var. *subaridum*
var. *versicolor*
Ranunculus andersonii
**Thalictrum fendleri*
**Potentilla propinqua*
Galium hilendiae
ssp. *nevadense*
**Galium parishii*
Heuchera rubescens
var. *pachypoda*
Castilleja linariaefolia
Castilleja martinii
var. *clokeyi*
Pedicularis semibarbata
Penstemon bridgesii
Penstemon eatonii
Penstemon palmeri
**Valeriana puberulenta*
Smilacina stellata
Agropyron spicatum
Bromus anomalus
Bromus carinatus
Muhlenbergia richardsonis
Poa fendleriana
Stipa lettermannii

Species of wet soils, at or near Clark Canyon Spring in the Spring Mountains, at 9000 ft:

Equisetum laevigatum
**Juniperus communis*
var. *depressa*
**Epilobium angustifolium*
ssp. *circumvagum*
Epilobium glandulosum
var. *tenue*
**Dodecatheon redolens*
**Kelloggia galioides*
**Habenaria sparsiflora*
Carex microptera
Carex praegracilis
Juncus longistylis
Agropyron trachycaulum
**Bromus ciliatus*
Deschampsia caespitosa

Artemisia—Cercocarpus

On the high mountain slopes and ridges of the Kawich Range, in the upper Eden Creek drainage of the Kawich Peak area, White Fir communities are absent; Limber Pine *(Pinus flexilis)*, an associate of White Fir elsewhere, is known from a few small, scattered trees on the slopes. Here, the high-elevation equivalent of the Fir communities is low *Artemisia tridentata* vegetation, with scattered groves of *Cercocarpus ledifolius* (Mountain Mahogany) beginning along an irregular boundary with *Artemisia*—Pinyon—Juniper at around 8000 ft. Although from a distance it has the aspect of a large fire scar over these steep slopes and high ridges at 8000 to 9300 ft, there is no evidence to indicate that this plant association is of fire origin. On the contrary, it appears to be a well-defined kind of vegetation, in equilibrium with the climate and soils of this part of the Kawich Range [bedrock is mapped as rhyolitic welded ash-flows of the tuff of White Blotch Spring (Cornwall, 1972)].

Plants of many of the species grow in and around the *Cercocarpus* thickets, but others are identified with the unbroken expanses of *Artemisia tridentata*. Most species occur elsewhere in upper *Artemisia*—Pinyon—Juniper, or especially the Fir communities of Bald Mountain, but some (marked with an asterisk) are known only from here or also from the Spring Mountains according to Clokey (1951), and the latter are designated by a dagger (†). A recently described endemic, *Penstemon pudicus*, occurs in profusion on these slopes, and is unknown outside of the *Artemisia—Cercocarpus* and peripheral *Artemisia*—Pinyon—Juniper of the Kawich Range. The following species characterize this plant association:

Trees and Shrubs

Artemisia tridentata
Chrysothamnus greenei
ssp. *filifolius*
Chrysothamnus viscidiflorus
ssp. *puberulus*
ssp. *viscidiflorus*
Gutierrezia sarothrae
Tetradymia canescens
Sambucus coerulea
Symphoricarpos vaccinioides
Eriogonum microthecum
var. *laxiflorum*
Eriogonum racemosum
Eriogonum umbellatum
var. *subaridum*
Amelanchier pallida
Cercocarpus ledifolius

Herbs

**Selaginella watsonii*
†*Cystopteris fragilis*
Crepis intermedia
†*Erigeron clokeyi*
Chenopodium atrovirens
Lesquerella kingii
ssp. *kingii*
Arenaria congesta
var. *subcongesta*
Silene verecunda
ssp. *andersonii*
Stellaria jamesiana

Astragalus calycosus
†*Astragalus platytropis*
Lupinus argenteus var. *tenellus*
Phacelia hastata
Monardella odoratissima
Gayophytum racemosum
Ipomopsis congesta
**Phlox pulvinata*
Polygonum douglasii var. *johnstonii*
**Androsace septentrionalis* var. *puberulenta*
Ivesia sabulosa
Heuchera rubescens var. *pachypoda*
Castilleja linariaefolia
**Castilleja viscidula*
**Penstemon pudicus*

DISTURBED VEGETATION

The region abounds in large and smaller areas where the original vegetation cover has been wholly or partially removed and the soils have been grossly modified. In the Mojave and transition desert areas, recovery of vegetation is an exceedingly slow process, to be considered in terms of centuries of time. At the higher elevations and higher rainfalls of the Great Basin Desert, recovery is a matter of at least several decades.

Characteristic of the whole region is the continuing occupancy of disturbed sites by species of the original vegetation of the area, with population expansions in some species and disappearance or reduced populations in other species. But essentially nowhere is there the intercalation of floristically discrete assemblages—seral stages between pioneer and climax communities—which occurs in orderly and predictable sequence in humid regions (plant succession).

Autosuccession, which characterizes the processes of recovery of desert vegetation, was pointed out many years ago by Muller (1940) and Shreve (1942). At the lower and middle elevations in this region, soils are scarcely modified by the soil-forming processes (Sec. 1); hence the substrates are usually not fundamentally different after disturbance than before. In the absence of climatic change, the only species able to occupy the soils after disturbance are still the original species, i.e., most native species are the pioneer species of the region.

These relationships are more complex at the higher elevations, where increases in rainfall and decreases in temperature are reflected by increased organic matter accumulations, and the whole complex of variables contributes to increased rates of the soil-forming processes at the higher elevations. Removal of the tree and shrub vegetation and physical disturbance of the soils present a greatly modified environment in which seedlings must become established. However, even here, although the species representation may differ markedly from that of the original community, it is usually species of the original community that become reestablished and survive in the modified environment.

A number of native species are known only from disturbed sites, but most of these are inferred to belong to the environment of washes, or seepage areas, since the populations are usually temporary. Among these are the local, more or less transient, roadside populations of most *Grindelia* and *Helianthus* species.

INTRODUCED SPECIES

Around 125 of the species known from the region (nearly 12% of the total flora) are introduced, and nearly all of them are Eurasian species. Most occur on disturbed soils that are either temporarily or continuously moist through irrigation practices, and fail to become established on the sites of their initial invasion.

Only a few species have successfully invaded undisturbed communities; these are all winter annuals—*Bromus rubens*, *Bromus trinii*, *Schismus arabicus*, and *Erodium cicutarium*. Of these, only *Bromus rubens* has become so integrated into the communities, on so many sites and over so large an area, that, were its history of introduction not known, it would be inferred to be one of the many native winter annual species. This is also the case with *Bromus trinii* and *Schismus arabicus*, on the local sites of their occurrence. *Erodium cicutarium* is widely distributed on disturbed sites, but also occurs in undisturbed communities, where it is usually restricted to microsites with soils in a disturbed condition. All are associated with the middle elevations of the region, especially in *Coleogyne* vegetation. At the higher elevations, *Bromus tectorum* is everywhere the *Bromus* of disturbed *Artemisia* or *Artemisia*–Pinyon–Juniper vegetation, rarely occurring (except sparingly) in the undisturbed communities.

In addition to *Bromus rubens* and *B. tectorum*, the successful invaders of disturbed soils of the region are the annual *Salsola* species—*S. paulsenii* at the lower elevations, *S. iberica* at the higher elevations, and both species and their hybrids at the middle elevations (Beatley, 1973a). In the Mojave and transition desert areas, their growing season, from time of germination in the spring until maturation in the autumn, coincides with the period of apparent dormancy of most native species with which they share the sites of their occurrence. The outstanding success of these species on disturbed grounds over all of southwestern United States is inferred to be related in part to their occupancy of an environmental niche that is often unoccupied by the native desert species (Beatley, 1974b). On the Test Site, large summer populations are the conspicuous plants in areas disturbed by testing activities, especially in Yucca Flat.

At the lower elevations, *Tamarix ramosissima* is a well established and reproducing species of disturbed, moist, usually heavy soils; it is particularly common in parts of Pahrump Valley, and occurs locally on the Test Site especially on or near the playas. At the higher elevations, *Bassia hyssopifolia*, *Kochia iranica*, and *Sisymbrium altissimum* are locally common persistent weeds of moist, disturbed soils (and *Bassia* is especially common also along irrigation channels in Ash Meadows). *Halogeton glomeratus* is most common on disturbed soils of lowland *Atriplex* communities in closed drainage basins, where it is especially predictable near the playas.

Polypogon monspeliensis, *Agrostis semiverticillata*, and *Veronica anagallis-aquatica* are predictable at most springs of the region; others less predictable, but well established at certain springs, are *Agrostis stolonifera* and *Poa pratensis.*

For most other species, introduced from humid regions, occurrence is local and survival is of short duration only.

ENDANGERED AND THREATENED SPECIES

As a result of exploitive utilization of the land by private interests and mission-oriented land use by government agencies—all of which is destructive to the natural biological/environmental systems—a number of plant species are today endangered or threatened in the region. These have been cited in a recent report to the Congress of the United States (Ripley, 1975).

Of the Test Site species placed on the current Endangered Species list, *Astragalus beatleyae* is presently the most critically endangered; it is known from but two local sites on northwestern and northeastern Pahute Mesa, both of which are in areas intensively affected by underground testing and construction activities. *Astragalus nyensis* and *Phacelia parishii* (to be added to the list), both extremely rare species, are perhaps now extinct on the Test Site, since neither was found during the 16-year period of collecting; both were collected in 1941 by Rupert Barneby and Dwight Ripley (Plant Collections, Part II) near the playa of Frenchman Flat, a site now occupied by a large gravel quarry.

The general disturbances associated with the testing program on Rainier Mesa and southern Pahute Mesa place in nearly continuous jeopardy the populations of *Frasera pahutensis*, *Penstemon pahutensis*, and *Trifolium andersonii* ssp. *beatleyae*, all recently described endemics. Yucca Flat is perhaps the home of the southern Nevada endemic *Mirabilis pudica*, and nowhere in the region has the landscape been more intensively and thoroughly modified than on the floor of this basin.

Camissonia megalantha, a narrow endemic of the slopes near Cane Spring and parts of the volcanic mountain mass marked by French Peak in northern Frenchman Flat, is endangered by the frequent visitors to Cane Spring, and on the mountain uplands (along with the rare *Astragalus funereus*) by an ammunition dump placed on the site of the principal population known; other local populations of this species, and the major population of *Phacelia beatleyae*, are much endangered by test operations in northern Frenchman Flat, nearby the mountain canyons where these two species occur.
species occur.

Other species endangered, and to be added to the list (some are currently designated as Threatened Species),, are *Arabis dispar* and *Arabis shockleyi*, both known in the whole region only from the northern Eleana Range, over or almost over tunnel excavations used for underground nuclear detonations. *Lupinus uncialis* is a rare species, and known from one plant on northwestern Pahute Mesa, on a site today highly disturbed by test operations. Endangered by any movements of rocks above drainage are cliff species, among which on the Test Site is especially *Haplopappus watsonii*, occurring in one of its few known localities in western United States in a canyon near the confluence of Rainier and Pahute mesas. *Thelypodium laxiflorum* and *Phacelia mustelina*, both rare species and restricted to the bases or faces of cliffs, are present today in the Pahute and Rainier mesa areas in a few isolated canyons, mostly amid rubble from the collapse of canyon walls.

Beyond the Test Site, every species of the spring areas, or nearby uplands, of northern and eastern Ash Meadows is gravely endangered. Here, a large agricultural operation utilizing the water from the springs has in recent years plowed the vegetation from many square miles (about half of Ash Meadows), without fencerows which might serve as refugia; other areas are heavily trampled and grazed by unconfined livestock of local residents. Plant species whose survival is most endangered are the endemics *Astragalus phoenix*, *Centaurium namophilum*, *Mentzelia leucophylla*, *Ivesia eremica*, and *Grindelia fraxino-pratensis*, all known from several sites of the spring areas, and *Nitrophila mohavensis*, known from only one site at the south end of the otherwise totally obliterated Carson Slough, which was the major drainage channel of western Ash Meadows.

Elsewhere in the region disappearance of *Penstemon arenarius* is imminent on the single site of its known occurrence in the region: The population, in southern Sarcobatus Flat below Tolicha Peak, is near a corral, where all plants are heavily browsed and trampled by livestock. *Lathyrus hitchcockianus*, known primarily from a particularly vulnerable site in the mining district of the Bullfrog Hills, and

Penstemon pudicus, an endemic of the high elevations of the Kawich Range, are endangered because of their rarity and potential for destruction by nearby mining operations. *Penstemon pudicus* and other species of the higher elevations of the Kawich Range are also more or less threatened or endangered by intensive grazing by the wild horses that live on the range.

Other species of the region designated Threatened Species on the official list are the following:

On the Test Site:

Haplopappus brickellioides
Perityle megalocephala var. *intricata*
Coryphantha vivipara var. *rosea*
Arctomecon merriamii
Gilia nyensis
Gilia ripleyi
Linanthus arenicola
Eriogonum concinnum
Agave utahensis var. *eborispina*

Elsewhere in the region:

Enceliopsis nudicaulis var. *corrugata* (Ash Meadows)
Machaeranthera ammophila (Ash Meadows)
Cordylanthus tecopensis (Ash Meadows)
Astragalus aequalis (Spring Mountains)
Astragalus oophorus var. *clokeyanus* (Spring Mountains)
Eriogonum bifurcatum (Pahrump Valley)
Salvia funerea (Stewart Valley)
Lupinus holmgrenanus (Sarcobatus Flat)

Unless the species are legally protected, or the prescriptions for southern Nevada land use are modified, it seems predictable that within the next decade some of the presently endangered species will become extinct in the region, and that many additional species will be endangered and threatened.

LITERATURE CITED

Abrams, LeRoy. 1923—1951. Illustrated flora of the Pacific states. Vols. 1-3. Stanford Univ. Press, Stanford, Calif. 538 pp., 635 pp., 866 pp.

——, and Roxanna S. Ferris. 1960. Illustrated flora of the Pacific states, Vol. 4. Stanford Univ. Press, Stanford, Calif. 732 pp.

Albers, J. P., and J. H. Stewart. 1972. Geology and mineral deposits of Esmeralda County, Nevada. Bull. 78, Nevada Bur. of Mines and Geology. Univ. of Nevada, Reno. 80 pp. (and plates).

Beatley, Janice C. 1962. Vascular plants of the U. S. Atomic Energy Commission's Nevada Test Site, Nye County, Nevada. UCLA 508. Lab. Nuclear Med. & Radiat. Biol., Univ. of California, Los Angeles. 33 pp.

——, 1965. Ecology of the Nevada Test Site I. Geographic and ecologic distributions of the vascular flora (annotated checklist). UCLA 12-553. Lab. Nuclear Med. & Radiat. Biol., Univ. of California, Los Angeles. 69 pp.

——, 1966. Ecological status of introduced brome grasses (*Bromus* spp.) in desert vegetation of southern Nevada. Ecology 47: 548-554.

——, 1967. Survival of winter annuals in the northern Mojave Desert. Ecology 48: 745-750.

——, 1969a. Vascular plants of the Nevada Test Site, Nellis Air Force Range, and Ash Meadows (northern Mojave and southern Great Basin Deserts, south-central Nevada). UCLA 12-705. Lab. Nuclear Med. & Radiat. Biol., Univ. of California, Los Angeles. 122 pp.

——, 1969b. Biomass of desert winter annual plant populations in southern Nevada. Oikos 20: 261-273.

——, 1970. Perennation in *Astragalus lentiginosus* and *Tridens pulchellus* in relation to rainfall. Madroño 20: 326-332.

——, 1971a. Ecologic and geographic distributions of the vascular plants of southern Nye County, and adjacent parts of Clark, Lincoln, and Esmeralda counties, Nevada. UCLA 12-800 (suppl. to 12-705). Lab. Nuclear Med. & Radiat. Biol., Univ. of California, Los Angeles. 49 pp.

——, 1971b. Vascular plants of Ash Meadows, Nevada. UCLA 12-845. Lab. Nuclear Med. & Radiat. Biol., Univ. of California, Los Angeles. 59 pp.

——, 1973a. Russian-thistle (*Salsola*) species in western United States. J. Range Manage. 26: 225-226.

——, 1973b. Check list of vascular plants of the Nevada Test Site and central-southern Nevada. COO-2307-4. Dept. of Biological Sciences, Univ. of Cincinnati, Cincinnati, Ohio. 42 pp.

——, 1974a. Effects of rainfall and temperature on the distribution and behavior of *Larrea tridentata* (Creosote-bush) in the Mojave desert of Nevada. Ecology 55: 245-261.

——, 1974b. Phenological events and their environmental triggers in Mojave Desert ecosystems. Ecology 55: 856-863.

——, 1975. Climates and vegetation pattern across the Mojave/Great Basin Desert transition of southern Nevada. Amer. Midl. Natur. 93: 53-70.

Blankennagel, Richard K., and J. E. Weir, Jr. 1973. Geohydrology of the eastern part of Pahute Mesa, Nevada Test Site, Nye County, Nevada. Geol. Surv. Prof. Paper 712-B. Govt. Printing Off., Washington. 35 pp. (and plates).

Bouyoucos, George J. 1962. Hydrometer method improved for making particle-size analyses of soils. Agron. J., 54: 464-465.

Brown, James H. 1971. The desert pupfish. Sci. Amer. 225: 104-110.

Clokey, Ira W. 1951. Flora of the Charleston Mountains, Clark County, Nevada. Univ. of California Press, Berkeley. 274 pp.

Cornwall, Henry R. 1972. Geology and mineral deposits of southern Nye County, Nevada. Bull. 77, Nevada Bur. of Mines and Geology. Univ. of Nevada, Reno. 49 pp. (and plates).

Coville, Frederick V. 1893. Botany of the Death Valley Expedition. Contr. U. S. Nat. Herb. 4. Govt. Printing Office, Washington. 363 pp.

Cronquist, Arthur, Arthur H. Holmgren, Noel H. Holmgren, and James L. Reveal. 1972. Intermountain flora. Vol. 1. Hafner Publ. Co., Inc., New York. 270 pp.

Denny, Charles S., and Harald Drewes. 1965. Geology of the Ash Meadows Quadrangle, Nevada—California. Geol. Surv. Bull. 1181-L. Govt. Printing Office, Washington. 56 pp. (and plate).

Ekren, E. B., 1968. Geologic setting of Nevada Test Site and Nellis Air Force Range. *In* Nevada Test Site. Geol. Soc. Amer., Memoir 110 (Edwin B. Eckel, Edit.). Geol. Soc. of Amer., Inc., Boulder, Colorado. pp. 5-19.

——, R. E. Anderson, C. L. Rogers, and D. C. Noble. 1971. Geology of northern Nellis Air Force Base Bombing and Gunnery Range, Nye County, Nevada.

Geol. Surv. Prof. Paper 651. Govt. Printing Office, Washington. 91 pp. (and plates).

Fenneman, Nevin H. 1931. Physiography of western United States. McGraw-Hill Book Co., New York. 534 pp.

Fernald, Arthur T., George S. Corchary, and William P. Williams. 1968. Surficial geologic map of Yucca Flat, Nye and Lincoln counties, Nevada. Misc. Geol. Invest., Map I-550. U. S. Geol. Surv., Washington.

Hunt, Charles B. 1966. Plant ecology of Death Valley, California, Geol. Surv. Prof. Paper 509. Govt. Printing Office, Washington. 68 pp. (and plates).

Johnson, Mike S., and Donald E. Hibbard. 1957. Geology of the Atomic Energy Commission Nevada proving grounds area, Nevada. Geol. Surv. Bull. 1021-K. Govt. Printing Office, Washington. pp. 333-384 (and plate).

Juhren, Marcella, F. W. Went, and Edwin Phillips. 1956. Ecology of desert plants. IV. Combined field and laboratory work on germination of annuals in the Joshua Tree National Monument, California. Ecology 37: 318-330.

Kearney, Thomas H., and Robert H. Peebles. 1960. Arizona flora. Univ. of California Press, Berkeley. 1085 pp.

Leavitt, V. D. 1970. Soil survey of Area 18, Nevada Test Site. Report SWRHL-74r, Southwest. Radiol. Health Lab., Las Vegas. 119 pp.

——, 1974. Soil surveys of five plutonium contaminated areas on the test range complex in Nevada. NERC-LV-539-28, Environmental Protection Agency, Las Vegas. 84 pp.

——, and B. J. Mason. 1971. Soil survey of Area 15, Nevada Test Site. Report SWRHL-106r, Southwest. Radiol. Health Lab., Las Vegas. 48 pp.

Lloyd, Robert M., and Richard S. Mitchell. 1973. A flora of the White Mountains, California and Nevada. Univ. of California Press, Berkeley. 202 pp.

Longwell, C. R., E. H. Pampeyan, Ben Bowyer, and R. J. Roberts. 1965. Geology and mineral deposits of Clark County, Nevada. Bull. 62, Nevada Bur. Mines. Univ. of Nevada, Reno. 218 pp. (and plates).

Miller, Robert R. 1948. Cyprinodont fishes in the Death Valley system of eastern California and southwestern Nevada. Univ. Mich. Mus. Zool. Misc. Publ. 68. 155 pp.

Minckley, W. L., and James E. Deacon. 1968. Southwestern fishes and the enigma of "Endangered Species." Science 159: 1424-1432.

Muller, Cornelius H. 1940. Plant succession in the *Larrea—Flourensia* climax. Ecology 21: 206-212.

Munz, Philip A. 1968. Supplement to a California Flora. Univ. of California Press, Berkeley. 224 pp.

——, 1974. A flora of southern California. Univ. of California Press, Berkeley. 1086 pp.

——, and David D. Keck. 1959. A California flora. Univ. of California Press, Berkeley. 1681 pp.

Ripley, S. Dillon. 1975. Report on endangered and threatened plant species of the United States, presented to the Congress of the United States of America by the Secretary, Smithsonian Institution. Comm. on Merchant Marine and Fisheries, Serial No. 94-A. House Document No. 94-51. Govt. Printing Office, Washington. 200 pp.

Romney, E. M., V. Q. Hale, A. Wallace, and others. 1973. Some characteristics of soil and perennial vegetation in northern Mojave Desert areas of the Nevada Test Site. UCLA 12-916. Lab. Nuclear Med. & Radiat. Biol., Univ. of California, Los Angeles. 340 pp.

Shreve, Forrest. 1936. The plant life of the Sonoran Desert. Sci. Monthly 42: 195-213.

——, 1942. The desert vegetation of North America. Bot. Rev. 8: 195-246.

——, 1951. Vegetation of the Sonoran Desert. Carnegie Institution of Washington, Publ. 591, Washington, D. C. 192 pp. (Later, *In* Shreve, Forrest, and Ira L. Wiggins. 1964. Vegetation and flora of the Sonoran Desert, Vol. 1. Stanford Univ. Press, Stanford, Calif.)

Spalding, Volney M. 1909. Distribution and movements of desert plants. Carnegie Institution of Washington, Publ. 113, Washington, D. C. 114 pp.

Tidestrom, Ivar. 1925. Flora of Utah and Nevada. Contr. U. S. Nat. Herb. 25. Govt. Printing Office, Washington. 665 pp.

Tschanz, C. M. and E. H. Pampeyan. 1970. Geology and mineral deposits of Lincoln County, Nevada. Bull. 73, Nevada Bur. Mines. Univ. of Nevada, Reno. 187 pp. (and plates).

Wallace, Arthur, and E. M. Romney. 1972. Radioecology and ecophysiology of desert plants at the Nevada Test Site. TID-25954. U. S. Atomic Energy Commission, Office of Information Services. Nat. Tech. Inform. Serv., Springfield, Va. 439 pp.

Went, F. W. 1948. Ecology of desert plants. I. Observations on germination in the Joshua Tree National Monument, California. Ecology 29: 242-253.

——, 1949. Ecology of desert plants. II. The effect of rain and temperature on germination and growth. Ecology 30: 1-13.

——, 1953. The effects of rain and temperature on plant distribution in the desert. Desert Research Proc., Res. Counc. Israel Spec. Publ. 2, pp. 230-240. Jerusalem.

——, 1955. The ecology of desert plants. Sci. Amer. 192: 68-75.

——, and M. Westergaard. 1949. Ecology of desert plants. III. Development of plants in the Death Valley National Monument, California. Ecology 30: 26-38.

Winograd, Isaac J. and William Thordarson. 1975. Hydrogeologic and hydrochemical framework, south-central Great Basin, Nevada-California, with special reference to the Nevada Test Site. Geol. Surv. Prof. Paper 712-C. Govt. Printing Office, Washington. 126 pp. (and plates).

Fig. 5. *Larrea—Ambrosia* vegetation of sandy soils of west Jackass Flats, from 3000 ft looking north (upslope) toward the south face of Shoshone Mountain. Dead stalks are those of winter annual plants, persisting into 1975 from the extraordinary growing season of 1973.

Fig. 6. *Larrea—Ambrosia,* with *Yucca schidigera* (Mojave Yucca), northern Mercury Valley, at 3300 ft, on the south-facing bajada below Red Mountain of the west Spotted Range. Looking southwest across Mercury Valley toward the Spring Mountains (background).

Fig. 7. *Larrea tridentata* (Creosote-bush), with codominant *Ambrosia dumosa* (Bur-Sage), *Grayia spinosa* (Hop-Sage), *Ceratoides lanata* (Winter-Fat), and *Lycium pallidum* (bottom foreground, left of center), at 3500 ft, on south-facing bajada below volcanic Skull Mountain (background), east Rock Valley. Soils are sand, with dark boulders and gravels throughout.

Fig. 8. *Larrea—Lycium—Grayia*, west-central Frenchman Flat. On this site, a claypan immediately underlies the soil pavement, and broad bare areas occur between the shrub-clumps. Looking northeast toward volcanic hills of the Halfpint Range.

Fig. 9. Cane Spring area at base of north slope of Skull Mountain, west Frenchman Flat drainage, 4000 ft. Foreground is *Typha domingensis* (Cat-tail), on the slope is *Atriplex canescens* (Four-winged Saltbush), and the trees are *Salix gooddingii* (Willow).

Fig. 10. Southern Frenchman Flat, west of playa at 3120 ft, looking south toward the limestone—dolomite Mercury Ridge (left) and Red Mountain (right) of the west Spotted Range, with the northern Spring Mountains in the far distance. *Larrea tridentata* (Creosote-bush) vegetation is in foreground and also below the mountains (midground), separated by a zone of *Lycium pallidum—Grayia* (and *Atriplex canescens*) marked by the east-west sharp boundaries of the *Larrea* vegetation.

Fig. 11. *Atriplex confertifolia* (Shadscale) vegetation and upslope sharp boundaries with *Larrea tridentata* (Creosote-bush) vegetation, north of Frenchman Flat playa. Photo is a composite, taken from 3100 ft near the playa, and includes on the left, the east side of French Peak mountain (volcanic tuffs), and center, Raysonde Buttes (between Scarp and Nye Canyons), also volcanic tuffs, all belonging to the Halfpint Range; at the right is the southwestern high peak of the limestone—dolomite Buried Hills.

Fig. 12. Southwestern Yucca Flat, at 4200 ft, looking east toward the playa (upper right) at 3900 ft, and the Halfpint Range (background). Vegetation is *Coleogyne—Grayia* (in the ecotone between *Coleogyne ramosissima* (Blackbrush) and *Grayia—Lycium*) with *Yucca brevifolia* (Joshua-Tree). Cactus, lower right, is *Opuntia echinocarpa*, the widely distributed cactus of the middle elevations.

Fig. 13. Southwestern Yucca Flat, at 4100 ft, looking northeast across *Grayia—Lycium* vegetation, with scattered *Yucca brevifolia* (Joshua-Tree) on the basin floor. Oak Spring Butte (6965 ft), at the south end of the Belted Range, is the flat-topped physiographic feature of upper left; the stratified limestone—dolomite hill (upper right) is Banded Mountain of the Halfpint Range. Valley floor is at mostly 4000 to 4500 ft.

Fig. 14. *Coleogyne ramosissima* (Blackbrush) vegetation on hills of Mine Mountain, at 5200 ft; looking northeast from near summit of Mine Mountain, across the lowlands of Yucca Flat. Oak Spring Butte, southern Belted Range, is at upper left; front range of hills, upper right, is Banded Mountain of the Halfpint Range; and background, upper right, is the Groom Range, including its high peak, Bald Mountain (9380 ft). See also Fig. 28.

Fig. 15. *Coleogyne ramosissima* (Blackbrush) vegetation (foreground) on valley floor of northern Mid Valley, at 4800 ft, below east face of Shoshone Mountain (background). Midground (formerly *Coleogyne*) and mountain slopes (formerly *Artemisia* and *Artemisia*—Pinyon—Juniper) were burned in 1959 and are now covered by dense stands of *Bromus rubens* and *B. tectorum*. Photo June 1972.

Fig. 16. Volcanic cliffs of the Sugar Loaves area, west Eleana Range and eastern Forty-Mile Canyon drainage, weathering forms of the Paintbrush Tuff capped by the Stockade Wash Member; base of cliffs at 5500 ft. On the light-colored loose sands from these tuffs, several regionally rare or endemic plant species occur (*Gilia nyensis, Lesquerella ludoviciana, Cymopterus ripleyi, Oenothera pallida*); *Atriplex canescens* and *Eriogonum kearneyi* var. *kearneyi* are the most common shrub and suffrutescent species.

Fig. 17. Volcanic buttes of the west side of the Eleana Range; *Artemisia tridentata* (Big Sagebrush) in foreground, butte slopes usually *Artemisia nova* (Black Sagebrush), with scattered Pinyon and Juniper trees. Base of buttes, around 5500 ft.

Fig. 18. Looking south from volcanic flatrock area (foreground and midground), at 6300 ft, northern Forty-Mile Canyon drainage below south rim of Pahute Mesa, where the flora contains a number of vernal species of restricted distributions. Bedrock is a densely welded member of the Timber Mountain Tuff. Background: Big Butte and Sugar Loaves (left half) and northern Shoshone Mountain (right half).

Fig. 19. Looking east from Fig. 18, showing continuation of flatrock area in relation to Rainier Mesa, the southwestern terminus of the Belted Range; at the north end (extreme left) is Doe Point, one of the two high-elevation (7679 ft) points on the Mesa. Vegetation is *Artemisia*—Pinyon—Juniper.

Fig. 20. Looking southwest from west face of Rainier Mesa (along Holmes Rd., at 7200 ft). From background to foreground: Grapevine Mountains (high peak, 8738 ft) 50 miles (80 km) distant (upper right); Timber Mountain (upper center and left), high peak 7445 ft; below Timber Mountain, volcanic cone and basaltic lava flow of Buckboard Mesa (around 5400 ft); south end of Twin Peaks (extreme right, center); foreground, *Artemisia*—Pinyon—Juniper of west slope of Rainier Mesa.

Fig. 21. Looking northwest from west face of Rainier Mesa (along Holmes Rd., at 7200 ft). South rim of Pahute Mesa (background), at around 7000 ft; Twin Peaks, midground; foreground, as in Fig. 20.

Fig. 22. Boulder-cliffs in the Timber Mountain Tuff, common feature of especially near south and north rims of Pahute Mesa. Photo at south rim, 7300 ft, where several plant species are identified with the cliff crevices and bases.

Fig. 23. Juniper Flats of north-central Pahute Mesa, at 6500 ft, a local area where trees are nearly all Juniper (*Juniperus osteosperma*), and shrub and herbaceous layers are either absent or scarcely represented. *Physaria chambersii* and *Polygala subspinosa* var. *subspinosa* are common on the otherwise largely bare soils of the area.

Fig. 24. *Artemisia tridentata* (Big Sagebrush) and *Artemisia*—Pinyon—Juniper vegetation of the gently rolling uplands over many square miles of central and northern Pahute Mesa. Photo looking northwest, at 6500 ft, from near headwaters (upper right) of Columbine Canyon, a tributary of South Silent Canyon, where there is the only known occurrence of *Aquilegia shockleyi* (Columbine) on the Test Site (see also Fig. 25).

Fig. 25. *Aquilegia shockleyi* (Columbine) in the recesses along boulder-cliffs of Columbine Canyon, northern Pahute Mesa, 6400 ft (see also Fig. 24).

Fig. 26. *Artemisia tridentata* (Big Sagebrush) vegetation, at 6200 ft, extreme northwestern Pahute Mesa, looking west toward Black Mountain (high point, 7235 ft), the center of a caldera.

Fig. 27. Foreground, *Artemisia tridentata* (Big Sagebrush), and midground, *Artemisia*—Pinyon—Juniper along a drainage course, at 6200 ft, extreme northwestern Pahute Mesa. Looking northwest toward Stonewall Mountain (left of center, background), and north (in background, right of center) toward the Cactus Range (high peak 7482 ft at extreme right) and basin floor of Gold Flat.

Fig. 28. Northwest side of Bald Mountain, central Groom Range. *Artemisia*—Pinyon—Juniper (foreground) extends about to level of saddle at base of mountain (upper left, at 8000 ft), and is over sedimentary rocks. Beginning near the base of the volcanic peak, open and discontinuous White Fir (*Abies concolor*) forests extend almost to the summit (9380 ft). A distance view of Bald Mountain and the Groom Range is shown in Fig. 14.

Part II The Vascular Plants

PLANT COLLECTIONS

The vascular flora known to date from the region delimited in Fig. 2 consists of 1093 taxa, as the taxa are here recognized (and excluding hybrids); these belong to 98 families. Around two-thirds of the taxa occur on the Test Site or near vicinity (Fig. 3), which is about one-fifth of the total area under consideration. That such a high proportion of the plants is known from a much smaller proportion of the total land area is due in part to the intensity and duration of the botanical collecting on the Test Site (1959–1975) as compared with the four years of collecting (1968–1971) in the mountains and drainage basins elsewhere in the region. It is due in part, also, to the particular combination of physiographic, geologic, and climatic features of the Test Site, as a result of which the area straddles two large deserts (Mojave and Great Basin) and the desert transition (Part I). In accord with all these features, there are two large floras represented, and species endemic to the transition, all within a 2000 sq. mile (5100 km^2) area. Many Mojave Desert species are at their northern limits, and Great Basin Desert species their southern limits, on the Nevada Test Site.

Only occasional or selected plant collections were made in this region prior to the late 1950s. Percy Train collected in the area in the 1930s and was on the present Test Site in August 1938, when he made the type collection for *Camissonia megalantha* from Cane Spring in west Frenchman Flat and collected the rare *Petradoria discoidea* from the Eleana Range in northwestern Yucca Flat. The only other collections known from the Test Site are those of the early 1940s by Rupert C. Barneby and H. Dwight Ripley, from Skull Mountain and southeastern Frenchman Flat below the west Spotted Range, including the type collections of *Astragalus nyensis*, *Polygala subspinosa* var. *heterorhyncha*, and *Cymopterus ripleyi* var. *saniculoides*, and from the Specter Range, the type collection of *Gilia ripleyi*.

Elsewhere in the region the pioneer collecting of Frederick Coville and Frederick Funston on the Death Valley Expedition of 1891 was the first (and apparently the last) large effort to sample comprehensively the plants over most of the Death Valley region, including Ash Meadows; many type collections were made on this expedition, including the Ash Meadows endemic *Ivesia* (*Potentilla*) *eremica*. Other notable efforts were the selected collections over much of the region by Rupert Barneby and Dwight Ripley in the 1940s and the intensive collecting by Ira W. Clokey, from 1935 to 1942, in the Spring Mountains, especially in the mesic canyons of the north slope and the Charleston Peak area, which culminated in his flora covering much of this mountain range; many new species were described from the collections of these botanical pioneers. Otherwise collecting has been occasional and selective through the years, mostly by California botanists, whose collections have been deposited in various California herbaria.

Plant collecting on the present project began in 1959, as an effort to attach correct names to Test Site plants for use in ecological studies. From 1959 through 1967, the collecting was an ancillary activity of the author, during which period 5500 collections were processed from the Test Site (and one sampling from Ash Meadows in 1965); 300 of these were made in the spring and summer of 1967 by Vernon Bostick, and around 100 were collections of William H. Rickard or Lora M. Shields.

Because regional understandings had become mandatory for understandings of the Test Site vegetation and flora, collecting beyond the boundaries of the Test Site was begun in 1968, especially on the Air Force Range and in Ash Meadows; over 2000 collections were made that season by James L. Reveal and the author, working separately and together. In 1969, another 2000 were added by the author, again mostly beyond the Test Site boundaries, especially in the mountain ranges in the southern part of the region. In 1970, 2000 more were processed, 600 of which were joint collections by James Reveal and the author; these were mostly from the mountain ranges in the northern part of the region and the northwest Spring Mountains of the southern part. In 1971, the final year of intensive collecting activity, around 1000 collections were made, both on the Test Site and elsewhere over the region. There were, in total, around 300 collections in 1972 and 1975. The Reveal collections carry his collection number, and the author's collection number is the accession number.

In addition to these collections, around 150, mostly from the mountains at the southern end of the Test Site, were contributed as distribution records by Thomas A. Ackerman. Around 200, mostly from beyond the Test Site, were occasional collections by William A. Rhoads and others during the years 1960 to 1972. In addition, there are around 400 miscellaneous collections from beyond the area of Fig. 2; these include 125 duplicates of collections by Vernon Bostick in the Hot Creek Range, made for the University of Nevada, Reno.

These research collections, made during the years 1959—1975, are the basis for the definition of the ecologic and geographic distributions of the vascular plants of the region, and hence most are voucher collections. These number in total around 13,600 specimens; around 5000 constitute the permanent Test Site herbarium (NTS), and the remainder will be deposited at the United States National Herbarium, Washington, D. C. In addition, the more than 13,000 voucher specimens, collected during the years 1962—1975 on all of the author's permanent study sites, are the basis for many of the distribution records on the Test Site. A few of the taxonomic and distribution records are necessarily based on collections of others made prior to 1950.

For most collections there were from 1 to 35 duplicate specimens, and these were given to various herbaria of the country through the years. Nearly all taxa are represented by one or more specimens in the herbarium of the Biology Department, University of Nevada, Reno, as the state repository for collections from this previously botanically unknown part of Nevada. Duplicates of many collections from the Mojave Desert and Test Site parts of the region were deposited also at the University of Nevada, Las Vegas, through the years. Dudley Herbarium of Stanford University, Stanford, Calif., was the chief out-of-state repository for duplicates from the whole region almost from the time of the beginning of the project. Beginning with the 1968 season, the New York Botanical Garden, Bronx, N. Y., became a repository for duplicates from the Intermountain Region, and, beginning in 1969, Rancho Santa Ana Botanic Garden, Claremont, Calif., became a principal repository for specimens from the Mojave Desert part of the region or for plants otherwise of interest to California botanists. In addition, duplicates restricted to certain families went largely to other herbaria: The Brassicaceae went to the Gray Herbarium of Harvard University, Cambridge, Mass., and many of the Scrophulariaceae went to the Jepson Herbarium of the University of California, Berkeley. In total, around 25,000 duplicates are in institutional herbaria of the country.

ACKNOWLEDGMENTS

The support and contributions of many people through the years have made possible this concluding report on the flora of central-southern Nevada. The problem of attempting field studies in a region not covered by any published flora and in a known critical floristic area, where neither reference collections nor the taxonomic literature were available and the flora was almost wholly unknown—but the vascular plants nevertheless requiring correct determinations—was solved by the generous and cordial cooperation of taxonomists of the country. To these who gave of their time and expertise on behalf of our needs at the Test Site, the author cannot adequately express her gratitude; so large and complex was the task that without their help and encouragement the project would surely have faltered and died somewhere along the way.

From almost the beginning, and continuing through 1967, the large assistance was by Peter H. Raven, who reviewed most of the collections of the Onagraceae and hundreds of other collections with uncertain or no determinations. To him for his sustained support, especially during these critical early years, there is a permanent large debt of gratitude.

Beginning in 1968, at the beginning of greatly increased intensity and areal coverage of the collecting efforts, came the total involvement in all aspects of the flora project by James L. Reveal. This has included the periodic sharing of the field work, description of new taxa, review of all uncertain determinations and problems of synonomy, and a thorough review of the taxonomic considerations of the distribution compilation (resulting in many last-minute manuscript changes). Through these years, the author has had continuous access to his judgments on all matters pertaining to the flora, and it is largely because of his supportive contributions that the work can be brought to a conclusion at this time. In addition, the largest genus in the flora, *Eriogonum*, is also the best known in the region, and the more than 1000 collections carry his annotations.

Each year, for most of the years of the project, the accumulated collections of two large groups received the critical attention of their authorities before the next collecting season began. Reed C. Rollins has reviewed nearly all the collections of the Brassicaceae, the fourth largest family in the flora. Rupert C. Barneby has seen nearly all the *Astragali* (and certain other collections) and, in addition, as one familiar with and having a deep interest in the region and a sympathetic understanding of the objectives of the project, has been a frequent source of information, encouragement, and uniquely thoughtful contributions. To these and to Lyman Benson, who reviewed all the Cactaceae, the author is especially grateful.

Other group specialists who have donated their time and expertise to the critical review of collections, or related matters, during the years and whose contributions in total are very large indeed are the following: Alva Day Whittingham, the genus *Gilia*; Lauramay T. Dempster, *Galium*; Rimo Bacigalupi and Noel H. Holmgren, Scrophulariaceae; Loran C. Anderson, *Chrysothamnus* and related genera; David B. Dunn, *Lupinus*; Herbert A. Wahl, *Chenopodium*; Henry J. Thompson and Joyce E. Roberts, Loasaceae; John M. Gillett, *Trifolium*; Roy L. Taylor, Saxifragaceae; Arthur H. Holmgren and Noel H. Holmgren, Poaceae; Lincoln Constance, certain of the Apiaceae and Hydrophyllaceae; T. M. Barkley, *Senecio*; Derek Burch, *Euphorbia*; Charles B. Heiser, Jr., *Helianthus*; Larry Higgins, perennial species of *Cryptantha*; Leslie Gottlieb and, more recently, Spencer Tomb, *Stephanomeria* and related genera; Walter Lewis, *Rosa*; Dieter Wilkin, *Hulsea*; Lawrence R. Heckard, Orobanchaceae; Donald W. Kyhos, *Chaenactis*; V. Botschantzev, *Salsola*; John Thomas Howell, *Carex occidentalis*; C. Leo Hitchcock, *Lycium shockleyi*; and Arthur Cronquist, certain problem collections in the Asteraceae.

Lastly, there is a special recognition of the contributions of Philip A. Munz. His California floras enabled determinations to be made, at least tentatively, for most taxa, and his cordial association with the project the last five years of his life was a personal privilege for the author and an asset of large significance to the work in its totality.

The work was supported by the U. S. Atomic Energy Commission (now Energy Research and Development Administration), with nearly all the field work performed under contract AT (04-1) Gen-12 between the Division of Biology and Medicine and the University of California, and the final manuscript prepared under contract AT (11-1) 2307 between the Division of Biomedical and Environmental Research and the University of Cincinnati. Base support at the Nevada Test Site was by the Civil Effects Test Organization, and all studies were conducted under CETO Project 61.5.4.

Certain personnel and organizations at the Test Site made possible or facilitated the field work and completion of the manuscript. Map reproduction and processing of photographs, including the composite of Fig. 11, are the products of the skills and special efforts of Hélène E. LaPlant, Chief, REECO Printing Plant, whose artistic judgments and cordial cooperation throughout were major contributions to the total production effort. For each period of access to the Nellis Air Force Range, arrangements were made by the AEC Operations Coordination Center, whose "bird-dogs" also were otherwise helpful to the needs of the field program. And it is with much personal gratitude that we acknowledge the vehicles and vehicle maintenance—especially the preventive maintenance—supplied by Arkey and Chester Huber and their associates of the REECO Motor Pool Department; this enabled us to travel in some of the most rugged and remote areas of southern Nevada with the maximum safety and minimum risk possible.

Lastly, it is a particular pleasure to acknowledge the contributions of Kathleen M. MacKay, who, as the herbarium assistant over a 13-year period, mounted all specimens, typed all duplicate labels, and shared the responsibility for the processing and distribution of the collections and the record-keeping; her services were an integral and indispensable part of the project and give permanent artistic expression to its tangible products.

Ecologic and Geographic Distributions

In the vascular plant compilation that follows, families are in alphabetical order under the major subdivisions, as are also the genera and species within each family.

The ecologic and geographic distributions of the taxa of the region of Fig. 2 are based entirely on record specimens, except as noted; the widely distributed and common species of the region were those, on the whole, least frequently collected, and their distributions, therefore, are known to be incomplete.

The vegetation types listed are those designated on the collections and described in Part I (The Vegetation); in general, the order of listing is from Mojave Desert, to transition desert, to Great Basin Desert, or from lower to higher elevations. *Atriplex* refers to *Atriplex confertifolia*, and whether the plant association is upland *Atriplex* (upper bajadas or mountains) or lowland *Atriplex* (lower bajadas of basin floors) is usually indicated by context within the distribution as described. Common names are given only for the woody plants, where such a name is in common use, and for the introduced herbaceous species and crop plants that are widely known by their common names.

Geographic distributions of bajada species are by drainage basins or by mountain ranges where the species' distribution is associated with the bajadas below certain mountain ranges. Distributions of all mountain species are by mountain, mountain range, or mesa. In general, the order of listing of mountain ranges and drainage basins is from south to north and west to east. NTS refers to the Nevada Test Site and areas outside its boundaries indicated in Fig. 3; locations of all physiographic features referred to in the NTS distributions are indicated on this map. Outside this area distributions are by the counties in alphabetical order, and the major physiographic and cultural features are indicated on the map of Fig. 2.

Ranges in elevation are from the record of the collections and do not necessarily define the actual elevational ranges of the taxa. Likewise, the flowering periods are those known from the collections, and indicated by the months appearing in the records. Perennials not designated as either shrubs or woody at the base are herbaceous perennials. For species designated as winter or summer annuals, the season of germination is known from field studies; for all others, referred to as annuals, either germination occurs in the spring or the time of germination is unknown in this region.

ABBREVIATIONS

Apr	April
Aug	August
auths.	authors
ca	around
cent., Cent.	central, Central
cm	centimeter(s)
CO.	county
coll(s).	(plant) collection(s)
Cyn(s), cyn(s)	Canyon(s), canyon(s)
Dec	December
diam.	diameter
e., E.	eastern, Eastern
E	east
elev(s).	elevation(s)
esp.	especially
f.	forma
Feb	February
fl.	flowering
fls.	flowers
fr.	fruiting
frs.	fruits
ft	feet
incl.	including
Jan	January
m	meter(s)
Mar	March
Mtn(s), mtn(s)	Mountain(s), mountain(s)
n., N.	northern, Northern
N	north
N. Amer.	North American
ne., Ne.	northeastern, Northeastern
NE	northeast
Nov	November
NTS	Nevada Test Site
nw., Nw.	northwestern, Northwestern
NW	northwest
Oct	October
s., S.	southern, Southern
S	south
se., Se.	southeastern, Southeastern
SE	southeast
Spg(s), spg(s)	Spring(s), spring(s)
sp., spp.	species (singular, plural)
ssp.	subspecies
sw., Sw.	southwestern, Southwestern
SW	southwest
var.(s)	variety; varities
w., W.	western, Western
W	west
yr(s)	year(s)

Division **LYCOPODIOPHYTA.** Club-Mosses

SELAGINELLACEAE. Selaginella Family

Selaginella

S. watsonii Underw. NYE CO.: Locally the ground cover in *Artemisia*—Pinyon—Juniper and *Artemisia—Cercocarpus*, n. Kawich Range (upper Eden Creek cyn). 7600—8400 ft. Perennial. May—June.

Division **EQUISETOPHYTA.** Horsetails

EQUISETACEAE. Horsetail Family

Equisetum

E. hyemale L. var. **affine** (Engelm.) A. A. Eat. CLARK CO.: Local on wet soils, *Artemisia*—Pinyon—Juniper, nw. Spring Mtns (Willow Creek); occurs with *E. laevigatum*. 6100 ft. Perennial. July.

E. laevigatum A. Br. Local on wet soils, *Artemisia*—Pinyon—Juniper. CLARK CO.: Nw. Spring Mtns (Willow Creek, Clark Cyn). NYE CO.: N. Kawich Range (Longstreet cyn). 6000—7000 ft. Perennial. May—July.

Division **POLYPODIOPHYTA.** Ferns

POLYPODIACEAE. Common Fern Family

Cheilanthes

C. covillei Maxon. NTS: Around boulders, *Coleogyne* below Yucca Mtn (nw. Jackass Flats). 5200 ft. Perennial. June.

C. feei T. Moore. Limestone crevices, *Artemisia*—Pinyon—Juniper. CLARK CO.: Nw. Spring Mtns (Clark Cyn). NYE CO.: Nw. Belted Range. 6000—6500 ft. Perennial. July—Aug.

Cystopteris

C. fragilis (L.) Bernh. NYE CO.: Local, *Artemisia—Cercocarpus*, n. Kawich Range (upper Eden Creek cyn). 7600 ft. Perennial. Summer.

Notholaena

N. jonesii Maxon. NTS: Limestone crevices in *Atriplex*, e. Specter Range. 4300 ft. Perennial. Appar. spring.

N. parryi D. C. Eat. Common, limestone crevices, *Larrea—Atriplex*, *Atriplex*, *Coleogyne*, *Artemisia nova*. NTS: Specter Range, w. Spotted Range (Red Mtn, Mercury Ridge), Ranger Mtns, Buried Hills and limestone butte to W, CP Hills, and dolomite hills of Eleana Range (at 6200 ft). CLARK CO.: N. Spotted Range, and below nw. Spring Mtns. NYE CO.: Mtn N end of Pahrump Valley, and Bare Mtn. 3200—5000 (—6200) ft. Perennial. Apr—May.

Pellaea

P. mucronata (D. C. Eat.) D. C. Eat. var. **mucronata.** NTS: Cliff crevices, bases of boulders, talus, or washes vicinity of volcanic rock outcrops, *Coleogyne* and *Artemisia*—Pinyon—Juniper; s. Shoshone Mtn (Topopah Valley); coll. from below Yucca Mtn to W (*Reveal & Beatley 6246*) consists of this and the following species from same site. 5200—5800 ft. Perennial. May—June.

P. truncata Goodd. [*P. longimucronata* Hook.]. Rock crevices and around boulders, *Coleogyne* and *Artemisia*—Pinyon—Juniper. NTS: Below Yucca Mtn (nw. Jackass Flats). NYE CO.: Nw. Spring Mtns (N slope below Mt. Stirling). 5800—5900 ft. Perennial. May—June.

Pityrogramma

P. triangularis (Kaulf.) Maxon var. **triangularis.** NTS: Occasional along cliffs or talus slopes, *Coleogyne* and *Artemisia*; w. Shoshone Mtn (near Forty-Mile Cyn wash), and below Oak Spring Butte (s. Belted Range). 3600—5500 ft. Perennial. May—Sept.

Division **PINOPHYTA.** Gymnosperms

CUPRESSACEAE. Cypress Family

Juniperus

J. communis L. var. **depressa** Pursh (DWARF JUNIPER). CLARK CO.: Nw. Spring Mtns (upper Clark Cyn), common locally as ground cover, moist soils along stream in Yellow Pine—Fir. 9000 ft. Shrub. June.

J. osteosperma (Torr.) Little (UTAH JUNIPER). Dominant tree in all mountain areas above 6000 ft, growing usually with *Pinus monophylla*, with which it forms a discontinuous tree layer in usually otherwise *Artemisia* communities. Lower and upper elevation limits are slightly lower than for *Pinus monophylla*. Essentially pure stands (i.e., without Pinyon Pine) occurring locally or over larger areas, are usually associated with calcareous bedrock. NTS: Eleana Range, s. Belted Range (Rainier Mesa, slopes, and uplands to E), Halfpint Range (Raysonde Buttes area). CLARK CO.: Nw. Spring Mtns (Wheeler Wash area). NYE CO.: Grapevine Mtns, Stonewall Mtn, n. Cactus Range, n. Kawich Range. 5500–*ca* 8000 ft. Small tree. May.

J. scopulorum Sarg. (ROCKY MTN JUNIPER). CLARK CO.: Nw. Spring Mtns (upper Clark Cyn), where it is the Juniper associated with *Pinus ponderosa* and *Abies concolor*. 8000–9000 ft. Small tree. June.

EPHEDRACEAE. Ephedra Family

Ephedra

E. fasciculata A. Nels. NYE CO.: The *Ephedra* of s. Stewart Valley, associated with *Atriplex hymenelytra*. 2500 ft. Shrub. Apr.

E. funerea Cov. & Mort. (DEATH VALLEY EPHEDRA). Common where various combinations of *Larrea*, *Atriplex*, *Ambrosia*, *Prosopis*, or *Coleogyne* are dominants, associated usually with limestone mountain ranges. NTS: In or below Specter Range, w. Spotted Range (Red Mtn), N side of Skull Mtn, Yucca Mtn (nw. Jackass Flats), Halfpint Range (Scarp Cyn). NYE CO.: E side of Stewart Valley, n. and e. Ash Meadows, below mtns N end of Pahrump Valley, e. Amargosa Valley. 2200–4000 ft. Shrub. Apr–May.

E. nevadensis S. Wats. (NEVADA EPHEDRA). The widely distributed and common *Ephedra* throughout most of region, associated with *Larrea*, *Grayia*, *Lycium*, *Coleogyne*, and *Artemisia;* locally a dominant. NTS: Mercury Valley, Rock Valley, Jackass Flats, Topopah Valley, Frenchman Flat, Mid Valley, Yucca Flat, Forty-Mile Cyn, s. Groom Lake. NYE CO.: Pahrump Valley; in or below Bullfrog Hills, Grapevine Mtns, Reveille Range, below cent. Belted Range. 2500–6000 ft. Shrub. Apr–May.

E. torreyana S. Wats. (MORMON TEA). Below certain parts of certain limestone mountain ranges, in *Larrea*, *Atriplex*, or *Coleogyne;* not known to occur with *E. funerea*. NTS: The common *Ephedra* in or below S face and E end of Specter Range, w. Spotted Range (incl.

Red Mtn, Mercury Ridge), Ranger Mtns, Buried Hills and limestone butte to W, Halfpint Range (upper Nye Cyn), CP Hills. CLARK CO.: Below N slope of nw. Spring Mtns, and N-S axis of Spotted Range. 3200—5000 ft. Shrub. Apr—May.

E. viridis Cov. (GREEN EPHEDRA). The *Ephedra* of upper *Artemisia* and *Artemisia*—Pinyon—Juniper throughout region; often also in *Coleogyne, Atriplex*. NTS: Hill and mountain areas of all drainages; colls. from Specter Range, Skull Mtn, Spotted Range (incl. Red Mtn), Mine Mtn, Eleana Range, Shoshone Mtn, s. Belted Range (Rainier Mesa), nw. Papoose Range. CLARK CO.: Nw. Spring Mtns (N slope below Mt. Stirling); N-S axis of Spotted Range. NYE CO.: Bullfrog Hills. 4500—7500 ft. Shrub. Apr—May.

PINACEAE. Pine Family

Abies

A. concolor (Gord. & Glend.) Lindl. ex Hildebr. var. **concolor** (WHITE FIR). CLARK CO.: Nw. Spring Mtns (upper Clark Cyn, Trough Spg), where it is a dominant tree with or without *Pinus monophylla* or *P. ponderosa*, 7000—9000 ft. LINCOLN CO.: Cent. Groom Range (N slope of Bald Mtn), *ca* 8000—9200 ft, in open forests with or without *Pinus monophylla* and locally with *P. flexilis*. Apparently absent from higher peaks (Wheelbarrow Peak, 8200 ft; Kawich Peak, 9400 ft) elsewhere in the region. 7000—9200 ft. Small to large tree. July into Aug.

Pinus

P. flexilis James (LIMBER PINE). LINCOLN CO.: Cent. Groom Range (Bald Mtn), where a common tree in certain areas of E slope above *ca* 7500 ft, with *P. monophylla* and *Abies concolor*. NYE CO.: Kawich Range (Kawich Peak area of upper Eden Creek cyn), occasional small trees (to 8 cm diam.) in *Cercocarpus* groves, at *ca* 8500 ft; larger trees probably in the Range. 7000—9000 ft. Small tree. June.

P. monophylla Torr. & Frém. (PINYON PINE). Dominant tree in all mountain areas, above *ca* 6000 ft, growing usually with *Juniperus osteosperma* (at highest elevations with *Abies concolor* and *P. flexilis*), forming a discontinuous tree layer in usually otherwise *Artemisia* communities. NTS: Eleana Range, s. Belted Range (Rainier Mesa), Pahute Mesa. CLARK CO.: Nw. Spring Mtns (Cold Creek area). NYE CO.: Grapevine Mtns, Stonewall Mtn, n. Kawich Range, cent. Belted Range. 5800—8500 ft. Small tree. June—July.

P. ponderosa Laws. var. **ponderosa** (WESTERN YELLOW PINE). CLARK CO.: Dominant tree, with *P. monophylla* and/or *Abies concolor*, certain cyns (incl. upper Clark Cyn, and Wheeler Pass area) of nw. Spring Mtns. 7000—9200 ft (colls. to 8100 ft). Large tree. Late June—July.

Division **MAGNOLIOPHYTA.** Angiosperms

Class MAGNOLIATAE. Dicotyledons

ACERACEAE. Maple Family

Acer

A. glabrum Torr. var. **diffusum** (Greene) Smiley. Rare in region. CLARK CO.: Nw. Spring Mtns (upper Clark Cyn), in Yellow Pine—Fir. NYE CO.: Cent. Belted Range (below Wheelbarrow Peak), in *Artemisia*—Pinyon—Juniper. 6600—9200 ft. Small tree. May.

AIZOACEAE. Carpet-Weed Family

Mollugo

M. cerviana (L.) Ser. NTS: Uncommon, local certain yrs in *Larrea—Ambrosia*, *Grayia—Lycium*; sands below Yucca—Shoshone Mtn (w. Jackass Flats), and below Eleana Range (nw. Yucca Flat). 3000—4500 ft. Summer annual, doubtfully introduced in this area. Aug—Oct.

AMARANTHACEAE. Amaranth Family

Amaranthus

A. albus L. Disturbed sites in *Atriplex*, *Grayia—Lycium*, *Coleogyne*, *Artemisia tridentata*, *Artemisia*—Pinyon—Juniper, uncommonly *Larrea*. NTS: Areas of Cane Spg, Tippipah Reservoir, Whiterock Spg; Mercury townsite and roadsides of w. Mercury Valley and nw. Yucca Flat. CLARK CO.: Indian Springs townsite. NYE CO.: Ash Meadows, Pahrump Valley; n. and cent. Belted Range, cent. Reveille Range, s. Kawich Range. 2200—6500 ft. Introduced annual. Aug—Oct.

A. blitoides S. Wats. [*A. graecizans* of N. Amer. auths.]. Frequent on moist disturbed soils, in *Larrea*, *Atriplex*, *Coleogyne*, *Artemisia*, esp. *Artemisia*—Pinyon—Juniper. NTS: Mercury Valley, se. Forty-Mile Cyn. LINCOLN CO.: Below n. Groom Range. NYE CO.: N. Ash Meadows, s. Oasis Valley; spgs or other moist sites in or below cent. Belted Range (Indian Spg), n. Reveille Range (in *Atriplex parryi—Sarcobatus* at Twin Spgs), s. Kawich Range (Cedar Spg), s. Monitor Range, s. San Antonio Mtns. 2300—6800 ft. Annual. June—Nov.

A. californicus (Moq.) S. Wats. Occasional, *Larrea—Ambrosia*, *Atriplex*. NTS: Forty-Mile Cyn wash (nw. Jackass Flats). NYE CO.: N. Cactus Range (Cactus Spg). 3600—6300 ft. Annual. May—Sept.

A. fimbriatus (Torr.) Benth. NTS: Locally common in *Coleogyne*, *Grayia—Lycium*, uncommon in *Larrea*; hills N of Cane Spg, s. and e. Shoshone Mtn, Mine Mtn, Eleana Range. 4000—4800 ft. Summer annual. Aug—Oct.

A. powellii S. Wats. NYE CO.: Disturbed moist soils, *Artemisia*—Pinyon—Juniper, cent. Belted Range (e. Kawich Valley). 6000—6800 ft. Annual. July—Aug.

A. retroflexus L. Uncommon weed of disturbed moist soils in *Larrea—Atriplex*, *Atriplex*, *Artemisia*. CLARK CO.: Indian Springs townsite. NYE CO.: N. Ash Meadows, n. Pahrump Valley; Tonopah townsite. 2300—6000 ft. Introduced annual. July—Oct.

Tidestromia

T. oblongifolia (S. Wats.) Standl. var. **oblongifolia.** Locally often abundant in *Atriplex* and *Larrea—Ambrosia*, in canyons of limestone mountain ranges, and bajadas below. NTS: Specter Range, w. Spotted Range. NYE CO.: Ash Meadows, Pahrump Valley; below Bare Mtn (n. Amargosa Valley). 2200—4200 ft. Perennial. July—Oct.

ANACARDIACEAE. Sumac Family

Rhus

R. trilobata Nutt. ex Torr. & Gray var. **anisophylla** (Greene) Jeps. (SQUAW BUSH). Occasional in probably all mountain areas in lower *Artemisia*—Pinyon—Juniper; toward lower elevational limits, scattered in washes of *Artemisia* and *Coleogyne/Artemisia*. NTS: W. Spotted Range (Red Mtn), Buried Hills, Yucca Mtn (nw. Jackass Flats), Shoshone Mtn, Eleana Range, s. Belted Range (below Rainier Mesa and Oak Spring Butte), Pahute Mesa; w. Forty-Mile Cyn, w. Emigrant Valley. CLARK CO.: Nw. Spring Mtns (Willow Creek area, Crystal Spg cyn). LINCOLN CO.: Cent. and s. Groom Range. NYE CO.: Nw. Spring Mtns (N slope below Mt. Stirling), n. Kawich Range (Longstreet cyn), cent. Belted Range (Johnnies Water cyn);

one large plant (probably planted), n. Ash Meadows. (2200—) 4800—6600 ft. Shrub. Apr, in advance of leaves.

APIACEAE (UMBELLIFERAE). Carrot Family

Angelica

A. kingii (S. Wats.) Coult. & Rose. CLARK CO.: Occasional, as large clumps in stream, in *Artemisia*—Pinyon—Juniper, nw. Spring Mtns (Cold Creek Spg). 6200 ft. Perennial. June—July.

Apium

A. graveolens L. (CELERY). NYE CO.: Common, moist soils near stream in Pinyon—Juniper, nw. Spring Mtns (Crystal Spg cyn). 4600—5200 ft. Introduced perennial. July—Aug.

Berula

B. erecta (Huds.) Cov. Locally common in year-round wet soils. NTS: Edge of Cane Spg pond, at one time common under cat-tails. CLARK CO.: Edge of small water impoundment, Cactus Spgs. NYE CO.: Local near spgs in Ash Meadows; common along drainage channels and swamp areas of Oasis Valley, and Crystal Spg cyn (nw. Spring Mtns). 2300—5000 ft. Perennial. June—Sept.

Cymopterus

C. aboriginum M. E. Jones. NTS: Crevices and ledges of limestone or dolomite hills and mountain ranges, in *Coleogyne*, *Artemisia nova*, lower *Artemisia*—Pinyon—Juniper; CP Hills, Mine Mtn, Eleana Range, Halfpint Range (Banded Mtn); summit of Skull Mtn [*Ripley & Barneby 3444* (CAS), 4800 ft]. CLARK CO.: Type locality near Indian Springs (Spring Mtns). 4400—6500 ft. Perennial. Apr—May.

C. gilmanii Morton. NTS: Locally common on ledges, lower talus slopes, and washes in certain cyns of limestone mountain ranges, or calcareous outcrops elsewhere, in *Atriplex* or *Coleogyne;* among rocks, summit of Skull Mtn; w. Spotted Range (Red Mtn, Mercury Ridge), Ranger Mtns. 3300—4500 (—5700) ft. Perennial. Apr—May.

C. globosus (S. Wats.) S. Wats. NTS: Known from only two sites in the region: *Atriplex—Kochia* N of Yucca Flat playa, where present in about one of four yrs; and *Coleogyne* of upper bajada below dolomite hill of Eleana Range, nw. Yucca Flat, where present most yrs. 3900—5000 ft. Perennial. Apr—May.

C. purpurascens (A. Gray) M. E. Jones. NTS: Locally common, *Coleogyne*, *Artemisia nova*, and *Artemisia*—Pinyon—Juniper; dolomite hill of Eleana Range, s. Belted Range (Doe Point of nw. Rainier Mesa). 5800—7500 ft. Perennial. Mar—Apr.

C. ripleyi Barneby. Widely distributed, esp. in sandy soils; *Atriplex*, *A. canescens*, *Larrea—Ambrosia*, *Larrea—Grayia—Lycium*, *Grayia—Lycium*, *Larrea/Coleogyne*, *Coleogyne—Grayia*, *Artemisia tridentata*, *Artemisia*—Pinyon—Juniper. Plants of the lower elevations are often referable to var. *saniculoides* Barneby, and higher-elevation plants to var. *ripleyi*, but characters distinguishing the two varieties become ill-defined and somewhat clinal with latitude and elevation when the populations as a whole are considered. NTS: Local in Rock Valley; abundant in sands of ne. Frenchman Flat (below Buried Hills); s. and cent. Frenchman Flat; more or less common locally in e. Forty-Mile Cyn (below light-colored tuffs of Eleana Range); most areas of Yucca Flat, S face of Rainier Mesa (s. Belted Range); white gravel talus in w. Emigrant Valley, sands of s. Groom Lake. NYE CO.: S. Penoyer Valley, cent. Kawich Valley. 3200—6700 ft. Perennial. Apr—June.

Daucus

D. carota L. (QUEEN ANNE'S LACE). CLARK CO.: Local, disturbed site, Indian Springs townsite. 3100 ft. Introduced biennial. June—Aug.

Hydrocotyle

H. verticillata Thunb. var. **verticillata.** NYE CO.: Locally abundant in wet soils, Ash thickets of n. Ash Meadows. 2300 ft. Perennial. July—Sept.

Lomatium

L. foeniculaceum (Nutt.) Coult. & Rose ssp. **fimbriatum** Theobald [*L. macdougalii* Coult. & Rose]. Locally common in *Artemisia nova* and *Artemisia*—Pinyon—Juniper. NTS: Flatrock areas of s. Belted Range (cent. and nw. Rainier Mesa) and s. Pahute Mesa. NYE CO.: S. Monitor Range. 6200—7500 ft. Perennial. Late Apr—early June.

ssp. **macdougalii** (Coult. & Rose) Theobald [*L. macdougalii* Coult. & Rose]. LINCOLN CO.: *Artemisia*—Pinyon—Juniper and Fir—Pinyon, cent. Groom Range (on and below Bald Mtn). 7000—8000 ft. Perennial. June.

L. nevadense (S. Wats.) Coult. & Rose var. **nevadense.** The widely distributed *Lomatium* of the region, occasional to sometimes locally common over several acres; *Coleogyne, Artemisia nova, Artemisia*—Pinyon—Juniper, uncommonly *Atriplex.* NTS: E. Specter Range, summit of Skull Mtn; in or below Shoshone Mtn, Mine Mtn, Eleana Range, s. Belted Range (below Rainier Mesa, Gold Meadows), Pahute Mesa. CLARK CO.: Nw. Spring Mtns (E of Wheeler Pass). NYE CO.: In or below nw. Spring Mtns (abundant in upland area of N slope below Mt. Stirling), nw. Yucca Mtn (Crater Flat), Bullfrog Hills (Sawtooth Mtn, where occurs with var. *parishii*), Stonewall Mtn. 4200—6700 ft. Perennial. Late Apr—May.

var. **parishii** (Coult. & Rose) Jeps. NYE CO.: Bullfrog Hills (with typical variety on Sawtooth Mtn), Grapevine Mtns. 6000—7800 ft. Perennial. Late Apr—May.

L. parryi (S. Wats.) Macbr. NYE CO.: Around boulders, on talus slopes, and in washes, *Artemisia*—Pinyon—Juniper; locally common in cyns of E and W slopes of cent. Belted Range (Johnnies Water and Cliff Spg), and n. Kawich Range (Eden Creek and Longstreet cyns). 6400—8200 ft. Perennial. May—June.

L. scabrum (Coult. & Rose) Mathias. Occasional to common locally, crevices and ledges of limestone and dolomite hills and mountains; *Atriplex, Coleogyne,* uncommonly *Larrea.* NTS: W. Spotted Range (Red Mtn), Ranger Mtns, Buried Hills and limestone butte to W. CLARK CO.: N-S axis of Spotted Range. 3500—5200 ft. Perennial. Apr—May.

Osmorhiza

O. occidentalis (Nutt.) Torr. NYE CO.: Locally common along stream, Willow thickets in *Artemisia*—Pinyon—Juniper, n. Kawich Range (Eden Creek cyn). 6700 ft. Perennial. June.

APOCYNACEAE. Dogbane Family

Amsonia

A. tomentosa Torr. & Frém. Includes tomentose and glabrous (*A. brevifolia* A. Gray) forms which occur in the same populations. Restricted to washes in or below limestone mountain ranges, in *Larrea—Atriplex, Atriplex.* NTS: Below w. Spotted Range (e. Mercury Ridge), and Ranger Mtns. NYE CO.: Washes of ne. Ash Meadows. 2200—3800 ft. Perennial. Apr—May.

Apocynum

A. androsaemifolium L. var. **androsaemifolium.** CLARK CO.: Local populations, esp. on disturbed sites, Yellow Pine—Fir, nw. Spring Mtns (upper Clark Cyn). 8000 ft. Perennial. July.

A. cannabinum L. var. **glaberrimum** A. DC. NYE CO.: Occasional, near spgs, n. Ash Meadows. 2300 ft. Perennial. July.

A. sibiricum Jacq. var. **salignum** (Greene) Fern. NTS: Washes in *Artemisia* or *Artemisia*—Pinyon—Juniper, n. and w. Pahute Mesa. NYE CO.: Hillside depression in *Atriplex*, e. Goldfield Hills (Wildhorse Spg). 5800—6800 ft. Perennial. June—July.

ASCLEPIADACEAE. Milkweed Family

Asclepias

A. erosa Torr. Scattered small populations throughout most areas of the region, restricted to usually sandy washes; in nearly all vegetation types of lower and middle elevations (incl. *Larrea*, *Atriplex*, *Grayia—Lycium*, *Coleogyne*, *Artemisia*, rarely *Artemisia*—Pinyon—Juniper). NTS: Rock Valley, Jackass Flats, Frenchman Flat, Mid Valley, Yucca Flat, Forty-Mile Cyn, s. Gold Flat. LINCOLN CO.: Desert Valley, NYE CO.: N. Ash Meadows, Pahrump Valley, Kawich Valley, cent. Groom Lake; n. Cactus Range, foothills of s. Quinn Canyon Range. 2300—6400 ft. Perennial. June—Sept.

A. fascicularis Dcne. in A. DC. NYE CO.: More or less common in moist soils, *Juncus—Distichlis*, *Atriplex*, Mesquite, and Ash—Screwbean; in agricultural areas along drainage channels and fencerows; Ash Meadows, Pahrump Valley, Oasis Valley. 2200—3400 ft. Perennial. June—July.

A. speciosa Torr. NYE CO.: Edge of *Phragmites* mound, Bole Spg, se. Ash Meadows; collected in Ash Meadows in March 1891 (Coville, F. V., *Contr. U.S. Nat. Herb.*, 4. 1893), but unknown today except for the Bole Spg population preserved within a long-time fenced area around a habitation. 2200 ft. Perennial. Apr—May.

Cynanchum

C. utahense (Engelm.) Woodson. NTS: Local with *Hymenoclea*, along edge of washes in *Larrea—Ambrosia* below w. Spotted Range (e. Mercury Valley). 3500 ft. Perennial, twining and woody at base.

Sarcostemma

S. hirtellum (A. Gray) R. Holm. Local, cove in *Atriplex*, N slope of cent. Specter Range (s. Rock Valley). 4000 ft. Perennial, twining and woody at base.

ASTERACEAE (COMPOSITAE). Sunflower Family

Acamptopappus

A. shockleyi A. Gray. More or less common associated shrub in *Larrea—Ambrosia*, *Larrea—Atriplex*, *Larrea—Lycium—Grayia*, *Larrea—Sarcobatus*, *Grayia—Lycium*, *Atriplex*; rare or local below certain mountain ranges (e.g., the limestone Specter Range), and abundant below others (e.g., bajada of w. Jackass Flats below Shoshone Mtn). NTS: Mercury Valley, Rock Valley, Jackass Flats, Topopah Valley, Frenchman Flat, Yucca Flat (incl. Plutonium Valley), s. Groom Lake (sw. Papoose Range). CLARK CO.: S. Spotted Range. NYE CO.: Oasis Valley; Bullfrog Hills, Goldfield Hills. 3000—5800 ft. Shrub. Apr—May.

Achillea

A. millefolium L. var. **lanulosa** (Nutt.) Piper. Disturbed moist sites in *Artemisia*—Pinyon—Juniper. NTS: Local, s. Pahute Mesa. NYE CO.: N. Kawich Range (Longstreet cyn). 6800—7300 ft. Perennial. June—July.

Agoseris

A. glauca (Pursh) Greene var. **laciniata** (D. C. Eat.) Smiley. Locally common, *Artemisia*—Pinyon—Juniper. NTS: S. Belted Range (Rainier Mesa, Gold Meadows), Pahute Mesa. 6800—7500 ft. Perennial. May—June.

Ambrosia

A. acanthicarpa Hook. [*Franseria a.* (Hook.) Cov.]. Locally abundant in washes, on sand dunes, or other disturbed sandy sites; *Larrea—Ambrosia*, *Grayia—Lycium*, *Atriplex*, *A. canescens*, *Artemisia tridentata*. NTS: N. Amargosa Valley, Frenchman Flat, Yucca Flat, sands of e. and s. Forty-Mile Cyn; Pahute Mesa. NYE CO.: E. Cactus Flat, Ralston Valley, Stone Cabin Valley, Kawich Valley, s. Penoyer Valley. 3400—6200 ft. Spring-germinating annual. July—Sept.

A. dumosa (A. Gray) Payne. [*Franseria d.* A. Gray] (BURSAGE). Either codominant or one of the prominent shrub species of many *Larrea* communities in the Mojave Desert part of the region; in s. Nye Co., reaches its n. limits in ne. Yucca Flat (local below Banded Mtn), on N slope of Jackass Flats, with *Atriplex* on Thirsty Cyn uplands and Obsidian Butte—Tolicha Peak area of n. Sarcobatus Flat. N. limits closely approximate, but are not precisely the same as, those of *Larrea*. Sometimes hybridizes with *Hymenoclea salsola* (*Beatley & Bostick 5121*, base of Ranger Mtns, se. Frenchman Flat, 14 June 1967). NTS: Mercury Valley, Rock Valley, Jackass Flats, Topopah Valley, Frenchman Flat, Yucca Flat, Thirsty Cyn. NYE CO.: Sparingly in Ash Meadows, n. Amargosa Valley, nw. Crater Flat, n. and e. Sarcobatus Flat; s. Bullfrog Hills. 2200—4500 (—5000) ft. Shrub. Apr—May, some yrs Aug—Oct.

A. eriocentra (A. Gray) Payne. [*Franseria e.* A. Gray]. Abundant locally, restricted to sandy washes in *Larrea*, *Atriplex*, *Coleogyne*, *Artemisia tridentata.* NTS: Sw. and extreme w. Frenchman Flat, nw. Jackass Flats (Forty-Mile Cyn wash), nw. Yucca Flat, Thirsty Cyn. LINCOLN CO.: Nw. Desert Valley. NYE CO.: Nw. Spring Mtns (N slope below Mt. Stirling), n. Crater Flat. 3200—5800 ft. Shrub. May—June.

A. psilostachya DC. var. **californica** (Rydb.) Blake. NYE CO.: Local in moist soils or on irrigated sites, Ash Meadows. 2200 ft. Perennial. Sept—Oct.

Amphipappus

A. fremontii Torr. & Gray. var. **fremontii**. (CHAFF-BUSH). Locally common, in crevices or esp. washes, in *Larrea* or *Atriplex* of limestone mountain ranges. NTS: Slopes of Specter Range, s. Striped Hills, washes below Yucca Mtn (w. Jackass Flats), w. Spotted Range (Red Mtn). CLARK CO.: S. Spotted Range. NYE CO.: Hills N of Ash Meadows, mtn E side of Stewart Valley, foothills below N end of Spring Mtns, cyns of Bare Mtn. 2600—4500 ft. Shrub. Apr—June.

Anisocoma

A. acaulis Torr. & Gray. Locally common on deep sands, often abundant on disturbed sites of same areas; *Larrea—Ambrosia*, *Larrea—Grayia—Lycium*, *Atriplex canescens*, *Artemisia tridentata*, rarely *Artemisia*—Pinyon—Juniper. NTS: W. Jackass Flats, sw. Frenchman Flat, sands of e. Forty-Mile Cyn; s. Belted Range (S slope of Rainier Mesa). NYE CO.: Foothills of Quinn Canyon Range. 3100—6700 ft. Winter annual. Apr—June.

Antennaria

A. dimorpha (Nutt.) Torr. & Gray. Locally common, talus slopes or esp. around volcanic flatrock areas, *Artemisia*—Pinyon—Juniper, Fir—Pinyon. NTS: S. Belted Range (Rainier Mesa, Gold Meadows), S rim of Pahute Mesa. LINCOLN CO.: Cent. Groom Range (Bald Mtn). NYE CO.: N. Reveille Range. 6500—8500 ft. Perennial. May—June.

A. rosea Greene. Locally common, *Artemisia*—Pinyon—Juniper, usually on talus at base of volcanic cliffs. NTS: Cyns of n. and e. Pahute Mesa and S rim. CLARK CO.: Nw. Spring Mtns (upper Clark Cyn). LINCOLN CO.: Cent. Groom Range (Bald Mtn). NYE CO.: Cent. Belted Range (Indian Spg of W slope), n. Kawich Range (Eden Creek cyn). 6400—9000 ft. Perennial. Late May—July.

Arnica

A. parryi A. Gray ssp. **sonnei** (Greene) Maguire. NYE CO.: Locally common in meadow near stream, edge of *Artemisia*—Pinyon—Juniper, n. Kawich Range (Longstreet cyn). 6800—7100 ft. Perennial. June.

Artemisia

A. biennis Willd. NYE CO.: Local, mountain meadow in *Artemisia*—Pinyon—Juniper, n. Kawich Range (Longstreet cyn). 7200 ft. Annual or biennial. Sept.

A. bigelovii A. Gray in Torr. The low-elevation, shrubby *Artemisia* of limestone mountain slopes, crevices, ledges, and washes; *Atriplex*, *Coleogyne*. NTS: E. Specter Range, w. Spotted Range (Red Mtn, Mercury Ridge), Ranger Mtns, Buried Hills and limestone butte to W, Halfpint Range (hills near head of Nye Cyn). CLARK CO.: N-S axis of Spotted Range. 3400—5200 ft. Shrub. Sept—Oct.

A. dracunculus L. Sandy soils in washes of shallow or deep canyons, around boulders, or bases of volcanic cliffs, usually above 5000 ft; *Atriplex—Ceratoides*, *Artemisia*, *Artemisia*—Pinyon—Juniper, Fir—Pinyon. NTS: Many local sites in Forty-Mile Cyn drainage; Pahute Mesa and below N rim (s. Gold Flat). CLARK CO.: Nw. Spring Mtns (Clark Cyn). LINCOLN CO.: Cent. Groom Range (Bald Mtn, to summit (9380 ft) in *Ceratoides*). NYE CO.: Cent. Belted Range (Indian Spg of W slope), foothills of s. Monitor Range, n. Kawich Range (Eden Creek cyn). 4700—9400 ft. Perennial. Aug—Sept.

A. ludoviciana Nutt. ssp. **ludoviciana.** NYE CO.: Local, mountain meadow in *Artemisia*—Pinyon—Juniper, n. Kawich Range (Longstreet cyn). 7200 ft. Perennial. Sept.

ssp. **incompta** (Nutt.) Keck. Widely distributed at middle elevations of region; locally common at bases of cliffs, around boulders, and in washes; *Atriplex*, *Coleogyne*, *Artemisia*, occasionally *Artemisia*—Pinyon—Juniper. NTS: E. Specter Range, summit of Skull Mtn, w. Spotted Range, Yucca Mtn (nw. Jackass Flats), in and near Shoshone Mtn, Mine Mtn, Timber Mtn, Eleana Range, s. Belted

Range, Halfpint Range (Banded Mtn); Thirsty Cyn. CLARK CO.: Nw. Spring Mtns (Clark Cyn area), N-S axis of Spotted Range. NYE CO.: Nw. Spring Mtns (Crystal Spg cyn), n. Kawich Range (Eden Creek cyn), Cactus Range. 4100—7600 ft. Perennial, woody at base. Aug—Oct.

A. nova A. Nels. [*A. arbuscula* Nutt. ssp. *n.* (A. Nels.) Ward]. (BLACK SAGEBRUSH). Dominant shrub of shallow stony soils above *ca* 5000 ft, with Pinyon Pine and Juniper above *ca* 6000 ft. Throughout the volcanic mountain and mesa areas, in mosaic with *A. tridentata* of the sandy and deeper soils. NTS: Common dominant over the rolling hills of e. Forty-Mile Cyn and dolomite hills of Eleana Range; Timber Mtn, s. Belted Range (Rainier Mesa, Gold Meadows), Pahute Mesa. NYE CO.: Dominant on slopes of n. Cactus Range. 5000—7500 ft. Shrub. Sept—Oct.

A. spinescens D. C. Eat. (BUD-SAGE). Common associated shrub of *Grayia—Lycium*, less common in *Coleogyne* and *Artemisia*; also frequent in lower canyons, esp. in washes, of both limestone and volcanic mountain ranges; in this region, a species of the middle elevations. NTS: In or below w. Spotted Range (Red Mtn, Mercury Ridge), Halfpint Range (incl. French Peak area), s. Shoshone Mtn, summit of Skull Mtn, foothills of se. Belted Range (s. Groom Lake); common on basin floor of Yucca Flat. 3400—5600 ft. Shrub. Apr—May.

A. tridentata Nutt. var. **tridentata.** (BIG SAGEBRUSH). Dominant shrub of sandy soils above *ca* 5000 ft, with Pinyon Pine and Juniper above *ca* 6000 ft; throughout the volcanic mountains and mesas of the region, in mosaic with *A. nova* of the shallow soils of mountain and hill slopes; the *Artemisia* of valley bottoms, depressions, and nearly level terrain. NTS: In or below Shoshone Mtn, s. Belted Range (Rainier Mesa), Pahute Mesa, Halfpint Range. LINCOLN CO.: Groom Range (Bald Mtn). NYE CO.: Cent. Belted Range, Stonewall Mtn; Kawich Range, where it occurs as low plants over large treeless areas, or with *Cercocarpus ledifolius*, from 8000 ft to the high ridges. (4000—) 5000—9200 ft. Shrub. Aug—Oct.

Aster

A. adscendens Lindl. in Hook. NYE CO.: Local, mountain meadows in *Artemisia*—Pinyon—Juniper, n. Kawich Range (Eden Creek cyn). 7000—7200 ft. Perennial. Aug—Sept.

A. exilis Ell. NYE CO.: Local populations frequent in wet alkaline soils in disturbed *Atriplex*, Ash Meadows. 2200—2300 ft. Annual. Aug—Oct.

A. frondosus (Nutt.) Torr. & Gray. NYE CO.: Locally common, moist soils in disturbed *Atriplex*, Warm Spgs (base of s. Hot Creek Range). 5400 ft. Annual. Sept.

A. intricatus (A. Gray) Blake. NYE CO.: Scattered populations frequent in *Atriplex* and *Distichlis* communities of Ash Meadows and Oasis Valley; in *Atriplex—Sarcobatus* in cent. Sarcobatus Flat, and *Sarcobatus—Atriplex parryi* in s. Hot Creek Valley (Twin Spgs area). 2200—5200 ft. Perennial. July—Oct.

A. pauciflorus Nutt. NYE CO.: Common in *Distichlis* meadows and *Juncus* swamps of Ash Meadows and Oasis Valley. 2200—3700 ft. Perennial. Apr—Sept.

Atrichoseris

A. platyphylla A. Gray. Uncommon, absent most years; esp. washes and on disturbed sites of bajadas below limestone mountain ranges; *Larrea*, *Atriplex*, and *Coleogyne*. NTS: Specter Range, w. Spotted Range (Red Mtn), Ranger Mtns. NYE CO.: S. Stewart Valley. 2500—4000 ft. Winter annual. Apr—May.

Baccharis

B. emoryi A. Gray. NTS: Occasional introduced roadside plants in e. Forty-Mile Cyn, Mercury Valley, and near pond in cent. Jackass Flats. NYE CO.: Abundant and native in many areas of moist to wet soils in Ash Meadows, where it is the most common tall shrub, usually with Ash and Screwbean or *Distichlis*; uncommon in Pahrump Valley [along stream of nw. Spring Mtns (Crystal Spg cyn)]. 2200—5000 (—6000) ft. Shrub. Sept—Oct.

Baileya

B. multiradiata Harv. & Gray. Washes and other disturbed sites, *Larrea*, *Atriplex*, *Coleogyne*, restricted to soils derived from limestones. NTS: Common, bajadas below w. Spotted Range (Red Mtn, Mercury Ridge), Ranger Mtns, Buried Hills; occasional below Specter Range and CP Hills. CLARK CO.: Below nw. Spring Mtns (Wheeler Wash cyn, foothills of N slope). 3300—4700 ft. Perennial. Apr—May, individuals at any season.

B. pleniradiata Harv. & Gray. Locally common to abundant in sandy soils, sometimes over extensive areas in *Atriplex canescens* and *Artemisia tridentata*; common in *Larrea—Ambrosia*, *Lycium pallidum—Grayia*, *Grayia—Lycium*, and washes in *Atriplex*. NTS: Jackass Flats, Frenchman Flat, Yucca Flat, throughout Forty-Mile Cyn drainage, Thirsty Cyn, s. Gold Flat. ESMERALDA CO.: E. Big

Smoky Valley. NYE CO.: Cent. Sarcobatus Flat, Stone Cabin Valley, sw. Penoyer Valley. 4000—6000 ft. Winter annual, sometimes biennial. Apr—Oct, mostly July—Aug.

Balsamorhiza

B. hookeri Nutt. var. **neglecta** (Sharp) Cronq. Locally common, *Coleogyne*, *Artemisia nova*, lower *Artemisia*—Pinyon—Juniper. NTS: Eleana Range (near Capt. Jack Spg). NYE CO.: Local, s. Groom Range; N slope of nw. Spring Mtns, where abundant on uplands of Mt. Stirling area. 5600—6200 ft. Perennial. Late Apr—June.

Bebbia

B. juncea (Benth.) Greene (SWEET BUSH). Local, washes in *Larrea*, below limestone mountain ranges. NTS: Below Specter Range, w. Spotted Range (Red Mtn), Ranger Mtns. NYE CO.: Below W end of Spring Mtns. 2500—4100 ft. Shrub. May—Aug.

Brickellia

B. arguta Rob. Common along limestone cliffs and outcrops, in *Atriplex*. NTS: Cyns of Specter Range, e. Skull Mtn. w. Spotted Range (Red Mtn), Ranger Mtns, Buried Hills and limestone butte to W, CP Hills. CLARK CO.: N-S axis of Spotted Range. NYE CO.: Mtn N end of Pahrump Valley, foothills W end of Spring Mtns (Johnnie Mine area), Bare Mtn. 2800—5000 ft. Shrub. May—June.

B. californica (Torr. & Gray) A. Gray. Widely distributed, bases of volcanic cliffs, or around boulders in canyon washes, *Artemisia tridentata*, *Artemisia*—Pinyon—Juniper. NTS: Cyns of Shoshone Mtn, Timber Mtn, Eleana Range, s. Belted Range (Rainier Mesa, Gold Meadows, Oak Spring Butte), Halfpint Range (Cockeyed Ridge), Thirsty Cyn, Pahute Mesa. CLARK CO.: Nw. Spring Mtns (Wheeler Wash area). NYE CO.: Nw. Spring Mtns (Crystal Spg cyn), Grapevine Mtns, Stonewall Mtn, Reveille Range, foothills of s. Quinn Canyon Range, cent. Belted Range. 5000—8400 ft. Perennial, woody at base. Aug—Oct.

B. desertorum Cov. Uncommon except locally, washes or cliff faces, in *Larrea*—*Ambrosia*, *Atriplex*, *Coleogyne*. NTS: Below Yucca Mtn (w. Jackass Flats), w. Spotted Range (Red Mtn). CLARK CO.: Nw. Spring Mtns (Crystal Spg cyn, Johnnie Mine area), N-S axis of Spotted Range. NYE CO.: Limestone mtns NE of Ash Meadows and N end of Pahrump Valley; Bare Mtn, Tolicha Peak area. 2700—5100 ft. Shrub. Aug—Nov.

B. incana A. Gray. Abundant in certain washes, in *Larrea—Ambrosia.* NTS: Below e. Specter Range and w. Spotted Range (Red Mtn). NYE CO.: N-cent. Pahrump Valley; below nw. Spring Mtns (n. Amargosa Valley). 2800—3700 ft. Shrub. June—July, Sept—Oct.

B. longifolia S. Wats. Rare, deep crevices or reentrants in limestone cliffs. NTS: W. Spotted Range (N and S slopes of Red Mtn), n. Ranger Mtns. 3500—4600 ft. Shrub. Sept—Oct.

B. microphylla (Nutt.) A. Gray. var. **scabra** A. Gray. Sandy soils derived usually from volcanic rocks, around boulders along washes or bases of cliffs; *Atriplex, A. canescens, Artemisia tridentata, Artemisia*—Pinyon—Juniper. NTS: Cyns of Shoshone Mtn, Eleana Range (incl. Sugar Loaves area of se. Forty-Mile Cyn), s. Belted Range (Rainier Mesa), Pahute Mesa. NYE CO.: Bare Mtn, N. Yucca Mtn (Beatty Wash), Stonewall Mtn, s. Monitor Range, Reveille Range, s. Hot Creek Range, cent. Belted Range. 3400—6800 ft. Shrub. Sept—Oct.

B. multiflora Kell. Canyon bottoms and bases of cliffs, *Larrea* and *Atriplex.* NTS: Buried Hills. NYE CO.: Cyns W face of cent. Bare Mtn, n. Yucca Mtn (Beatty Wash); one extremely large plant along irrigation channel, cent. Pahrump Valley. 2800—4200 ft. Shrub. Sept—Oct.

B. oblongifolia Nutt. var. **linifolia** (D. C. Eat.) Rob. Widely distributed and often common in washes at middle elevations, esp. around certain white rock outcrops weathering to gravels; esp. in *Artemisia tridentata,* but also *Atriplex, A. canescens, Grayia—Lycium, Coleogyne, Artemisia*—Pinyon—Juniper, and Yellow Pine—Pinyon. NTS: In or below w. Spotted Range (Red Mtn, Mercury Ridge) Ranger Mtns, Halfpint Range (incl. Nye Cyn, hills of e. Yucca Flat, s. Emigrant Valley), CP Hills, Eleana Range, s. Belted Range (n. Yucca Flat, s. Groom Lake), Pahute Mesa, nw. Papoose Range; throughout Forty-Mile Cyn drainage, w. Emigrant Valley. CLARK CO.: Nw. Spring Mtns (upper Clark Cyn and Wheeler Wash area). NYE CO.: Gold Flat, Cactus Flat, Stone Cabin Valley, Kawich Valley, s. Penoyer Valley, nw. Desert Valley. 3400—6400 (—7600) ft. Perennial, woody at base. May—June, some yrs Aug—Sept.

B. watsonii Rob. Widely distributed and common at lower elevations, usually associated with limestone or dolomite outcrops, sometimes calcareous volcanic rocks, esp. cliff crevices and rock outcrops along washes; *Atriplex, Coleogyne, Artemisia tridentata,* rarely *Artemisia*—Pinyon—Juniper. NTS: Specter Range, Skull Mtn, w. Spotted Range (incl. Red Mtn), Ranger Mtns, Buried Hills, Halfpint Range (French Peak mtn, Banded Mtn), calcareous outcrops of Shoshone Mtn, Mine Mtn, dolomite hills of Eleana Range, uplands of Thirsty Cyn, below N rim of Pahute Mesa (s. Gold Flat). NYE CO.: Ne. Bare Mtn, n. Yucca Mtn. 3100—5000 (—6100) ft. Shrub. Sept—Oct.

Calycoseris

C. **parryi** A. Gray. Local, *Coleogyne*, *Artemisia*, and lower *Artemisia*—Pinyon—Juniper. NTS: N slope of Jackass Flats, n. Topopah Valley (esp. on burned sites), s. and w. Mid Valley, nw. Yucca Flat. NYE CO.: Nw. Spring Mtns (Crystal Spg cyn). 4600—6000 ft. Winter annual. Apr—May.

C. **wrightii** A. Gray. Widely distributed, occasional to common locally, *Larrea* (esp. *Larrea—Lycium—Grayia*), *Grayia—Lycium*, *Coleogyne*. NTS: Mercury Valley, Rock Valley, Jackass Flats, Frenchman Flat, Yucca Flat. NYE CO.: Beatty Mtn, Bullfrog Hills. 3000—4200 ft. Winter annual. Apr—May.

Centaurea

C. **maculosa** Lam. NYE CO.: Occasional roadside plants, disturbed *Atriplex* or *Artemisia*; Ralston Valley, s. Hot Creek Valley. 5500—5800 ft. Introduced biennial. Aug—Sept.

C. **melitensis** L. NYE CO.: Occasional, fencerows and disturbed sites, Ash Meadows and Pahrump Valley. 2200—2600 ft. Introduced annual. June—July.

C. **repens** L. NYE CO.: Local, disturbed moist soils, Ash Meadows and Pahrump Valley. 2200—2600 ft. Introduced perennial. May—Oct.

Chaenactis

C. **carphoclinia** A. Gray. Occasional to often the dominant annual, identified with calcareous soils below limestone mountain ranges; *Larrea*, *Atriplex*, *A. hymenelytra*, *Grayia—Lycium*, *Coleogyne*. NTS: Most parts of Mercury Valley, Rock Valley, Jackass Flats, Frenchman Flat, and Yucca Flat. NYE CO.; S. Stewart Valley; below Bare Mtn, Bullfrog Hills. 2500—4500 ft. Winter annual. Apr—May.

C. **douglasii** (Hook.) Hook. & Arn. Occasional to common, *Artemisia*—Pinyon—Juniper, Fir, or Fir—Pinyon. NTS: Shoshone Mtn, Eleana Range, s. Belted Range (Rainier Mesa), Pahute Mesa. CLARK CO.: Nw. Spring Mtns (Wheeler Cyn, Clark Cyn). LINCOLN CO.: Cent. Groom Range (Bald Mtn). NYE CO.: Tolicha Peak area, s. Monitor Range, n. and s. Kawich Range (Eden Creek cyn and Cedar

Pass), cyns of both slopes of cent. Belted Range. 5500—9000 ft. Biennial or perennial. Late May—Oct, mostly June.

C. fremontii A. Gray. Sandy soils, when present often the dominant annual; *Larrea—Ambrosia.* NTS: Below Yucca Mtn (extreme w. Jackass Flats) and red cinder cone near S end of Yucca Mtn (n. Amargosa Valley); below Skull Mtn (cent. and s. Jackass Flats, sw. Frenchman Flat, N slope of Rock Valley) and N and S slopes of Specter Range (S slope of Rock Valley, n. Amargosa Valley). NYE CO.: Below Bare Mtn (Amargosa Valley and Crater Flat), Bullfrog Hills. 2600—3800 ft. Winter annual. Apr—May.

C. macrantha D. C. Eat. Rather widely distributed, but usually not common except locally, esp. on calcareous soils; *Larrea, Lycium pallidum—Grayia, Grayia—Lycium, Coleogyne.* NTS: Mercury Valley, Rock Valley, Jackass Flats, Frenchman Flat, Mid Valley, Yucca Flat, s. Gold Flat. NYE CO.: Bullfrog Hills; s. Sarcobatus Flat, Stonewall Flat. 3000—5200 ft. Winter annual. Apr—May.

C. stevioides Hook. & Arn. The most widely distributed and common *Chaenactis* of the region, often the dominant annual in *Larrea, Lycium pallidum—Grayia, Grayia—Lycium,* less frequent in *Atriplex, Coleogyne,* and *Artemisia,* rare in *Artemisia*—Pinyon—Juniper. NTS: Mercury Valley, Rock Valley, Jackass Flats, Topopah Valley, Frenchman Flat, Mid Valley, Yucca Flat, ne. Forty-Mile Cyn, s. Groom Lake, s. Gold Flat. NYE CO.: S. Stewart Valley (in *Atriplex hymenelytra*), n. Amargosa Valley (Bullfrog Hills, near saddle of Fluorspar Cyn/Tates Wash), Railroad Valley. 2500—6000 ft. Winter annual. Late Mar—June.

C. xantiana A. Gray. Occasional to common, esp. in sandy soils; *Larrea—Ambrosia, Larrea—Grayia—Lycium, Lycium pallidum—Grayia, Grayia—Lycium, Larrea/Coleogyne* (rarely *Coleogyne*), *Atriplex, Artemisia tridentata,* rarely *Artemisia*—Pinyon—Juniper. NTS: Mercury Valley, Rock Valley, w. Jackass Flats (sands below Shoshone Mtn), Frenchman Flat, Mid Valley, Yucca Flat, Forty-Mile Cyn, s. Gold Flat, sw. Groom Lake; n. and nw. Pahute Mesa. NYE CO.: Below n. Yucca Mtn (Beatty Wash); s. Penoyer Valley. 3100—6000 ft. Winter annual. Apr—July.

Chaetadelphia

C. wheeleri A. Gray. Usually on loose sands, *Larrea—Ambrosia, Atriplex, A. canescens, Grayia—Lycium,* Mesquite. NTS: W. Spotted Range (sw. Frenchman Flat), sands below Buried Hills, se. Belted Range and nw. Papoose Range (s. Groom Lake), and below NW rim of Pahute Mesa (s. Gold Flat). ESMERALDA CO.: E. Big Smoky Valley. NYE CO.: Low dunes of Ash Meadows, playa N of Stonewall

Mtn (Stonewall Flat), below n. Belted Range (s. Penoyer Valley). 2200—5500 ft. Perennial. Apr—June, some yrs Sept.

Chrysothamnus

C. albidus (M. E. Jones) Greene. NYE CO.: Occasional to common in several areas of Ash Meadows and Oasis Valley, in *Atriplex—Haplopappus*, *Juncus*, or *Distichlis*; in *Atriplex parryi—Sarcobatus* in s. Hot Creek Valley (near Twin Spgs). In n. Ash Meadows, may hybridize with the hybrid of *C. nauseosus* ssp. *mohavenis* and *C. n.* ssp. *hololeucus* (Anderson, L. C., *Bull. Torr. Bot. Club*, **100**: 171. 1973). 2200—5200 ft. Shrub. Aug—Oct.

C. greenei (A. Gray) Greene ssp. **filifolius** (Rydb.) Hall & Clem. NTS: Local, in *Artemisia*—Pinyon—Juniper, e. Pahute Mesa. NYE CO.: Dominant shrub over large areas of basin floor of Kawich Valley; common in *Atriplex* and *Atriplex—Ceratoides* in Gold Flat and e. Cactus Flat; *Atriplex canescens*, *Artemisia tridentata*, and *Grayia—Lycium* in Penoyer Valley. 5000—6800 ft. Shrub. Sept—Oct.

C. nauseosus (Pall.) Britton ssp. **albicaulis** (Nutt.) Hall & Clem. *Artemisia tridentata* and *Artemisia*—Pinyon—Juniper. CLARK CO.: Common locally, nw. Spring Mtns (upper Clark Cyn, and along Willow Creek). NYE CO.: Locally abundant, n. Kawich Range (Longstreet cyn). 6000—7400 ft. Shrub. Sept—Oct.

ssp. **consimilis** (Greene) Hall & Clem. NTS: Known from one population on a flat sandy area of e. Pahute Mesa, in *Artemisia tridentata*. 6900 ft. Shrub. Aug—Oct.

ssp. **hololeucus** (A. Gray) Hall & Clem. Common in sandy soils, esp. in washes, throughout most of the region above 5000 ft; *Artemisia tridentata*, *Artemisia*—Pinyon—Juniper, uncommon in *Atriplex*. NTS: In or near Shoshone Mtn, Timber Mtn, Eleana Range, s. Belted Range (Rainier Mesa, Gold Meadows), Pahute Mesa; along washes in *Atriplex—Ceratoides*, s. Gold Flat; scattered plants in washes in *Larrea*, Mercury Valley and s. Frenchman Flat. NYE CO.: Washes in or near Bare Mtn, nw. Yucca Mtn, Grapevine Mtns, Goldfield Hills, Cactus Range, s. San Antonio Mtns, foothills of Monitor Range, Kawich Range, Reveille Range; with *Atriplex torreyi*, as hybrids with ssp. *mohavensis* in w. Pahrump Valley (Anderson, L. C., *Bull. Torr. Bot. Club*, **100**: 171. 1973). (3400—) 4500—7600 ft. Shrub. Sept—Oct.

ssp. **leiospermus** (A. Gray) Hall & Clem. NTS: Widely distributed and more or less common in washes in *Coleogyne*, *Atriplex*, rarely *Larrea*, in Mercury Valley, e. Rock Valley, Topopah Valley, n. and w. Frenchman Flat, Mid Valley, ne. Yucca Flat, Thirsty Cyn; common in *Artemisia* and lower *Artemisia*—Pinyon—Juniper in or below Shoshone Mtn, Timber Mtn, Mine Mtn, Eleana Range, s. Belted Range (slopes of Rainier Mesa), Pahute Mesa. CLARK CO.: Nw. Spring Mtns (Wheeler Wash). NYE CO.: In or below Bare Mtn, n. Yucca Mtn, Cactus Range, Kawich Range, Reveille Range, cent. Belted Range. 3500—6500 ft. Shrub. Sept—Oct.

ssp. **mohavensis** (Greene) Hall & Clem. CLARK CO.: W. Indian Springs Valley. NYE CO.: The common subspecies in *Atriplex* communities of Pahrump Valley (where atypical), Oasis Valley, and s. and cent. Sarcobatus Flat, extending along canyon washes into *Artemisia*—Pinyon—Juniper in the Grapevine Mtns; occas. plants in lowlands of e. Amargosa Valley. In n. and e. Ash Meadows, populations are a hybrid swarm, with ssp. *hololeucus* the other inferred parental taxon, and with characters varying between the extremes of the two subspecies (Anderson, L. C., *Bull. Torr. Bot. Club*, **100**: 171. 1973). 2200—6800 ft. Shrub. Sept—Oct.

C. paniculatus (A. Gray) Hall. Essentially restricted to washes in *Larrea*, where if present it is often the dominant shrub, usually at around 3500 ft. NTS: Below both slopes of Specter Range, both slopes of w. Spotted Range, Ranger Mtns, Halfpint Range (below French Peak mtn, and esp. along Nye Cyn wash), abundant in Forty-Mile Cyn wash below w. Shoshone Mtn; Thirsty Cyn. NYE CO.: Nw. Spring Mtns (Crystal Spg cyn), below Bare Mtn (Chuckwalla Cyn, Fluorspar Cyn), uncommon in Oasis Valley, cent. and n. Sarcobatus Flat. 3200—4600 ft. Shrub. Sept—Oct.

C. parryi (A. Gray) Greene ssp. **asper** (Greene) Hall & Clem. CLARK CO.: Nw. Spring Mtns (upper Clark Cyn) in Yellow Pine—Fir and Yellow Pine—Pinyon. LINCOLN CO.: With Limber Pine, cent. Groom Range (Bald Mtn). NYE CO.: *Artemisia—Cercocarpus* in n. Kawich Range (Eden Creek cyn). In Pinyon—Fir—*Cercocarpus* in Quinn Canyon Range to the N. 7400—9000 ft. Shrub. July—Sept.

ssp. **nevadensis** (A. Gray) Hall & Clem. Uncommon except locally, *Artemisia*—Pinyon—Juniper. NTS: Eleana Range, s. Belted Range (length of E rim and sw. Rainier Mesa, Gold Meadows), ne. Pahute Mesa. NYE CO.: N. Kawich Range (Eden Creek cyn). 6000—8000 ft. Shrub. Late Aug—Oct.

C. teretifolius (Dur. & Hilg.) Hall. Widely distributed species of the foothills of the middle elevations, around volcanic or limestone outcrops, boulders, and cliff crevices, esp. in *Coleogyne*, less frequent in *Larrea—Atriplex*, *Atriplex*, *Grayia—Tetradymia*, *Artemisia*. NTS: Mercury Valley, Rock Valley, Jackass Flats, Topopah Valley, Frenchman Flat, Mid Valley, Forty-Mile Cyn, s. Groom Lake, Emigrant Valley, and Thirsty Cyn. NYE CO.: Mtns E of Ash Meadows and n. Pahrump Valley; nw. Spring Mtns (N slope below

Mt. Stirling), Bare Mtn, Grapevine Mtns. 2700—5700 ft. Shrub. Aug—Oct.

C. viscidiflorus (Hook.) Nutt. ssp. **puberulus** (D. C. Eat.) Hall & Clem. The most widely distributed and common *Chrysothamnus* in the volcanic mountain and mesa areas, usually above 5500 ft; *Artemisia*, *Artemisia*—Pinyon—Juniper; occasional in *Coleogyne*, *Atriplex—Ceratoides*, *Grayia—Lycium*. NTS: In or below Shoshone Mtn, Buckboard Mesa, Eleana Range, s. Belted Range (Rainier Mesa), Pahute Mesa, Halfpint Range. LINCOLN CO.: Groom Range (to summit of Bald Mtn in *Ceratoides*). NYE CO.: In or below nw. Spring Mtns (N slope below Mt. Stirling), Grapevine Mtns, Tolicha Peak, Stonewall Mtn, Goldfield Hills, Cactus Range, n. Kawich Range (in *Artemisia—Cercocarpus* to the high ridges of Kawich Peak area), cent. Belted Range; n. Groom Lake, sw. Penoyer Valley. (4800—) 5500—9400 ft. Shrub. June—Aug.

ssp. **stenophyllus** (A. Gray) Hall & Clem. Widely distributed, the common *Chrysothamnus* of basin floors and foothills, esp. in volcanic areas and on disturbed sites, usually below 5500 ft; *Atriplex*, *A. canescens*, *Grayia—Lycium*, *Coleogyne*, *Artemisia*, uncommon in *Artemisia*—Pinyon—Juniper or *Larrea*. NTS: Washes of n. and w. Mercury Valley, e. Rock Valley, Topopah Valley, n. Frenchman Flat, Mid Valley, Yucca Flat, throughout Forty-Mile Cyn drainage, Thirsty Cyn. CLARK CO.: Nw. Spring Mtns (area of lower Clark Cyn). NYE CO.: Crater Flat, Sarcobatus Flat, Cactus Flat, Gold Flat, cent. Groom Lake. 3600—6500 ft. Shrub. Aug—Oct.

ssp. **viscidiflorus.** *Artemisia tridentata* and *Artemisia*—Pinyon—Juniper, in washes and on sandy uplands, usually above 6000 ft; absent over large discrete areas. NTS: Shoshone Mtn, both slopes of Eleana Range, s. Belted Range (abundant on and below Rainier Mesa, where plants are 2n, 4n, or 6n (Anderson, L. C., *Amer. J. Bot.*, 53: 204. 1966)), n. and e. Pahute Mesa, Thirsty Cyn. CLARK CO.: Nw. Spring Mtns (cent. and upper Clark Cyn). LINCOLN CO.: Cent. Groom Range (to summit of Bald Mtn, in *Ceratoides*). NYE CO.. Stonewall Mtn, n. Kawich Range (lower Eden Creek cyn, and the common *Chrysothamnus* of *Artemisia—Cercocarpus* to the high ridges of Kawich Peak area), cent. and n. Belted Range. 4900—9400 ft. Shrub. Aug—Oct.

Cirsium

C. mohavense (Greene) Petr. NYE CO.: Common in *Atriplex* communities throughout Ash Meadows. 2200—2300 ft. Biennial or perennial. June—Sept.

C. neomexicanum A. Gray. Scattered along washes and lower canyon walls throughout most of region, associated esp. with limestone mountain ranges; *Larrea, Atriplex, Coleogyne, Artemisia,* lower *Artemisia*—Pinyon—Juniper. NTS: In or below both slopes of w. Spotted Range (Red Mtn), Buried Hills, w. Shoshone Mtn, E slope of Eleana Range, Halfpint Range (s. Banded Mtn), nw. Papoose Range. CLARK CO.: N-S axis of Spotted Range. NYE CO.: Cent. Belted Range. 3800—6500 ft. Biennial. May—July.

C. **nidulum** (M. E. Jones) Petr. CLARK CO.: Nw. Spring Mtns (common, upper Clark Cyn) in Yellow Pine. 7000—7600 ft. Perennial. July—Aug.

C. **vulgare** (Savi) Airy-Shaw. NYE CO.: Moist to wet soils, in Ash—Screwbean of n. Ash Meadows; disturbed mountain meadows in *Artemisia*—Pinyon—Juniper, n. Kawich Range (Longstreet cyn). 2300—7200 ft. Introduced biennial. June—Nov.

Conyza

C. **canadensis** (L.) Cronq. Disturbed moist soils in *Larrea, Atriplex, Coleogyne, Artemisia*—Pinyon—Juniper. NTS: Near ponds of n. Mercury Valley and cent. Jackass Flats; Mercury townsite; Whiterock Spg (nw. Yucca Flat). CLARK CO.: Nw. Spring Mtns (common along Cold Creek); Indian Springs townsite. NYE CO.: Common along irrigation ditches of n. Pahrump Valley, local in Ash Meadows; n. Kawich Range (common locally in Eden Creek cyn). 2200—7200 ft. Annual. July—Nov.

C. **coulteri** A. Gray. Occasional plants or locally common, disturbed moist soils in *Larrea, Atriplex.* NTS: Near ponds of n. Mercury and cent. Jackass Flats. NYE CO.: Pahrump Valley and Ash Meadows. 2200—3700 ft. Annual. Sept—Nov.

Crepis

C. **intermedia** A. Gray. Widely distributed in *Artemisia*—Pinyon—Juniper; Fir—Pinyon, *Cercocarpus* thickets. NTS: Shoshone Mtn, Timber Mtn, Eleana Range, s. Belted Range (Rainier Mesa, Gold Meadows), Pahute Mesa. LINCOLN CO.: Cent. Groom Range (Bald Mtn). NYE CO.: Nw. Spring Mtns (N slope below Mt. Stirling), n. Kawich Range (upper Eden Creek cyn). 6000—8500 ft. Perennial. June—July.

C. **occidentalis** Nutt. ssp. **occidentalis.** Uncommon except locally, *Artemisia*—Pinyon—Juniper. NTS: S rim, nw. and e. Pahute Mesa, s. Shoshone Mtn. LINCOLN CO.: Foothills N end of Groom Range. NYE CO.: Bullfrog Hills (summit of Sawtooth Mtn), s. Kawich Range (Cedar Pass). 5800—7000 ft. Perennial. Apr—July.

C. runcinata (James) Torr. & Gray ssp. **hallii** Babc. & Steb. NYE CO.: Seasonally water-saturated soils of a *Distichlis* meadow in cent. Ash Meadows, habitat now destroyed. 2200 ft. Perennial. May.

Dyssodia

D. cooperi A. Gray. Occasional to common locally, usually in washes, *Larrea*, *Grayia—Lycium*, *Coleogyne*, *Artemisia*. NTS: Mercury Valley, cent. Jackass Flats, Frenchman Flat, w. Mid Valley, nw. Yucca Flat. CLARK CO.: N-S axis of Spotted Range. LINCOLN CO.: Wash below N end of Groom Range. 3000—5600 ft. Perennial. May—June, some yrs Aug—Oct.

D. pentachaeta (DC.) Rob. var. **belenidium** (DC.) Strother. [*D. thurberi* (A. Gray) Rob.]. Rock crevices and lower canyon walls, restricted to limestone mountain ranges and foothills of lower elevations; *Larrea*, *Atriplex*, *Coleogyne*. NTS: W. Spotted Range (Red Mtn, Mercury Ridge), Buried Hills. CLARK CO.: Nw. Spring Mtns (limestone hill below N slope), N-S and E-W axes of Spotted Range; Indian Springs townsite. 3200—5200 ft. Perennial. Apr—May, some yrs Aug—Sept.

Encelia

E. farinosa A. Gray (BRITTLE BUSH). NYE CO.: Common in *Atriplex—Ambrosia* and *Larrea—Ambrosia*, ledges of talus slopes, mtns E side of Stewart Valley and N end of Pahrump Valley; scattered roadside plants in n. Amargosa and Pahrump Valley drainages. 2600—3500 ft. Shrub. Apr.

E. frutescens A. Gray. NTS: Occasional plant on disturbed sites, Mercury Valley and n. Amargosa Valley. NYE CO.: Common in *Atriplex* communities of Ash Meadows. 2200—3700 ft. Shrub. May, some yrs Sept.

E. virginensis A. Nels. ssp. **actoni** (Elmer) Keck. NTS: *Larrea—Ambrosia* below w. Specter Range (n. Amargosa Valley), and Forty-Mile Cyn wash (nw. Jackass Flats). NYE CO.: N. Bare Mtn (Amargosa Valley, w. Crater Flat). 3300—4100 ft. Shrub. Apr—July.

ssp. **virginensis.** Widely distributed, common in washes and on lower canyon walls in and below limestone mountain ranges, or calcareous outcrops elsewhere; *Larrea* (esp. *Larrea—Atriplex*), *Grayia—Lycium*, *Coleogyne*. NTS: N. Amargosa Valley, Mercury Valley, Rock Valley, Jackass Flats, Topopah Valley, Frenchman Flat, Yucca Flat, Thirsty Cyn. NYE CO.: Cent. Reveille Range. 2600—5600 ft. Shrub. Apr—Sept, mostly May—June.

Enceliopsis

E. nudicaulis (A. Gray) A. Nels. var. **corrugata** Cronq. (Cronquist, A., *Bull. Torr. Bot. Club*, 99: 246. 1972). NYE CO.: Populations in several localities of Ash Meadows, in *Atriplex*; type locality, n. Ash Meadows. 2200—2300 ft. Perennial. Apr—May.

var. **nudicaulis.** Local populations in or near limestone mountain ranges or limestone outcrops elsewhere, in *Atriplex*, *Artemisia nova.* NTS: E. Specter Range, w. Spotted Range (incl. Mercury Ridge), Ranger Mtns. NYE CO.: S. Monitor Range, n. Belted Range, foothills of Quinn Canyon Range. Several colls. approaching var. *corrugata* (fidė Cronquist) occur without geographic definition. 3300—6400 ft. Perennial. Apr—June.

Erigeron

E. aphanactis (A. Gray) Greene. NTS: Uncommon, nw. Pahute Mesa in *Artemisia tridentata* and *A. nova.* NYE CO.: N. Kawich Range (Longstreet cyn) in *Artemisia*—Pinyon—Juniper. 5600—8000 ft. Perennial. May—June.

E. breweri A. Gray var. **porphyreticus** (M. E. Jones) Cronq. NTS: Rim of East Thirsty Cyn, in *Artemisia tridentata.* 5600 ft. Perennial. July—Aug.

E. clokeyi Cronq. NYE CO.: Common on steep mountain slopes, in *Artemisia—Cercocarpus* of n. Kawich Range (upper Eden Creek cyn). 8000—9100 ft. Perennial. June—July.

E. divergens Torr. & Gray. Occasional small populations, moist soil at base of cliffs, lower talus slopes of mountain canyons, or near springs; *Artemisia*, *Artemisia*—Pinyon—Juniper, or rarely *Coleogyne.* NTS: Shoshone Mtn, s. Belted Range (Rainier Mesa), esp. Pahute Mesa; Topopah Spg, Tippipah Reservoir, Whiterock Spg. NYE CO.: White Blotch Spg W of n. Groom Range. 5200—7500 ft. Annual or biennial. May—Sept.

E. pumilus Nutt. ssp. **concinnoides** Cronq. The common *Erigeron* of the region, usually in *Atriplex*, *Coleogyne*, *Artemisia nova*, or *Artemisia*—Pinyon—Juniper. NTS: Cyns of Shoshone Mtn, Timber Mtn, Eleana Range, Yucca Mtn (nw. Jackass Flats), w. Spotted Range (Red Mtn, Mercury Ridge), CP Hills, s. Belted Range (Rainier Mesa, Gold Meadows), Pahute Mesa, nw. Papoose Range. CLARK CO.: N-S axis of Spotted Range. NYE CO.: Nw. Spring Mtns (N slope below Mt. Stirling), Stonewall Mtn, Goldfield Hills, n. and s. Kawich Range, Reveille Range, cent. and n. Belted Range. 4000—8200 ft. Perennial. Apr—May at lower elevs., June—July at higher elevs.

Eriophyllum

E. ambiguum (A. Gray) A. Gray. NYE CO.: Common, *Larrea—Ambrosia*, washes below nw. Spring Mtns. 3000—4000 ft. Winter annual. Apr—May.

E. pringlei A. Gray. Widely distributed and common locally in *Larrea*, *Lycium pallidum—Grayia*, *Grayia—Lycium*, *Coleogyne*, lower *Artemisia*, *Atriplex*, *Atriplex—Ceratoides*. NTS: Mercury Valley, Rock Valley, Jackass Flats, Topopah Valley, cent. and w. Frenchman Flat, ne., cent. and s. Yucca Flat, sw. Groom Lake, s. Gold Flat. NYE CO.: Below se. Bullfrog Hills, Stonewall Mtn; Oasis Valley. 3300—5000 ft. Winter annual. Apr—May.

Gaillardia

G. pulchella Foug. CLARK CO.: Escape from cultivation, base of nw. Spring Mtns (Cold Creek Field Sta.). 6000 ft. Introduced perennial. June—Sept.

Geraea

G. canescens Torr. & Gray. NTS: Common locally in n. Amargosa Valley (esp. below red cinder cone near S end of Yucca Mtn); occasional plants in *Larrea* of e. Jackass Flats, and disturbed *Larrea* of s. Jackass Flats, e. Rock Valley. NYE CO.: Occasional in *Atriplex*, cent. and s. Ash Meadows. 2200—3700 ft. Winter annual. Mar—June.

Glyptopleura

G. marginata D. C. Eat. NTS: Widely distributed in sandy soils at lower elevs., usually as scattered plants; *Larrea*, *Lycium pallidum—Grayia*, *Grayia—Lycium*, *Atriplex—Ceratoides*, *Coleogyne;* Mercury Valley, Rock Valley, Jackass Flats, Frenchman Flat, Yucca Flat. 3000—4500 ft. Winter annual. Apr.

G. setulosa A. Gray. NYE CO.: Local in *Atriplex* and *Larrea—Sarcobatus*; Oasis Valley; Bullfrog Hills and below Stonewall Mtn. 3400—5000 ft. Winter annual. Apr—May.

Gnaphalium

G. luteo-album L. NYE CO.: Local small populations in wet soils, disturbed *Atriplex*; Ash Meadows, Pahrump Valley; nw. Spring Mtns (Crystal Spg cyn). 2300—5000 ft. Introduced annual. May—July.

G. palustre Nutt. NTS: Disturbed *Artemisia tridentata*, near Tippipah Reservoir (se. Forty-Mile Cyn). NYE CO.: Locally common in disturbed *Artemisia*—Pinyon—Juniper, s. Kawich Range (Cedar Pass). 5200—7000 ft. Annual. July—Aug.

Grindelia

G. fraxino-pratensis Reveal & Beatley (Reveal, J. L., and J. C. Beatley, *Bull. Torr. Bot. Club*, 98: 332. 1971). NYE CO.: Endemic species occurring as usually small populations in *Atriplex*, Ash—Screwbean, and *Distichlis* meadows, in n. and e. Ash Meadows; on moist to wet clay soils, esp. in spg areas; type locality, N of Lodge. 2100—2300 ft. Perennial or biennial. June—Oct.

G. nana Nutt. NYE CO.: Common, disturbed *Atriplex* of moist soils near Warm Spgs (base of s. Hot Creek Range). 5400 ft. Perennial. Aug—Sept.

G. squarrosa (Pursh) Dunal var. **serrulata** (Rydb.) Steyerm. Occasional plants in disturbed *Larrea*, *Atriplex*, *Artemisia*, esp. on road shoulders. NTS: W. Mercury Valley, s. Frenchman Flat. CLARK CO.: Below nw. Spring Mtns (near Cold Creek). ESMERALDA CO.: E. Big Smoky Valley. NYE CO.: Oasis Valley, Sarcobatus Flat, Ralston Valley. 3200—6000 ft. Perennial. June—Sept.

var. **squarrosa.** NYE CO.: Large population in abandoned field, cent. Pahrump Valley. 2600 ft. Perennial. July—Oct.

Gutierrezia

G. microcephala (DC.) A. Gray. Widely distributed and common to abundant, canyon walls, rock outcrops and washes, in essentially all mountain, hill, and upland areas, esp. common and predictable where soils are derived from calcareous rocks, and in *Atriplex*, *Coleogyne*, and *Artemisia*; less common in *Larrea*, *Atriplex*—*Ceratoides*, *Grayia*—*Lycium*, and *Artemisia*—Pinyon—Juniper. NTS: In or below Specter Range, Yucca Mtn, Skull Mtn, w. Spotted Range (Red Mtn), Ranger Mtns, Halfpint Range (French Peak mtn, Banded Mtn), Mine Mtn, Eleana Range, Shoshone Mtn, Timber Mtn, Buckboard Mesa, uplands of Thirsty Cyn, s. Belted Range, Pahute Mesa. LINCOLN CO.: Groom Range. NYE CO.: Ash Meadows; nw. Spring Mtns, Bare Mtn, Stonewall Mtn, Cactus Range, Reveille Range. 2200—8500 ft. Perennial, woody at base. July—Oct.

G. sarothrae (Pursh) Britt. & Rusby. (MATCHWEED). Common, on lower canyon walls, washes, cliffs and ledges, in *Larrea*—*Atriplex*, *Atriplex*, and *Artemisia*—Pinyon—Juniper. NTS: In two dissimilar areas: Common in the limestone w. Spotted Range and Ranger Mtns at 3500—4000 ft; and in the volcanic areas of s. Belted Range (Rainier Mesa) and Pahute Mesa at 5500—7000 ft. CLARK CO.: Nw.

Spring Mtns (Clark Cyn). NYE CO.: In *Artemisia—Cercocarpus*, n. Kawich Range (upper Eden Creek cyn), to 8800 ft. Of more restricted distribution than *G. microcephala*, but growing with it in some areas, e.g., certain cyns of Pahute Mesa and in the limestone ranges. 3500—8800 ft. Perennial, woody at base. July—Oct.

Haplopappus

H. acaulis (Nutt.) A. Gray. NYE CO.: Locally common on gravel talus in *Artemisia*—Pinyon—Juniper of s. Kawich Range (Cedar Pass) and n. Reveille Range (slopes above Reveille). 6500—7000 ft. Perennial. May—June.

H. acradenius (Greene) Blake ssp. **acradenius.** NTS: Roadside plants in n. Amargosa Valley, and s. and w. Mercury Valley, introduced from Ash Meadows. NYE CO.: Widely distributed in Ash Meadows in *Atriplex* and Mesquite, frequently a dominant species and occurring on same sites as ssp. *eremophilus*; roadside plants in Pahrump Valley and w. Indian Springs Valley. 2200—3500 ft. Shrub. Aug—Oct.

ssp. **eremophilus** (Greene) Hall. Distribution as in ssp. *acradenius*, somewhat less common than the typical subspecies, and with no apparent geographic or site distinction in Ash Meadows. 2100—3400 ft. Shrub. Aug—Oct.

H. brickellioides Blake. Common locally, *Atriplex*, rarely *Coleogyne*; around rock outcrops and cliffs, limestone mountain ranges. NTS: Specter Range, w. Spotted Range (incl. Red Mtn), Ranger Mtns, CP Hills. NYE CO.: Limestone mtns and hills NE of Ash Meadows and nw. Pahrump Valley; Bare Mtn. Suspected hybrid plant (*Beatley 7041, 7043*) of this species with *H. acradenius* ssp. *eremophilus*, from roadside of cent. Mercury Valley. 2600—4400 ft. Shrub. June—Oct.

H. cooperi (A. Gray) Hall. Locally common on bajadas in *Larrea*, *Grayia—Lycium*, *Coleogyne*, *Artemisia*. NTS: Nw. Rock Valley, n. and e. Jackass Flats, s. Topopah Valley, w. and cent. Frenchman Flat, s. Mid Valley, cent. and e. Forty-Mile Cyn, sw. and ne. Yucca Flat, s. Groom Lake, w. Emigrant Valley. LINCOLN CO.: Nw. Desert Valley. NYE CO.: W. Crater Flat, s. Sarcobatus Flat, n. Stonewall Flat; Bullfrog Hills. 3600—5600 ft. Shrub. Apr—June.

H. cuneatus A. Gray. Locally common, washes and bases of cliffs in *Coleogyne*, *Artemisia*, lower *Artemisia*—Pinyon—Juniper. NTS: Foothills of s. Shoshone Mtn (n. Topopah Valley), and cliffs along Forty-Mile Cyn wash of w. Shoshone Mtn (nw. Jackass Flats). NYE CO.: Nw. Spring Mtns (near Gold Spg and Rock Spg of N slope), Bullfrog Hills. 3500—6000 ft. Shrub. Sept—Oct.

H. gooddingii (A. Nels.) Munz & Jtn. Locally common, limestone cliff crevices and ledges, in *Atriplex*. NTS: Buried Hills, Halfpint Range (Banded Mtn). 4200—4800 ft. Perennial, woody at base. Apr—May.

H. gracilis (Nutt.) A. Gray. NTS: Occasional plant in disturbed wash below e. Specter Range (ne. Amargosa Valley). 2800 ft. Annual, introduced in this area. Aug—Sept.

H. linearifolius DC. Washes, less common on canyon slopes, in *Coleogyne*, *Artemisia*, and lower *Artemisia*—Pinyon—Juniper. NTS: In or below Yucca Mtn (nw. Jackass Flats), s. and e. Shoshone Mtn (nw. Jackass Flats, n. Topopah Valley, cent. and w. Mid Valley), Eleana Range, s. Belted Range (Oak Spring Butte). NYE CO.: Nw. Spring Mtns (N slope below Mt. Stirling); around rock outcrops or talus slopes, n. Bare Mtn and n. Yucca Mtn (Beatty Wash). 3500—6600 ft. Shrub. Apr—May.

H. nanus (Nutt.) D. C. Eat. On cliffs, around rock outcrops on canyon slopes, or flatrock areas of mesas; *Larrea/Coleogyne*, *Coleogyne*, *Artemisia*, esp. *Artemisia*—Pinyon—Juniper; the common *Haplopappus* of volcanic mountain and hill areas of the middle and higher elevations; occasional in limestone ranges. NTS: W. Spotted Range (Mercury Ridge), Halfpint Range (cyns of French Peak mtn, Banded Mtn), in or below Shoshone Mtn, Timber Mtn, Eleana Range, s. Belted Range (Rainier Mesa), Pahute Mesa and below N rim (s. Gold Flat). CLARK CO.: N-S axis of Spotted Range. NYE CO.: Cactus Range, cent. Reveille Range, n. and cent. Belted Range; around low rock outcrops of cent. Gold Flat. 4200—7500 ft. Shrub. Aug—Sept.

H. racemosus (Nutt.) Torr. ssp. **glomeratus** (Nutt.) Hall. Locally common in *Distichlis*—*Juncus* meadow, s. Oasis Valley. 3400 ft. Perennial. Aug—Oct.

ssp. **sessiliflorus** (Greene) Hall. Locally common in *Distichlis* and *Atriplex*—*Haplopappus* of clay soils, n. and e. Ash Meadows. 2200—2300 ft. Perennial. Aug—Oct.

H. watsonii A. Gray. NTS: Restricted to crevices in volcanic cliffs, *Artemisia*—Pinyon—Juniper, cyn below W slope of Rainier Mesa (s. Belted Range) and SE rim of Pahute Mesa; occurs in large population with two other autumn-flowering species it superficially resembles (*Heterotheca villosa* and *Haplopappus nanus*). 6400—6600 ft. Perennial, woody at base. Sept—Oct.

Hecastocleis

H. shockleyi A. Gray. Locally common, limestone cliff crevices and ledges, *Atriplex* or *Coleogyne*. NTS: W. Spotted Range (Mercury

Ridge), Ranger Mtns, Buried Hills and limestone butte to W, Halfpint Range (French Peak mtn), CP Hills. 3700—5000 ft. Shrub. Apr—May.

Helianthus

H. annuus L. ssp. **jaegeri** (Heiser) Heiser. Occasional individuals or small populations on moist soils; Ash—Screwbean, *Atriplex*, or *Distichlis*, sites usually disturbed. CLARK CO.: Indian Springs townsite. NYE CO.: Spring areas of Ash Meadows, and a number of areas of cent. Pahrump Valley. 2200—3100 ft. Perennial. July—Sept.

spp. **lenticularis** (Dougl.) Ckll. Widely scattered small populations on moist disturbed sites, in *Artemisia*, *Atriplex*. NTS: Disturbed site, Mercury Valley; Pahute Mesa. CLARK CO.: W. Indian Spgs Valley; common near Cold Creek (below nw. Spring Mtns). LINCOLN CO.: Nw. Desert Valley. NYE CO.: Occasional plants in Ash Meadows, Oasis Valley, and n. Groom Lake. 2200—6600 ft. Perennial. July—Sept.

H. nuttallii Torr. & Gray. NYE CO.: Local small population in *Distichlis* wet meadow, n. Ash Meadows. 2300 ft. Perennial. Sept—Oct.

H. petiolaris Nutt. ssp. **fallax** Heiser. NTS: Disturbed *Larrea*, cent. Jackass Flats. 3700 ft. Introduced perennial. Sept—Nov.

ssp. **petiolaris.** NTS: Small persistent population on disturbed moist site in *Artemisia tridentata*, nw. Pahute Mesa. 6200 ft. Perennial. July—Sept.

Heterotheca

H. grandiflora Nutt. NYE CO.: Occasional plants along roadside, disturbed *Larrea—Atriplex*, n. Pahrump Valley. 2800 ft. Annual, introduced in this area. Sept—Oct.

H. villosa (Pursh) Shinners var. **hispida** (Hook.) Harms. [*Chrysopsis v.* (Pursh) Nutt. ex DC, var. *h.* (Hook.) A. Gray ex D. C. Eat.]. NTS: Around rock outcrops, boulders, and crevices, in *Artemisia*—Pinyon—Juniper; s. Belted Range (W slope of Rainier Mesa), cyns of Pahute Mesa, washes of Thirsty Cyn. (4800—) 6000—7000 ft. Perennial. July—Sept.

Hulsea

H. vestita A. Gray ssp. **inyoensis** (Keck) Wilken. Locally common, usually on steep sandy slopes, occasionally in sandy washes or valley bottoms, soils usually derived from white tuff bedrock; *Artemisia*—Pinyon—Juniper, less commonly *Artemisia tridentata* and

Atriplex canescens. NTS: Forty-Mile Cyn wash below Timber Mtn, Eleana Range, s. Belted Range (slopes of Rainier Mesa and uplands to E), below S rim of Pahute Mesa, and Kawich Valley below ne. Pahute Mesa. NYE CO.: N. and cent. Belted Range (Cliff Spg, Johnnies Water cyn). (4600—) 6000—7200 ft. Perennial. May—July, some yrs Sept—Oct.

Hymenoclea

H. salsola Torr. & Gray. Widely distributed and common in washes of the lower and middle elevations, often the common shrub of washes in *Larrea*, *Atriplex*, *Grayia—Lycium*, and *Coleogyne.* Sometimes hybridizes with *Ambrosia dumosa (Beatley & Bostick 5121).* NTS: N. Amargosa Valley, Mercury Valley, Rock Valley, Jackass Flats, Topopah Valley, Frenchman Flat, Mid Valley, Yucca Flat, s. Groom Lake. CLARK CO.: Below s. Spotted Range. NYE CO.: N. Pahrump Valley; Bullfrog Hills. 3000—5000 ft. Shrub. Apr—May.

Hymenopappus

H. filifolius Hook. var. **eriopodus** (A. Nels.) Turner. CLARK CO.: Occasional to common locally, Yellow Pine—Pinyon; nw. Spring Mtns (upper Clark Cyn). 7200 ft. Perennial. June—Sept.

var. **megacephalus** Turner. Occasional to locally common, *Artemisia*—Pinyon—Juniper. NTS: S. Belted Range (Rainier Mesa, Gold Meadows), Pahute Mesa. NYE CO.: Cent. Belted Range (Johnnies Water cyn). 6200—7500 ft. Perennial. June—July.

Hymenothrix

H. loomisii Blake. CLARK CO.: Scattered roadside plants, disturbed *Larrea—Atriplex*, w. Indian Springs Valley. 3400 ft. Perennial, introduced from Arizona where native. Sept—Oct.

Hymenoxys

H. cooperi (A. Gray) Ckll. var. **cooperi.** Occasional to locally common, shallow soils, *Artemisia nova* and *Artemisia*—Pinyon—Juniper. NTS: In or below Timber Mtn, s. Belted Range (Rainier Mesa), Pahute Mesa. CLARK CO.: Nw. Spring Mtns (Wheeler Cyn, Trough Spg). LINCOLN CO.: Foothills of n. Groom Range. NYE CO.: N. and cent. Belted Range, cent. and s. Kawich Range (Stinking Spg and Cedar Pass), n. Reveille Range. 5600—8000 ft. Perennial. June—July.

H. lemmonii (Greene) Ckll. CLARK CO.: Yellow Pine—Fir, nw. Spring Mtns (lowlands of upper Clark Cyn). 8000—9000 ft. Perennial. July.

Iva

I. acerosa (Nutt.) Jacks. [*Oxytenia a.* Nutt.]. NYE CO.: Common in many areas of Ash Meadows, with Mesquite and *Atriplex.* 2200—2300 ft. Perennial, woody at base. July—Oct.

I. axillaris Pursh var. **robustior** Hook. CLARK CO.: *Larrea—Atriplex* and low dunes with Mesquite, Cactus Spgs. NYE CO.: Moist soils, esp. near springs, in *Atriplex*, *Artemisia*, *Artemisia*—Pinyon—Juniper; Oasis Valley; Goldfield Hills, Cactus Range, n. and s. Kawich Range (mouth of Eden Creek cyn and Cedar Spg), s. Monitor Range, n. Groom Range and White Blotch Spg to W. 3200—6600 ft. Perennial. June—Sept.

I. nevadensis M. E. Jones. Locally common, *Atriplex*, *A. canescens*, *Atriplex—Ceratoides.* NTS: Sandy soils near playa, Frenchman Flat; common in sands of e. and s. Forty-Mile Cyn. LINCOLN CO.: Below n. Groom Range. NYE CO.: Widely distributed over Gold Flat, Kawich Valley, and e. Groom Lake. 3100—6300 ft. Annual. June—Sept.

Lactuca

L. oblongifolia Nutt. in Fras. [*L. pulchella* (Pursh) DC.; *L. tartarica* (L.) C. A. Mey. ssp. *pulchella* Steb.]. CLARK CO.: Common locally along Willow Creek of nw. Spring Mtns, in *Artemisia*—Pinyon—Juniper. 6000 ft. Perennial. Sept—Oct.

L. serriola L. Scattered plants or small populations on usually disturbed and moist soils, esp. along irrigation channels and seepage areas near springs; in *Larrea*, *Atriplex*, *Coleogyne*, *Artemisia tridentata.* NTS: Near pond, Mercury townsite, and Whiterock Spg (nw. Yucca Flat); small populations on s. Pahute Mesa. CLARK CO.: Indian Springs townsite. NYE CO.: Common as weed along irrigation channels of Pahrump Valley, and locally in undisturbed Mesquite; common at Willow Spg (e. Goldfield Hills), and Tonopah townsite. Most populations are f. *integrifolia* Bogenh. (with unlobed leaves). 2600—7000 ft. Introduced annual. June—Oct.

Lepidospartum

L. latisquamum S. Wats. Essentially restricted to washes, where it may be occasional or the dominant shrub; washes in *Atriplex*, *Artemisia*, *Artemisia*—Pinyon—Juniper. CLARK CO.: Nw. Spring

Mtns (dominant shrub of Wheeler Wash; Cold Creek—Willow Creek area). NYE CO.: Grapevine Mtns, occasional in washes of Tolicha Peak—Obsidian Butte area, common in washes of s. Cactus Range, occasional s. San Antonio Mtns, below N end of Kawich Range, and cyns of E face of Reveille Range. 5000—6200 ft. Shrub. Sept—Oct.

Leucelene

L. ericoides (Torr.) Greene. In washes or shallow soils, usually common locally, esp. on disturbed sites; *Atriplex—Ceratoides*, *Grayia—Lycium*, *Coleogyne*, *Artemisia*, and *Artemisia*—Pinyon—Juniper. NTS: In or near Shoshone Mtn, Buckboard Mesa, Eleana Range, s. Belted Range (Rainier Mesa, Oak Spring Butte), Pahute Mesa (Juniper Flats), nw. Papoose Range. CLARK CO.: Nw. Spring Mtns (Clark Cyn). LINCOLN CO.: N. and s. Groom Range. NYE CO.: Cactus Range, cent. and n. Belted Range, foothills of s. Quinn Canyon Range. 4500—7800 ft. Perennial. May—Aug.

Lygodesmia

L. dianthopsis (D. C. Eat.) Tomb [*L. grandiflora* (Nutt.) Torr. & Gray]. Uncommon and local, apparently absent most years, sandy disturbed sites or washes; in *Artemisia tridentata*, *Artemisia*—Pinyon—Juniper. NTS: S. Belted Range (S and W slopes of Rainier Mesa, Gold Meadows). NYE CO.: Cent. Belted Range (Johnnies Water cyn). 6300—6700 ft. Perennial. Early to mid-June.

Machaeranthera

M. ammophila Reveal (Reveal, J. L., *Bull. Torr. Bot. Club*, 97: 171. 1970). NYE CO.: Endemic to Ash Meadows, on sandy soils, incl. low dunes, in scattered localities over the area, in *Atriplex* communities, type locality W of Carson Slough; continuing into Inyo Co., Calif. 2100—2200 ft. Annual. May—Sept.

M. canescens (Pursh) A. Gray var. **canescens.** The *Machaeranthera* of the middle and higher elevations, widely distributed and common, esp. in sandy soils; *Atriplex*, *A. canescens*, *Atriplex—Ceratoides*, *Grayia—Lycium*, *Coleogyne*, *Artemisia*, and *Artemisia*—Pinyon—Juniper, uncommon in *Larrea*. NTS: Rock Valley, n. Jackass Flats, Topopah Valley, Frenchman Flat (*Atriplex hymenelytra* of French Peak mtn, *Lycium pallidum—Grayia* and *Larrea* of lowlands), Mid Valley, Yucca Flat, abundant on sands of e. Forty-Mile Cyn, Thirsty Cyn, s. Gold Flat; Eleana Range, s. Belted Range (Rainier Mesa), Pahute Mesa. CLARK CO.: Nw. Spring Mtns (Clark Cyn). LINCOLN CO.: Cent. Groom Range (Bald Mtn). NYE CO.: Nw. Yucca Mtn,

Grapevine Mtns, Stonewall Mtn, n. Kawich Range (upper Eden Creek cyn in *Artemisia—Cercocarpus*), foothills of s. Monitor Range, Reveille Range; Cactus Flat, common in many areas of Kawich Valley. 3200—9000 ft. Biennial (sometimes winter annual) at lower and middle elevs., perennial at higher elevs. Apr—Oct.

M. tortifolia (A. Gray) Cronq. & Keck var. **imberbis** Cronq. NTS: The variety of w. Spotted Range and Ranger Mtns, esp. in washes or locally on white gravels, in *Atriplex*. 3600—4800 ft. Perennial, woody at base. Apr—June.

var. **tortifolia.** The widely distributed *Machaeranthera* of the lower elevations, not known to occur with *M. canescens*; rather consistently present in *Larrea* communities (esp. *Larrea—Lycium—Grayia*), *Grayia—Lycium*, and *Coleogyne*. NTS: N. Amargosa Valley (below Specter Range), Mercury Valley, Rock Valley, Jackass Flats, s., w., and nw. Frenchman Flat, cent. and s. Yucca Flat. NYE CO.: N. Ash Meadows, below N end of Spring Mtns, Beatty Mtn, Bare Mtn, se. Bullfrog Hills, Tolicha Peak area. 2300—4500 ft. Perennial, woody at base. Apr—June.

Malacothrix

M. coulteri A. Gray. NTS: Occasional plants, *Larrea—Ambrosia*, *Larrea—Lycium—Grayia*, and *Larrea/Coleogyne* of e. Rock Valley and N and E slopes of Jackass Flats. NYE CO.: Common, *Larrea—Ambrosia—Atriplex* of se. Bullfrog Hills. 3500—4300 ft. Winter annual. Apr.

M. glabrata A. Gray. Widely distributed, usually below 5000 ft; common to abundant locally in *Larrea—Ambrosia*, *Larrea—Atriplex*, *Larrea—Lycium—Grayia*, *Lycium pallidum—Grayia*, *Grayia—Lycium*, *Atriplex*, *Atriplex—Ceratoides*, rarely *Artemisia* or *Artemisia*—Pinyon—Juniper. Does not occur with *M. sonchoides*. NTS: Mercury Valley, Rock Valley, Jackass Flats, Topopah Valley, Frenchman Flat, Mid Valley, Yucca Flat, Forty-Mile Cyn. NYE CO.: Below nw. Yucca Mtn (Beatty Wash), common in e. Bullfrog Hills, n. Groom Lake (rare in this basin). 3100—5800 ft. Winter annual. Apr—May.

M. sonchoides (Nutt.) Torr. & Gray. Common in local or extensive areas of sandy soils, esp. *Larrea—Ambrosia*, *Atriplex canescens*, and *Artemisia tridentata*; uncommon in *Grayia—Lycium* and *Atriplex*. NTS: Mercury Valley, s. and w. Jackass Flats, sw. and e. Frenchman Flat, Yucca Flat, e. and s. Forty-Mile Cyn, sw. Groom Lake. NYE CO.: Beatty Wash (below Yucca Mtn); Stone Cabin Valley, s. Penoyer Valley. 3000—6000 ft. Winter annual. Apr—June.

Microseris

M. linearifolia (Nutt.) Sch.-Bip. [*M. lindleyi* (DC.) A. Gray]. Uncommon, usually growing in and under shrubs; *Coleogyne*, *Artemisia*, *Artemisia*—Pinyon—Juniper. NTS: Shoshone Mtn, Timber Mtn, below CP Hills, s. Belted Range (W face of Rainier Mesa), and below S rim of Pahute Mesa. 4200—6500 ft. Annual. Apr—June .

Monoptilon

M. bellidiforme Torr. and Gray ex Gray. Common in usually sandy soils, present to some extent in most kinds of *Larrea* communities, but esp. *Larrea—Ambrosia*; less frequent in *Lycium pallidum—Grayia* and lowland *Atriplex*. NTS: N. Amargosa Valley, w. Mercury Valley, Rock Valley, Jackass Flats, Frenchman Flat, in Yucca Flat only in NE part where *Ambrosia dumosa* occurs locally (below Banded Mtn). NYE CO.: Amargosa Valley S of Bullfrog Hills, Oasis Valley. 2600—3400 ft. Winter annual. Late Mar—May.

M. bellioides (A. Gray) Hall. The less common *Monoptilon* of the region, usually in *Larrea—Ambrosia*. NTS: Ne. Amargosa Valley, e. Rock Valley, n. and sw. Mercury Valley. NYE CO.: Stewart Valley (in *Atriplex hymenelytra*), Ash Meadows (in *Atriplex*), n. Pahrump Valley, ne. Amargosa Valley. 2200—3500 ft. Winter annual. Late Mar—June.

Pectis

P. papposa Harv. & Gray ex Gray. When present, usually locally abundant; *Larrea*, *Coleogyne*, esp. *Grayia—Lycium*, *Artemisia*. NTS: N. Amargosa Valley, Frenchman Flat, many areas of Yucca Flat, Thirsty Cyn. NYE CO.: N. Pahrump Valley, w. Railroad Valley, s. Hot Creek Valley, cent. and n. Groom Lake. 2800—5800 ft. Summer annual. July—Oct, depending on time of germination.

Perityle

P. intricata (Brandeg.) Shinners [*Laphamia megalocephala* S. Wats. ssp. *i.* (Brandeg.) Keck]. Restricted to limestone areas, locally common on lower slopes and washes of canyons (rarely on upper bajadas) in *Atriplex*. In the field, this plant is grossly distinct from *P. megalocephala*, and is here accorded its former status of species. NTS: Number of parts of w. Spotted Range (incl. Red Mtn, Mercury Ridge), Specter Range. NYE CO.: Low limestone hill of n. Pahrump Valley; Bare Mtn. 2600—4800 ft. Shrub. June—Aug.

P. megalocephala (S. Wats.) Macbr. var. **megalocephala** [*Laphamia m.* S. Wats.]. Local and infrequent, usually at bases of volcanic cliffs in *Atriplex*, *Artemisia*, or most often *Artemisia*—Pinyon—Juniper. NTS: Shoshone Mtn, Pahute Mesa. NYE CO.: Tolicha Peak—Obsidian Butte area, Stonewall Mtn. 4200—6800 ft. Shrub. July—Sept.

Petradoria

P. discoidea L. C. Anderson [*Chrysothamnus gramineus* Hall]. NTS: Common locally in *Artemisia*—Pinyon—Juniper, often around outcrops of white volcanic tuff; n. Shoshone Mtn, several areas of n. Eleana Range, s. Belted Range (NE rim of Rainier Mesa). CLARK CO.: Nw. Spring Mtns (common on slopes of upper Clark Cyn, in Yellow Pine—Pinyon and Yellow Pine—Fir). 5900—7600 ft. Perennial, woody at base. Late July—Sept.

P. pumila (Nutt.) Greene. Common around boulders, cliffs, talus slopes, in *Artemisia*—Pinyon—Juniper, *Artemisia*—*Cercocarpus*, and Yellow Pine—Fir. NTS: Shoshone Mtn, s. Belted Range (Rainier Mesa), Pahute Mesa. CLARK CO.: Nw. Spring Mtns (upper Clark Cyn, where less common than *P. discoidea*). LINCOLN CO.: Cent. Groom Range (Bald Mtn). NYE CO.: S. Monitor Range, n. Kawich Range (upper Eden Creek cyn). 5800—9100 ft. Perennial, woody at base. June—Aug.

Peucephyllum

P. schottii (A. Gray) A. Gray (PYGMY-CEDAR). Crevices and cliffs of limestone mountain ranges, in *Atriplex*, *A. hymenelytra*, *Larrea*—*Ambrosia*. NTS: S. Striped Hills, Specter Range, w. Spotted Range. CLARK CO.: S. Spotted Range. NYE CO.: Mtns N end of Pahrump Valley and E side of Stewart Valley. 2600—4200 ft. Shrub. Late Apr—May.

Pleurocoronis

P. pluriseta (A. Gray) King & Robinson [*Hofmeisteria p.* A. Gray]. Uncommon, limestone mountain canyons, in *Larrea*—*Ambrosia*, *Atriplex*. NTS: Cyn wash, S face of Specter Range. NYE CO.: Occasional to abundant, mtns at N end of Pahrump Valley and E of Ash Meadows, and on ledges and in crevices of E face of Bare Mtn. 2700—3400 ft. Shrub. Apr—May.

Pluchea

P. purpurascens (Sw.) DC. NYE CO.: Local in wet meadows, with *Baccharis*, *Scirpus*, or *Distichlis*, in spring areas of n. and e. Ash Meadows. 2200—2300 ft. Perennial. July—Oct.

P. sericea (Nutt.) Cov. NYE CO.: Occasional to common locally, usually on disturbed soils (often along fencerows) near certain springs of ne. Ash Meadows, with *Baccharis* or *Scirpus*. 2200—2300 ft. Shrub. May—June.

Porophyllum

P. gracile Benth. Local populations in washes and roadside depressions, *Larrea—Ambrosia*. NTS: Disturbed site, cent. Mercury Valley. NYE CO.: Below N end of Spring Mtns (Amargosa Valley, Pahrump Valley). 2500—3500 ft. Perennial. Apr—Sept.

Prenanthella

P. exigua (A. Gray) Rydb. [*Lygodesmia e.* A. Gray]. Widely distributed, but seldom common; *Larrea*, *Atriplex*, *A. canescens*, *Atriplex—Ceratoides*, *Grayia—Lycium*, *Coleogyne*. NTS: Mercury Valley, Rock Valley, Jackass Flats, Frenchman Flat, Mid Valley, s. and e. Yucca Flat, sw. Forty-Mile Cyn. CLARK CO.: N-S axis of Spotted Range (e. Frenchman Flat). NYE CO.: Stewart Valley, n. Pahrump Valley, Amargosa Valley N of Ash Meadows, sw. Gold Flat. 2600—5400 ft. Winter annual. Apr—July.

Psathyrotes

P. annua (Nutt.) A. Gray. More or less common in washes and on other disturbed sites in *Larrea*, *Lycium pallidum—Grayia*, *Atriplex canescens*, *Artemisia tridentata*, but belonging to *Atriplex* types of floors of closed drainage basins, and washes and slopes in *Atriplex* of limestone mountain canyons and bajadas. NTS: Mercury Valley, Rock Valley, Frenchman Flat, s. Forty-Mile Cyn, Thirsty Cyn, s. Gold Flat. LINCOLN CO.: S. Penoyer Valley. NYE CO.: Ash Meadows, cent. Pahrump Valley, w. Indian Springs Valley, cent. Sarcobatus Flat, w. Cactus Flat, many areas of Kawich Valley, s. Hot Creek Valley. 2200—6000 ft. Annual. May—Oct.

P. ramosissima (Torr.) A. Gray. NTS: Wash in *Larrea—Ambrosia* below cent. Specter Range (ne. Amargosa Valley). 3000 ft. Perennial (*Beatley 10025*). Apr.

Psilostrophe

P. cooperi (A. Gray) Greene. Occasional, usually in washes, *Larrea* or *Coleogyne*. NTS: N. and se. Mercury Valley. CLARK CO.: Nw. Spring Mtns (common on uplands and in washes of upper S-facing bajada). 3500—4500 ft. Shrub. May—June, Sept—Oct.

Rafinesquia

R. neomexicana A. Gray. Widely distributed, but usually not common locally; *Larrea*, *Grayia—Lycium*, *Coleogyne*. NTS: Mercury Valley, e. Rock Valley, Jackass Flats, w. and s. Frenchman Flat, cent. and s. Yucca Flat. NYE CO.: Occasional, s. Crater Flat. 3000—4200 ft. Winter annual. Late Mar—May.

Senecio

S. douglasii DC. Occasional in washes or other disturbed sites, or bases of cliffs, *Atriplex* or *Larrea—Ambrosia*, canyons of limestone mountain ranges and bajadas below. NTS: Specter Range, w. Spotted Range (incl. Red Mtn, Mercury Ridge), Ranger Mtns, Halfpint Range (French Peak mtn, Raysonde Buttes area), Yucca Mtn (w. Jackass Flats), w. Shoshone Mtn (limestone outcrops along Forty-Mile Cyn wash), and Thirsty Cyn. NYE CO.: Talus slopes of mtn E side of Stewart Valley, washes below nw. Spring Mtns (Amargosa Valley). 2600—4800 ft. Shrub. May—June, Sept—Oct.

S. integerrimus Nutt. var. **exaltatus** (Nutt.) Cronq. *Artemisia*—Pinyon—Juniper. NTS: Rare, s. Belted Range (Rainier Mesa). NYE CO.: n. Belted Range (common on moist soil, Eden Creek cyn). 7200—7600 ft. Perennial. May—June.

S. multilobatus Torr. & Gray ex Gray. Following Barkley (Barkley, T. M., *Brittonia*, **20**: 267. 1968), *S. multilobatus* is interpreted to include *S. stygius* Greene, and *S. lynceus* Greene (of the Spring Mtns). As treated, the species occurs in most parts of the region, in *Coleogyne*, *Artemisia*, *Artemisia*—Pinyon—Juniper. The *stygius* phase is mostly from (3700—) 4000—6000 ft, the *lynceus* phase at 7200—7800 ft (in Yellow Pine—Fir and Fir), and typical *multilobatus* is common from 6000—8000 ft in most areas of the region. NTS: Yucca Mtn (nw. Jackass Flats), Shoshone Mtn, Eleana Range, s. Belted Range (Rainier Mesa, incl. slopes), Pahute Mesa, nw. Papoose Range. CLARK CO.: Nw. Spring Mtns (common, upper Clark Cyn in Yellow Pine—Fir, Wheeler Wash). LINCOLN CO.: N. and cent. Groom Range (incl. Bald Mtn, where occurs in Fir). NYE CO.: Below nw. Yucca Mtn (Beatty Wash), cent. and n. Belted Range, Reveille Range. 3700—9000 ft. Perennial. May—July.

S. spartioides Torr. & Gray. Uncommon and local, base of cliffs, lower talus slopes, and washes, in *Artemisia*—Pinyon—Juniper, Yellow Pine—Fir. NTS: S. Belted Range (ne. Rainier Mesa, and cyn to NE), certain cyns of n. Pahute Mesa. CLARK CO.: Nw. Spring Mtns (upper Clark Cyn). LINCOLN CO.: Cent. Groom Range (with Limber Pine on Bald Mtn). 6000—9000 ft. Perennial. Late July—Oct.

Solidago

S. sparsiflora A. Gray. CLARK CO.: Local in Yellow Pine—Fir, nw. Spring Mtns (upper Clark Cyn). 7400 ft. Perennial. Aug—Sept.

S. spectabilis (D. C. Eat.) A. Gray. CLARK CO.: With Mesquite on low dunes, Cactus Spgs and Indian Spgs. NYE CO.: Local populations in Ash—Screwbean and *Atriplex*, n. and e. Ash Meadows, *Juncus* in Oasis Valley. 2300—3700 ft. Perennial. Sept—Oct.

Sonchus

S. arvensis L. NYE CO.: Occasional along wash in Pinyon—Juniper, nw. Spring Mtns (Crystal Spg cyn). 5000 ft. Introduced perennial. May.

S. asper L. Disturbed and irrigated soils. NTS: Mercury townsite. CLARK CO.: Indian Springs townsite. NYE CO.: Occasional, along irrigation channels or near springs, Ash Meadows. 2200—3800 ft. Introduced annual. May—Aug.

S. oleraceus L. Locally abundant weed of disturbed moist soils. CLARK CO.: Indian Springs townsite; nw. Spring Mtns (Willow Creek). NYE CO.: N. Pahrump Valley, Beatty townsite, local in n. Kawich Range. 2700—7200 ft. Introduced annual. May—Sept.

Stephanomeria

S. exigua Nutt. ssp. **exigua.** Widely distributed throughout the region in sandy soils; *Larrea* (esp. *Larrea—Ambrosia), Atriplex, A. canescens, Atriplex—Ceratoides, Lycium pallidum—Grayia, Artemisia tridentata*, and *Artemisia*—Pinyon—Juniper, uncommon in *Coleogyne*. Pappus bristles vary from commonly 5 [as in ssp. *pentachaeta* (D. C. Eat.) Hall] to 18, without geographic definition. NTS: Rock Valley, w. Jackass Flats, n. Topopah Valley, nw. and ne. Frenchman Flat, cent. and n. Yucca Flat, s. and e. Forty-Mile Cyn, s. Gold Flat; s. Belted Range (Rainier Mesa, Oak Spring Butte), n. and nw. Pahute Mesa. NYE CO.: Tolicha Peak area of Gold Flat, cent. and w. Groom Lake, s. Penoyer Valley. 3000—6500 ft. Winter annual, at least at lower elevs. May—July.

S. **parryi** A. Gray. Locally common, *Larrea—Grayia—Lycium*, *Larrea—Sarcobatus*, *Grayia—Lycium*, *Coleogyne*, and *Coleogyne/Artemisia*. NTS: N. and cent. Jackass Flats, Topopah Valley, w. Frenchman Flat, Mid Valley, w. half of Yucca Flat, s. Groom Lake; rarely ne. Yucca Flat and w. Mercury Valley; associated primarily with soils derived from s. and e. Shoshone Mtn, CP Hills, Mine Mtn, and Eleana Range. NYE CO.: Oasis Valley; nw. Spring Mtns (N slope below Mt. Stirling), Bare Mtn, nw. Yucca Mtn, Bullfrog Hills. 3400—5500 ft. Perennial. Late Apr—June.

S. **pauciflora** (Torr.) Nutt. Common to abundant, lower canyon walls and washes of limestone mountain ranges and bajadas below, calcareous outcrops elsewhere; *Larrea*, *Atriplex*, *Coleogyne*. NTS: In or below Specter Range, w. Spotted Range (Red Mtn, Mercury Ridge), e. Skull Mtn, Ranger Mtns, Halfpint Range (French Peak mtn, Raysonde Buttes area), CP Hills, Mine Mtn, Shoshone Mtn, Yucca Mtn, s. Belted Range (Oak Spring Butte). NYE CO.: S. Stewart Valley, n. Ash Meadows, cent. Pahrump Valley, n. Amargosa Valley. Most plants are glabrous, but the tomentulose form [var. *parishii* (Jeps.) Munz] and the glabrous form occur in Ash Meadows and Pahrump Valley; in the French Peak area of Frenchman Flat, both occur in the same local population (*Beatley*, *12848*, *12849*). 2200—5200 ft. Perennial, woody at base. June—Oct.

S. **spinosa** (Nutt.) Tomb [*Lygodesmia s.* Nutt.]. Occasional, to locally common esp. on disturbed sites, *Artemisia*, *Artemisia—Cercocarpus*, or usually *Artemisia*—Pinyon—Juniper. NTS: Shoshone Mtn, uplands E of Buckboard Mesa, s. Belted Range (Rainier Mesa, Gold Meadows), Pahute Mesa and below S rim. NYE CO.: S. Monitor Range, n. and s. Kawich Range (Eden Creek cyn and Cedar Pass area), cent. Belted Range (Johnnies Water cyn). 5900—9000 ft. Perennial, woody at base. July—Aug.

S. **tenuifolia** (Torr.) Hall. CLARK CO.: Nw. Spring Mtns (local populations in upper Clark Cyn in Yellow Pine—Pinyon, and near Cold Creek Spg in *Artemisia*—Pinyon—Juniper). 6200—7400 ft. Perennial. July—Sept.

Stylocline

S. **micropoides** A. Gray. Locally common, *Larrea*, *Coleogyne*, *Atriplex*, less common in *Artemisia nova*, *Artemisia*—Pinyon—Juniper. NTS: Bajadas of n. Amargosa Valley, Mercury Valley, Rock Valley, n. Jackass Flats, Mid Valley, w. and ne. Frenchman Flat, w. Yucca Flat; nw. and e. Pahute Mesa, and s. Belted Range (sw. Rainier Mesa). CLARK CO.: S. Spotted Range (w. Indian Springs Valley). NYE CO.: N. Crater Flat; below Bare Mtn and se. Bullfrog Hills

(Amargosa Valley). 3100—7400 ft. Winter annual (at middle and lower elevs.). Apr—May.

Syntrichopappus

S. fremontii A. Gray. Often locally abundant in *Atriplex*, *A. hymenelytra*, *Coleogyne*, *Artemisia nova*, and lower *Artemisia*—Pinyon—Juniper, uncommon in *Larrea*. NTS: E. Rock Valley, n. and e. Jackass Flats, Topopah Valley, n., s. and w. Frenchman Flat, esp. Mid Valley, w. and ne. Yucca Flat, e. Forty-Mile Cyn. NYE CO.: Nw. Spring Mtns (N slope below Mt. Stirling), nw. Yucca Mtn, Bullfrog Hills, Grapevine Mtns, Tolicha Peak area, Stonewall Mtn. 3100—7200 ft. Winter annual. Late Mar—May.

Taraxacum

T. officinale Wiggers. Predictable in moist soils of spring areas at higher elevations throughout the region, usually in *Artemisia*—Pinyon—Juniper; also irrigated sites at lower elevations. CLARK CO.: Indian Springs townsite; nw. Spring Mtns (Cold Creek area). NYE CO.: Stonewall Mtn Spg, wet meadows of n. Kawich Range. 3100—7200 ft. Introduced perennial. Apr—July.

Tetradymia

T. axillaris A. Nels. var. **axillaris** (COTTONTHORN). Associated shrub species on bajadas (usually as occasional plants) in *Larrea—Grayia—Lycium*, *Grayia—Lycium*, *Lycium pallidum—Grayia*, *Coleogyne*, and lower *Artemisia*, most common in *Atriplex* of lower canyons of limestone mountain ranges. NTS: Mercury Valley, Jackass Flats, Frenchman Flat, Mid Valley, Yucca Flat (incl. Plutonium Valley), Forty-Mile Cyn, s. Groom Lake. CLARK CO.: N-S axis of Spotted Range. NYE CO.: Nw. Yucca Mtn; Stonewall Flat. 3600—5400 ft. Shrub. Apr—May.

T. canescens DC. (GRAY HORSEBRUSH). Widely distributed in *Artemisia*—Pinyon—Juniper as occasional plants. NTS: Shoshone Mtn, Timber Mtn, Eleana Range, esp. s. Belted Range (Rainier Mesa), se. Pahute Mesa. LINCOLN CO.: Cent. Groom Range (with Limber Pine on Bald Mtn). NYE CO.: N. Kawich Range (upper Eden Creek cyn, in *Artemisia—Cercocarpus)*. 5500—9000 ft. Shrub. July—Aug.

T. glabrata A. Gray. (HORSEBRUSH). An associated or sometimes locally dominant species of *Atriplex*, *Grayia—Lycium*, *Coleogyne*, *Artemisia tridentata*, and *Artemisia*—Pinyon—Juniper. NTS: N. Frenchman Flat (French Peak area and near playa in *Atriplex canescens)*, Mid Valley, Yucca Flat and Forty-Mile Cyn where frequently occurs with *T. axillaris*, s. Groom Lake; cyns of n. Pahute Mesa. NYE CO.: Oasis Valley, s. Sarcobatus Flat, cent. Groom Lake. 3100—6400 ft. Shrub. May—June.

Townsendia

T. scapigera D. C. Eat. Uncommon, small populations occurring as scattered plants in *Artemisia nova* and *Artemisia*—Pinyon—Juniper (esp. around flatrock outcrops). NTS: Nw. Pahute Mesa. LINCOLN CO.: N. Groom Range (Desert Valley). NYE CO.: NE of n. Belted Range (sw. Penoyer Valley), n. Reveille Range (Railroad Valley), below s. Monitor Range (Ralston Valley). 6100—6500 ft. Perennial. May—June.

Tragopogon

T. dubius Scop. NYE CO.: Occasional as a weed along roadsides, e. Ralston Valley. 5900 ft. Introduced perennial. June—July.

T. pratensis L. CLARK CO.: Occasional plants, disturbed *Artemisia* of Cold Creek area, below nw. Spring Mtns. 6000 ft. Introduced perennial. June—July.

Viguiera

V. multiflora (Nutt.) Blake var. **nevadensis** (A. Nels.) Blake. Widely distributed and common, esp. on disturbed sites at middle and higher elevations; *Grayia—Lycium*, *Coleogyne*, *Artemisia nova*, *Artemisia*—Pinyon—Juniper, Yellow Pine. NTS: In or below Shoshone Mtn, CP Hills, Eleana Range, s. Belted Range (Rainier Mesa), nw. Papoose Range. CLARK CO.: Nw. Spring Mtns (common in middle and upper Clark Cyn). LINCOLN CO.: Common along washes of foothills, n. and cent. Groom Range. NYE CO.: Nw. Spring Mtns (around rock outcrops near Gold Spg); cent. Sarcobatus Flat and s. Penoyer Valley. 4000—7500 ft. Perennial, woody at base. Apr—Oct.

V. reticulata S. Wats. NYE CO.: Along washes in *Larrea—Ambrosia* and *Coleogyne*, below certain limestone mountain ranges; mtn E side of Stewart Valley, N end of Spring Mtns (Johnnie Mine area, Crystal Spg cyn), n. Bare Mtn—Yucca Mtn area (Tungsten Cyn and Fluorspar Cyn). 2600—4600 ft. Perennial, woody at base. May—June.

Xanthium

X. strumarium L. var. **canadense** (Mill.) Torr. & Gray. Moist, usually disturbed sites in Mesquite, *Distichlis*, *Atriplex*, *Larrea*,

Coleogyne, or *Artemisia*. NTS: Along Forty-Mile Cyn wash (nw. Jackass Flats), Whiterock Spg (nw. Yucca Flat). NYE CO.: Ash Meadows, cent. Pahrump Valley, Oasis Valley; W of Warm Springs (base of s. Hot Creek Range). 2300—5800 ft. Annual. July—Aug.

BERBERIDACEAE. Barberry Family

Berberis

B. fremontii Torr. Scattered plants or common locally in *Artemisia*—Pinyon—Juniper, Pinyon—Fir. CLARK CO.: Nw. Spring Mtns (Cold Creek—Willow Creek area). LINCOLN CO.: Cent. Groom Range (Bald Mtn). 6000—7000 ft. Shrub. June.

B. repens Lindl. CLARK CO.: Locally common in Yellow Pine—Fir, nw. Spring Mtns (head of Clark Cyn). 9400 ft. Shrub. May—June.

BORAGINACEAE. Borage Family

Amsinckia

A. tessellata A. Gray. Common, to the dominant annual over large areas, occurring to some extent on at least the upper bajadas of most drainage basins of the region; *Larrea*, *Atriplex*, *Lycium pallidum—Grayia*, *Grayia—Lycium*, *Coleogyne*, *Artemisia*. NTS: Mercury Valley, Rock Valley, esp. Jackass Flats, Topopah Valley, Frenchman Flat, Mid Valley, Yucca Flat, s. Groom Lake, w. Emigrant Valley. CLARK CO.: Boulders below cliffs, N-S axis of Spotted Range. NYE CO.: N. Amargosa Valley (below N end of Spring Mtns), Crater Flat (below Bare Mtn and nw. Yucca Mtn); Bullfrog Hills, Grapevine Mtns, e. Reveille Range. 2600—6000 ft. Winter annual. Apr—May.

Coldenia

C. canescens DC. var. **canescens.** NTS: Local around limestone outcrops, in *Atriplex* of upper bajadas below cent. and s. Ranger Mtns. 3600—3700 ft. Shrub. May—June.

C. nuttallii Hook. Locally common in extensive or small areas of loose sands; *Larrea—Ambrosia*, *Atriplex canescens*, *Atriplex—Ceratoides*, *Artemisia tridentata*, uncommonly *Artemisia*—Pinyon—Juniper. NTS: Sands of nw. Jackass Flats, ne. and sw. Frenchman Flat, esp. s. and e. Forty-Mile Cyn, red cinder cone (s. Yucca Mtn) of n. Amargosa Valley, s. Gold Flat, s. Groom Lake; sandy washes of n.

and e. Pahute Mesa. NYE CO.: N. Kawich Valley, s. Penoyer Valley. 2600—6500 ft. Spring-germinating annual. June—Sept.

C. plicata (Torr.) Cov. Locally common on loose sands. NTS: *Larrea—Ambrosia*, *Atriplex canescens*, usually growing with *C. nuttallii*; nw. Jackass Flats, ne. and sw. Frenchman Flat, s. and e. Forty-Mile Cyn. NYE CO.: N. Amargosa Valley (Big Dune). 2400—5500 ft. Perennial. May—Oct.

Cryptantha

C. ambigua (A. Gray) Greene. Common locally, *Artemisia*—Pinyon—Juniper. NTS: S. Shoshone Mtn, s. Belted Range (W rim and cent. Rainier Mesa), Pahute Mesa. NYE CO.: Cent. Belted Range (Johnnies Water cyn). 6000—7500 ft. Annual. Apr—July.

C. angustifolia (Torr.) Greene. Common in *Larrea—Ambrosia* and *Atriplex* below limestone mountain ranges of the n. Amargosa drainage. NTS: In or below Striped Hills, Specter Range, w. Spotted Range (Red Mtn), sw. Skull Mtn, Yucca Mtn (Mercury Valley, Rock Valley, Jackass Flats, and n. Amargosa Valley). NYE CO.: Stewart Valley, Ash Meadows, n. and e. Amargosa Valley below N end of Spring Mtns, Bare Mtn, area of n. Yucca Mtn. 2200—4000 ft. Winter annual. Late Mar—May.

C. barbigera (A. Gray) Greene. Local and uncommon; *Larrea/ Coleogyne*, *Artemisia*. NTS: Base of cliffs, s. Shoshone Mtn. CLARK CO.: Base of cliffs, N-S axis of Spotted Range. NYE CO.: Banks of wash, nw. Spring Mtns (Crystal Spg cyn); volcanic cliff ledge, Tolicha Peak area; Grapevine Mtns. 4600—6000 ft. Winter annual. May—June.

C. circumscissa (Hook. & Arn.) Jtn. Abundant and widely distributed throughout the region; *Larrea*, *Atriplex*, *Lycium pallidum—Grayia*, *Grayia—Lycium*, *Coleogyne*, *Artemisia*, rarely *Artemisia*—Pinyon—Juniper. NTS: Mercury Valley, Rock Valley, Jackass Flats, Topopah Valley, Frenchman Flat, Mid Valley, Yucca Flat, Forty-Mile Cyn, sw. Groom Lake; s. Belted Range (Rainier Mesa, Gold Meadows), Pahute Mesa. NYE CO.: Below n. Yucca Mtn (Beatty Wash), Stonewall Mtn, n. Kawich Range (lower Eden Creek cyn), Reveille Range; cent. Sarcobatus Flat, ne. Groom Lake. 3000—7500 ft. Winter annual. Apr—June.

C. confertiflora (Greene) Pays. Occasional to common in mountain ranges, and washes below, in *Atriplex* and *Coleogyne* of limestone mountains, and *Artemisia* and *Artemisia*—Pinyon—Juniper of usually volcanic ranges. There are two geographic and ecologic segments in the region, with gross differences in aspect, and two taxa may be distinguishable with further study. NTS: Specter Range, w.

Spotted Range (incl. Red Mtn, Mercury Ridge), Ranger Mtns, Buried Hills, Halfpint Range (incl. French Peak mtn, Banded Mtn), CP Hills, Shoshone Mtn, Eleana Range (incl. margin of sand dune), s. Belted Range (Rainier Mesa, Gold Meadows), Pahute Mesa. CLARK CO.: Nw. Spring Mtns (Wheeler Wash); N-S axis of Spotted Range. NYE CO.: Ash Meadows (incl. low dunes); Tolicha Peak—Obsidian Butte area, s. Monitor Range, cent. Belted Range. 2200—8000 ft. Perennial. Apr—June.

C. **decipiens** (M. E. Jones) Heller. Uncommon; *Larrea—Ambrosia* and *Atriplex*. NTS: Below Specter Range and outliers; limestone butte W of Buried Hills; base of cliff at edge of Forty-Mile Cyn wash, in area of limestone outcrop (w. Shoshone Mtn); along washes below Yucca Mtn (w. Jackass Flats). NYE CO.: Below nw. Yucca Mtn (Beatty Wash). 3400—4000 ft. Winter annual. Apr—May.

C. **dumetorum** (Greene ex Gray) Greene. Locally common, nearly always growing in shrubs; *Larrea—Ambrosia*, *Larrea—Lycium—Grayia*. NTS: Below w. Spotted Range (e. Mercury Valley); s., cent. and e. Jackass Flats, and w.-cent. Frenchman Flat, on soils derived from N slope of Skull Mtn. NYE CO.: Oasis Valley; below nw. Yucca Mtn (Beatty Wash) and se. Bullfrog Hills. 3100—3600 ft. Winter annual. Apr—May.

C. **flavoculata** (A. Nels.) Pays. Common in *Artemisia* and *Artemisia*—Pinyon—Juniper of most volcanic mountain ranges in the region. NTS: Shoshone Mtn, Eleana Range, s. Belted Range (Rainier Mesa, slopes and nearby areas), Pahute Mesa. CLARK CO.: Nw. Spring Mtns (Wheeler Cyn area). LINCOLN CO.: N. Groom Range. NYE CO.: Nw. Spring Mtns (N slope below Mt. Stirling), s. Monitor Range, n. and s. Kawich Range, n. Reveille Range, n. Belted Range, White Blotch Spg W of n. Groom Range. 5600—8100 ft. Perennial. Apr—June.

C. **gracilis** Osterh. The most common annual *Cryptantha* of volcanic mountain areas, *Artemisia*—Pinyon—Juniper, often growing under Pinyon Pine. NTS: W. Spotted Range, Shoshone Mtn, Eleana Range, s. Belted Range (Rainier Mesa, Gold Meadows), Pahute Mesa. NYE CO.: Nw. Spring Mtns (N slope below Mt. Stirling), Tolicha Peak area, cent. Belted Range, White Blotch Spg W of n. Groom Range. 4700—7500 ft. Annual. May—June.

C. **humilis** (Greene) Pays. var. **ovina** (Pays.) Higgins. Common locally in *Artemisia nova*, on shallow soils usually derived from white tuff bedrock. NTS: S. Belted Range (Gold Meadows). NYE CO.: Below n. Belted Range, cent. Kawich Range, s. Monitor Range. 6000—6700 ft. Perennial. May—June.

C. jamesii (Torr.) Pays. var. **abortiva** (Greene) Pays. Locally common to abundant, *Artemisia*—Pinyon—Juniper and Yellow Pine—Fir. CLARK CO.: Nw. Spring Mtns (throughout upper Clark Cyn). NYE CO.: Nw. Spring Mtns (N slope below Mt. Stirling), n. Kawich Range (Eden Creek and Longstreet cyns). 5400—9200 ft. Perennial. May—June.

C. maritima (Greene) Greene. Locally common on cliff ledges or talus of canyons and washes of usually limestone mountain ranges; *Atriplex.* Some colls. are referable to var. *pilosa* Jtn. NTS: Specter Range, w. Spotted Range, Ranger Mtns, Buried Hills, Yucca Mtn (w. Jackass Flats). NYE CO.: Bare Mtn—n. Yucca Mtn area (n. Amargosa Valley). 2800—4400 ft. Winter annual. Apr—May.

C. micrantha (Torr.) Jtn. Occasional to common locally on sandy soils (usually loose sands); *Larrea—Ambrosia, Larrea—Lycium—Grayia, Atriplex canescens, Lycium pallidum—Grayia, Grayia—Lycium, Artemisia tridentata,* and local in *Coleogyne;* often growing with *C. circumscissa,* which also may have red roots. NTS: Mercury Valley, Rock Valley, Jackass Flats, Frenchman Flat, e. Forty-Mile Cyn, uncommon in Yucca Flat. NYE CO.: Bullfrog Hills. 3000—5500 ft. Winter annual. Apr—May.

C. nevadensis Nels & Kenn. var. **nevadensis.** Widely distributed and common, usually growing in shrubs; *Larrea—Ambrosia, Larrea—Lycium—Grayia, Lycium pallidum—Grayia, Grayia—Lycium, Atriplex, Atriplex—Ceratoides, Coleogyne.* NTS: N. Amargosa Valley, Mercury Valley, Rock Valley, Jackass Flats, Topopah Valley, Frenchman Flat, Mid Valley, Yucca Flat, Forty-Mile Cyn. NYE CO.: S. Stewart Valley (in *Atriplex hymenelytra*), Oasis Valley (in *Larrea—Sarcobatus*), sw. Sarcobatus Flat, Railroad Valley, cent. Groom Lake. 2500—6000 ft. Winter annual. Apr—May.

C. pterocarya (Torr.) Greene. Widely distributed, common to abundant in *Larrea—Ambrosia, Larrea—Lycium—Grayia, Lycium pallidum—Grayia, Grayia—Lycium, Atriplex, Coleogyne, Artemisia,* uncommon in *Artemisia*—Pinyon—Juniper. NTS: N. Amargosa Valley, Mercury Valley, Rock Valley, Jackass Flats, Topopah Valley, Frenchman Flat, Mid Valley, Forty-Mile Cyn; S face of Rainier Mesa (s. Belted Range). NYE CO.: Stewart Valley, Stonewall Flat, Railroad Valley; nw. Spring Mtns (N slope below Mt. Stirling, and n. Amargosa Valley), Bullfrog Hills. 3100—7000 ft. Winter annual. Late Mar—May.

C. racemosa (S. Wats.) Greene. Locally common, *Larrea—Ambrosia, Atriplex, A. hymenelytra,* cliff crevices and talus slopes of limestone mountain ranges. NTS: Cent. and w. Specter Range. NYE

CO.: Mtns E side of Stewart Valley and N end of Pahrump Valley, Bare Mtn. 2600—4300 ft. Winter annual. Apr—May.

C. recurvata Cov. Frequent in *Larrea*, *Lycium pallidum—Grayia*, *Grayia—Lycium*, *Atriplex*, *A. canescens*, less commonly *Coleogyne* and *Artemisia*. NTS: Mercury Valley, Rock Valley, Jackass Flats, many areas of Frenchman Flat, Mid Valley, Yucca Flat, Forty-Mile Cyn, s. Groom Lake; Pahute Mesa. CLARK CO.: N-S axis of Spotted Range. NYE CO.: Crater Flat, Oasis Valley, Stonewall Flat, Cactus Flat, Railroad Valley, sw. Penoyer Valley; nw. Spring Mtns (N slope below Mt. Stirling), Bullfrog Hills. 3000—6100 ft. Winter annual. Apr—May.

C. scoparia A. Nels. Locally common in *Artemisia tridentata* and *Artemisia*—Pinyon—Juniper. NTS: S., e., and n. Pahute Mesa. NYE CO.: N. Kawich Range (upper Eden Creek cyn). 6400—7600 ft. Annual. June—July.

C. torreyana (A. Gray) Greene. NYE CO.: N. Kawich Range (local in *Artemisia*—Pinyon—Juniper, base of volcanic cliffs near mouth of Eden Creek cyn). 6700 ft. Annual. May—June.

C. utahensis (A. Gray) Greene. Washes in *Larrea*, *Coleogyne*, and *Atriplex* of limestone mountain areas. NTS: Often abundant in washes of upper bajadas below Specter Range, w. Spotted Range (Red Mtn, Mercury Ridge), Ranger Mtns, Buried Hills, French Peak/CP Hills area of n. Frenchman Flat and s. Yucca Flat, Mine Mtn, local in washes of Topopah Valley and below N end of Shoshone Mtn; Emigrant Valley. NYE CO.: Uncommon below n. Yucca Mtn (Beatty Wash), Bare Mtn, Bullfrog Hills, Grapevine Mtns. 3300—6000 ft. Winter annual. Apr—May.

C. virginensis (M. E. Jones) Pays. In two geographic and ecologic segments, the plants differing in gross aspect and perhaps worthy of taxonomic distinction: Common below 4000 ft in washes and limestone crevices of *Larrea—Atriplex*, and mostly above 6000 ft in *Artemisia* or *Artemisia*—Pinyon—Juniper of the volcanic mountains and mesas, seldom at the middle elevations. NTS: Specter Range, w. Spotted Range (incl. Red Mtn, Mercury Ridge), Ranger Mtns, Halfpint Range (areas of French Peak mtn and Raysonde Buttes), Shoshone Mtn, Timber Mtn, Eleana Range, s. Belted Range (incl. slopes of Rainier Mesa), n. Pahute Mesa and below N face in Gold Flat; w. Emigrant Valley. CLARK CO.: Nw. Spring Mtns (Wheeler Cyn area, and E of Cold Creek). ESMERALDA CO.: Goldfield Hills. LINCOLN CO.: Cent. Groom Range (below Bald Mtn). NYE CO.: Ash Meadows; in or below nw. Spring Mtns (N slope below Mt. Stirling, Crystal Spg cyn, and Johnnie Mine area), Goldfield Hills, Stonewall Mtn, Cactus Range, s. Monitor Range, n. and s. Kawich

Range, n. and cent. Reveille Range, s. Quinn Cyn Range, cent. Belted Range; cent. Kawich Valley. 2200—7200 ft. Perennial. Apr—June.

C. watsonii (A. Gray) Greene. Common, esp. under Pinyon trees, in *Artemisia*—Pinyon—Juniper of certain areas. NTS: S. Belted Range (Rainier Mesa, Gold Meadows). LINCOLN CO.: Cent. Groom Range (Bald Mtn). NYE CO.: Cyns of E and W slopes of cent. Belted Range, n. Kawich Range (Eden Creek and Longstreet cyns). 6400—7800 ft. Annual. May—July.

Heliotropium

H. curassavicum L. var. **oculatum** (Heller) Jtn. Moist soils, disturbed *Distichlis* or *Atriplex*. CLARK CO.: Indian Springs and Cactus Springs townsites. NYE CO.: Common in n. and e. Ash Meadows and s. Oasis Valley, local in Pahrump Valley. 2200—3300 ft. Perennial. May—Oct.

Lappula

L. occidentalis (S. Wats.) Greene var. **occidentalis** [*L. redowskii* of N. Amer. auths.]. Common locally in canyons, washes, and esp. flatrock areas, *Artemisia nova*, *Artemisia*—Pinyon—Juniper, Yellow Pine—Fir. Most colls. are assignable to var. *cupulata* (A. Gray) Higgins (*L. redowskii* var. *desertorum* (Greene) Jtn.). NTS: In or below s. Belted Range (Rainier Mesa, Gold Meadows, Oak Spring Butte), Pahute Mesa; *Atriplex* near E and W margins of playa, Frenchman Flat. CLARK CO.: Nw. Spring Mtns (upper Clark Cyn, Wheeler Cyn, Willow Creek—Cold Creek area). LINCOLN CO.: Cent. Groom Range (Bald Mtn). NYE CO.: In or below n. and cent. Belted Range; edge of sedge meadow, n. Kawich Range (Longstreet cyn); n. Reveille Range. 3100—7800 ft. Winter annual at lower elevs. Apr—June.

Lithospermum

L. ruderale Dougl. ex Lehm. Locally common, esp. around boulder-cliffs, *Artemisia tridentata* and *Artemisia*—Pinyon—Juniper. NTS: SE rim of Pahute Mesa. CLARK CO.: Nw. Spring Mtns (Oak thickets on burned area of S face below Wheeler Pass). NYE CO.: N. Kawich Range (Eden Creek cyn). 7200—8100 ft. Perennial. May—June.

Pectocarya

P. heterocarpa (Jtn.) Jtn. Common to abundant locally, *Larrea*—*Ambrosia*, *Larrea*—*Atriplex*, *Atriplex*, *A. hymenelytra*. NTS: Bajadas below Specter Range (Rock Valley and Amargosa Valley), w. Spotted Range (incl. Red Mtn), Ranger Mtns. NYE CO.: N. Stewart Valley, n. Amargosa Valley (below W end of Spring Mtns), Oasis Valley; below Bare Mtn, Beatty Mtn. 2500—3900 ft. Winter annual. Mar—Apr.

P. platycarpa (Munz & Jtn.) Munz & Jtn. Common to abundant, *Larrea—Ambrosia—Atriplex.* NTS: Bajadas below Specter Range, w. Spotted Range (incl. Red Mtn). NYE CO.: Below W end of Spring Mtns (n. Amargosa Valley), Bare Mtn, Beatty Mtn, nw. Yucca Mtn (Beatty Wash), se. Bullfrog Hills. 2600—4000 ft. Winter annual. Mar—Apr.

P. recurvata Jtn. Locally common to abundant, *Larrea—Ambrosia*, *Larrea—Atriplex.* NTS: Below w. Specter Range, w. Spotted Range (Red Mtn), limestone butte W of Buried Hills, Yucca Mtn (w. Jackass Flats). NYE CO.: Below Resting Spring Range (n. Stewart Valley), nw. Spring Mtns (n. Amargosa Valley), Bullfrog Hills, Bare Mtn—n. Yucca Mtn area (n. Amargosa Valley). 2600—3700 ft. Winter annual. Mar—May.

P. setosa A. Gray. Locally common, *Larrea—Grayia—Lycium*, esp. *Coleogyne*, *Artemisia nova*, uncommonly *Artemisia*—Pinyon—Juniper. NTS: Bajadas below Specter Range (w. Mercury Valley, e. Rock Valley), N slope of Skull Mtn, s. and e. Shoshone Mtn, Mine Mtn, Eleana Range; e. Forty-Mile Cyn, Emigrant Valley. NYE CO.: Nw. Spring Mtns (N slope below Mt. Stirling, and below W end in Amargosa Valley), n. Yucca Mtn, Bare Mtn, Bullfrog Hills, Grapevine Mtns. 3500—6000 ft. Winter annual. Mar—May.

Plagiobothrys

P. arizonicus (A. Gray) Greene ex Gray. NTS: Local, *Coleogyne* and *Artemisia* below S face of Shoshone Mtn (n. Topopah Valley), *Coleogyne* below Eleana Range (nw. Yucca Flat). 5000—5800 ft. Winter annual. Apr—May.

P. jonesii A. Gray. Frequent, esp. in washes, bases of cliffs, rock ledges, or lower slopes of limestone hills and mountains, in *Larrea—Atriplex*, *Atriplex*, *Coleogyne.* NTS: Striped Hills, Specter Range, below sw. Skull Mtn, w. Spotted Range (Red Mtn), Ranger Mtns, Buried Hills and limestone butte to W, Halfpint Range (French Peak mtn, Scarp Cyn), CP Hills. CLARK CO.: Nw. Spring Mtns (Cold Creek area); N-S axis of Spotted Range. NYE CO.: Below W end of Spring Mtns (Crystal Spg cyn; ne. Amargosa Valley), Bare Mtn, Beatty Mtn, n. Yucca Mtn (Beatty Wash). 3100—5200 ft. Winter annual. Apr—May.

P. kingii (S. Wats.) A. Gray. Local populations in *Artemisia* and *Artemisia*—Pinyon—Juniper. NTS: S rim of Pahute Mesa. ESMERALDA CO.: Goldfield Hills. NYE CO.: Areas of Quartz Mtn, Tolicha Peak, e. Goldfield Hills, s. Monitor Range, Reveille Range, s. Kawich Range (Rose Spg cyn). 5600—7000 ft. Annual. Apr—June.

P. scouleri (Hook. & Arn.) Jtn. var. **penicillatus** (Greene) Cronq. [*P. cognata* (Greene) Jtn.]. NYE CO.: Locally common, near springs in *Artemisia*—Pinyon—Juniper, cent. and n. Kawich Range (Stinking Spg, Eden Creek cyn). 6600—8100 ft. Annual. June.

P. stipitatus (Greene) Jtn. NYE CO.: Scattered plants or locally common in *Juncus* and *Distichlis* meadows, Ash Meadows. 2100—2300 ft. Perennial. Apr—May.

BRASSICACEAE (CRUCIFERAE). Mustard Family

Arabis

A. dispar M. E. Jones. NTS: Locally common in Eleana Range, red-brown volcanic talus of hill near Capt. Jack Spg, in *Artemisia nova*—Pinyon—Juniper. 5800—6200 ft. Perennial. Apr—June.

A. fendleri (S. Wats.) Greene. Locally common around cliffs and on talus slopes, *Coleogyne/Artemisia*, *Artemisia*—Pinyon—Juniper, Fir—Pinyon, Yellow Pine—Pinyon, Yellow Pine—Fir. CLARK CO.: Nw. Spring Mtns (upper Clark Cyn, Trough Spg, Wheeler Cyn and Pass area). LINCOLN CO.: Cent. Groom Range (Bald Mtn). NYE CO.: Nw. Spring Mtns (N slope below Mt. Stirling). 5800—9200 ft. Perennial. Apr—May.

A. fernaldiana Roll. var. **stylosa** (S. Wats.) Roll. LINCOLN CO.: Pinyon—Fir, cent. Groom Range (Bald Mtn). 8800 ft. Perennial. June.

A. glaucovalvula M. E. Jones. Occasional in or under shrubs, *Larrea—Ambrosia—Atriplex*, *Larrea—Lycium—Grayia*, *Grayia—Lycium*. NTS: W. Spotted Range (Red Mtn); W half of Frenchman Flat (below N and E slopes of Skull Mtn and CP Hills); cent.—w. Yucca Flat (below Mine Mtn). NYE CO.: S. Bullfrog Hills. 3300—4400 ft. Perennial. Apr—May.

A. hirsuta (L.) Scop. var. **pycnocarpa** (Hopkins) Roll. Local under shrubs, moist soils in *Artemisia*—Pinyon—Juniper. CLARK CO.: Nw. Spring Mtns (Cold Creek). NYE CO.: N. Kawich Range (Eden Creek cyn). 6200—7200 ft. Perennial. June—July.

A. holboellii Hornem. var. **pinetorum** (Tidestr.) Roll. Common and widely distributed in *Artemisia*—Pinyon—Juniper, Yellow Pine—Pinyon, nearly always growing in open; characteristic of shallow soils, esp. flatrock areas of mesas, and *Artemisia nova*. NTS:

Shoshone Mtn, Eleana Range, s. Belted Range (Rainier Mesa, Gold Meadows), Pahute Mesa. CLARK CO.: Nw. Spring Mtns (lower Clark Cyn area, Trough Spg). LINCOLN CO.: Cent. Groom Range (Bald Mtn). NYE CO.: Grapevine Mtns, n. and s. Kawich Range, Reveille Range, cent. Belted Range. 5500—8200 ft. Perennial. Apr—June.

var. **retrofracta** (Grah.) Rydb. NYE CO.: *Artemisia*—Pinyon—Juniper, n. Kawich Range (Longstreet cyn). 6700 ft. Perennial. May—June.

A. inyoensis Roll. Uncommon in the region. NTS: Volcanic rock crevices in *Coleogyne*, s. Shoshone Mtn. LINCOLN CO.: Cent. Groom Range (Bald Mtn), in Pinyon—Fir. NYE CO.: *Artemisia*—Pinyon—Juniper, s. Kawich Range (Rose Spg cyn). 4600—8500 ft. Perennial. Apr—June.

A. microphylla Nutt. NYE CO.: *Artemisia*—Pinyon—Juniper, n. Kawich Range (occasional plants, upper Longstreet cyn). 7400—7800 ft. Perennial. May—June.

A. pendulina Greene. *Artemisia*—Pinyon—Juniper, Yellow Pine—Pinyon, Yellow Pine—Fir; usually local at bases of cliffs, or in crevices, sometimes common on disturbed sites. NTS: S face of Pahute Mesa. CLARK CO.: Nw. Spring Mtns (upper Clark Cyn). LINCOLN CO.: Cent. Groom Range (Bald Mtn). NYE CO.: Cent. Belted Range (Wheelbarrow Peak). 7000—9200 ft. Perennial. June.

A. perennans S. Wats. Limestone cliff crevices and ledges, in *Larrea—Ambrosia*, *Atriplex*, *Artemisia nova*, *Artemisia*—Pinyon—Juniper. NTS: Ranger Mtns. LINCOLN CO.: Cent. Groom Range (Bald Mtn). NYE CO.: Nw. Spring Mtns (N slope below Mt. Stirling), N side of Stonewall Mtn, s. Kawich Range (Rose Spg cyn). 3600—8000 ft. Perennial. Apr—June.

A. pulchra M. E. Jones var. **gracilis** M. E. Jones. Common and widely distributed, nearly always growing in shrubs, often on sites with *A. holboellii* var. *pinetorum* at the higher elevations; the variety occurring most frequently in volcanic areas and at the higher elevations in *Coleogyne*, *Artemisia*, *Artemisia*—Pinyon—Juniper, uncommonly in *Larrea* and *Atriplex*. NTS: Shoshone Mtn, Eleana Range, s. Belted Range (Rainier Mesa), Yucca Mtn (w. Jackass Flats), Skull Mtn, Buried Hills, Halfpint Range (French Peak mtn), and nw. Papoose Range at 5000—7500 ft, but also in Specter Range, Little Skull Mtn, and Ranger Mtns at 3400—3800 ft. NYE CO.: Nw. Spring Mtns (N slope below Mt. Stirling), Bullfrog Hills, Tolicha Peak area, cent. Grapevine Mtns, s. Kawich Range, n. Reveille Range, at 5200—7000 ft, and below n. Yucca Mtn at 3600—4200 ft. 3500—7500 ft. Perennial. Apr—May.

var. **munciensis** M. E. Jones. Common and widely distributed, nearly always growing in shrubs, most predictable at lower elevations and in limestone mountain ranges or areas of calcareous outcrops, in *Larrea*, *Atriplex*, *Grayia—Lycium*, *Coleogyne*, *Artemisia*. NTS: Cent. Specter Range, Little Skull Mtn, w. Spotted Range (Red Mtn, Mercury Ridge), Ranger Mtns, Halfpint Range (areas of French Peak mtn, Scarp Cyn, and Banded Mtn), Shoshone Mtn, dolomite hills of Eleana Range, CP Hills, Mine Mtn, nw. Pahute Mesa. CLARK CO.: N-S axis of Spotted Range. NYE CO.: N. Yucca Mtn, Bullfrog Hills, below Stonewall Mtn. 3500—6400 ft. Perennial. Apr—May.

A. shockleyi Munz. NTS: Locally common, *Artemisia nova*—Pinyon—Juniper, near or at summit of dolomite hills of Eleana Range. 6000—6400 ft. Biennial, sometimes perennial. Late Mar—May.

Brassica

B. geniculata (Desf.) J. Ball. Disturbed moist soils. NTS: Se. Mercury Valley, n. Frenchman Flat, w. Yucca Flat. NYE CO.: Pahrump Valley; Beatty townsite. 2600—4500 ft. Introduced biennial or perennial. Apr—June.

B. kaber (DC.) Wheeler. NYE CO.: Scattered small populations in cultivated fields, Pahrump Valley. 2700 ft. Introduced annual. May—July.

Capsella

C. bursa-pastoris (L.) Medic. (SHEPHERD'S PURSE). Disturbed moist soils in *Artemisia*—Pinyon—Juniper. CLARK CO.: Nw. Spring Mtns (Willow Creek). NYE CO.: N. Kawich Range (Eden Creek cyn). 6000—7200 ft. Introduced annual or biennial. June—July.

Cardaria

C. pubescens (C. A. Mey.) Jarm. var. **elongata** Roll. NYE CO.: Moist mountain meadows and *Salix* thickets in *Artemisia*—Pinyon—Juniper, n. Kawich Range (Longstreet cyn). 6800—7200 ft. Introduced perennial. June—July.

Caulanthus

C. cooperi (S. Wats.) Pays. Occasional to sometimes abundant, nearly always growing in shrubs; *Larrea* (usually *Larrea—Lycium—Grayia*), *Grayia—Lycium*, and *Coleogyne*. NTS: Bajadas of Mercury Valley, Rock Valley, e. half of Jackass Flats, Mid Valley, Frenchman

Flat, s., cent. and ne. Yucca Flat. NYE CO.: Crater Flat, n. Amargosa Valley (Bullfrog Hills, Beatty Wash). 3000—4300 ft. Winter annual. Late Mar—Apr.

C. crassicaulis (Torr.) S. Wats. Common, *Artemisia*, *Artemisia*—Pinyon—Juniper of certain areas of certain basin floors. LINCOLN CO.: N. and cent. Groom Range (incl. Bald Mtn). NYE CO.: Upper bajadas and foothills of Belted Range (s. Penoyer Valley, e. Kawich Valley), Kawich Range (e. Cactus Flat, nw. Kawich Valley), s. Monitor Range (Stone Cabin Valley). 5800—8000 ft. Perennial. May—June.

C. glaber (M. E. Jones) Rydb. Local, *Artemisia tridentata*, *Artemisia*—Pinyon—Juniper, lower canyon walls and around rock outcrops on talus slopes. NTS: Sw. Shoshone Mtn, Timber Mtn (length of Cat Cyn), Eleana Range, s. Belted Range (S face of Rainier Mesa). CLARK CO.: Nw. Spring Mtns (uplands near Cold Creek). NYE CO.: Nw. Spring Mtns (N slope below Mt. Stirling). 4800—6700 ft. Herbaceous perennial. May—June.

C. lasiophyllus (Hook. & Arn.) Pays. var. **utahensis** (Rydb.) Pays. [*Thelypodium l.* (Hook. & Arn.) Greene var. *u.* (Rydb.) Jeps.]. Common member of *Larrea*, *Grayia—Lycium*, and some *Coleogyne* communities. NTS: Mercury Valley, Rock Valley, Jackass Flats, Frenchman Flat, Mid Valley, Yucca Flat. NYE CO.: Stewart Valley (in *Atriplex hymenelytra*), n. Amargosa Valley (below N end of Spring Mtns, Bare Mtn), Crater Flat; Beatty Wash, Bullfrog Hills. 2600—4500 ft. Winter annual. Late Mar—Apr.

C. pilosus S. Wats. Where present, usually growing in shrubs and occurring over extensive areas; *Atriplex*, *A. canescens*, *Atriplex—Ceratoides*, *Artemisia*. NTS: Nw. Pahute Mesa, and below N face in s. Gold Flat. LINCOLN CO.: Below n. Groom Range. NYE CO.: Nw. Yucca Mtn (Beatty Wash), Bullfrog Hills, Tolicha Peak area, Cactus Range, s. Kawich Range, n. Reveille Range, cent. and n. Belted Range (sw. Penoyer Valley, Kawich Valley), n. Groom Range; n. Groom Lake. 3500—6400 ft. Perennial. Late Apr—early June.

Chorispora

C. tenella (Pall.) DC. NYE CO.: Locally common, mountain meadow in *Artemisia*—Pinyon—Juniper, n. Kawich Range (Longstreet cyn). 7100 ft. Introduced annual. May—June.

Descurainia

D. californica (A. Gray) O. E. Schulz. NYE CO.: Common, under trees and around shrubs, on moist soils in *Artemisia*—Pinyon—

Juniper, n. Kawich Range (Eden Creek cyn). 7000—7600 ft. Biennial. June—July.

D. obtusa (Greene) O. E. Schulz. Occasional, under trees and shrubs, moist soils in *Artemisia*—Pinyon—Juniper or mountain meadows, n. Kawich Range (Eden Creek and Longstreet cyns). 7100—7200 ft. Biennial. June—July.

D. pinnata (Walt.) Britt. ssp. **glabra** (Woot. & Standl.) Detl. Often large and abundant on loose soil over rodent burrows, perhaps the most consistently present herbaceous species of *Larrea* communities (usually absent in *Larrea—Ambrosia* of sandy soils); *Grayia—Lycium*, *Lycium pallidum—Grayia*, *Coleogyne*, *Artemisia tridentata*. NTS: Mercury Valley, Rock Valley, E half of Jackass Flats, Frenchman Flat, Mid Valley, most parts of Yucca Flat. NYE CO.: N. Ash Meadows (in Ash—Screwbean); nw. Spring Mtns (N slope below Mt. Stirling), below Grapevine Mtns (sw. Sarcobatus Flat), cent. Reveille Range (Railroad Valley), s. Kawich Range (e. Cactus Flat). 2200—7000 ft. Winter annual at lower elevs. Mar—May.

ssp. **halictorum** (Ckll.) Detl. The subspecies in *Artemisia* and *Artemisia*—Pinyon—Juniper, in sandy soils of *Artemisia tridentata* or disturbed loose soils over rodent burrows in *A. nova; Atriplex*, Yellow Pine—Pinyon, Yellow Pine—Fir. NTS: Common and locally the dominant ground cover species some years in ne. Forty-Mile Cyn drainage; Pahute Mesa. CLARK CO.: Nw. Spring Mtns (upper Clark Cyn, Trough Spg). LINCOLN CO.: Cent. Groom Range (Bald Mtn). NYE CO.: Below n. Yucca Mtn (Beatty Wash), Goldfield Hills, Reveille Range. 3900—9000 ft. Winter annual at lower elevs. May—June.

D. sophia (L.) Webb. Occasional local populations on disturbed, often temporarily wet soils as in roadside depressions, usually in *Artemisia*, *Artemisia*—Pinyon—Juniper, also in *Larrea* and *Atriplex* at lower elevations. NTS: Mercury townsite; s. and w. Frenchman Flat; s. Groom Lake. CLARK CO.: Nw. Spring Mtns (Cold Creek, upper Clark Cyn). NYE CO.: Ash Meadows; n. Kawich Range, near spgs in cent. Belted Range. 2200—9000 ft. Introduced winter annual. Apr—June.

Draba

D. cuneifolia Nutt. ex Torr. & Gray var. **cuneifolia**. NTS: Local around volcanic flatrock outcrops, *Artemisia nova*, below S rim of Pahute Mesa. 6000—6200 ft. Annual. Apr—May.

var. **integrifolia** S. Wats. Uncommon or abundant locally, around limestone outcrops or ledges, *Larrea*, *Atriplex*, *Coleogyne*. NTS:

Specter Range, w. Spotted Range (e. Mercury Valley). CLARK CO.: N-S axis of Spotted Range. 3600—5000 ft. Winter annual. Apr.

Erysimum

E. asperum (Nutt.) DC. var. **purshii** Dur. [*E. capitatum* (Dougl.) Greene]. CLARK CO.: Common, *Artemisia*—Pinyon—Juniper and Yellow Pine—Pinyon, nw. Spring Mtns (Clark Cyn area, Trough Spg). 5000—8200 ft. Biennial. May—June.

Halimolobos

H. diffusa (A. Gray) O. E. Schulz var. **jaegeri** (Munz) Roll. NYE CO.: Common locally in crevices of columnar volcanic cliffs, in *Artemisia nova*, cent. Belted Range (Johnnies Water cyn); uncommon on grotto cliff face, *Artemisia*—Pinyon—Juniper, Stonewall Mtn (Stonewall Spg). 5600—6200 ft. Perennial. May—June.

Hutchinsia

H. procumbens (L.) Desv. Local to common on moist soils of spring and seepage sites in *Artemisia*, *Artemisia*—Pinyon—Juniper. NYE CO.: Ash Meadows (in Ash—Screwbean); n. and cent. Kawich Range (Eden Creek cyn, Stinking Spg). 2300—7000 ft. Annual. Apr—June.

Lepidium

L. flavum Torr. var. **flavum.** Locally common on bajadas in areas of limestone mountain ranges, *Larrea*, *Atriplex*, *A. hymenelytra*, Mesquite and Screwbean, rarely *Coleogyne*. NTS: Below Specter Range (Rock Valley), w. Spotted Range. CLARK CO.: Below s. Spotted Range. NYE CO.: Ash Meadows, Pahrump Valley, n. and e. Amargosa Valley. 2200—4000 ft. Winter annual. Mar—Apr.

L. fremontii S. Wats. Widely distributed, most common and predictable in *Atriplex* of limestone mountain canyons and washes in *Larrea—Atriplex* of bajadas below; occasional in *Larrea/Coleogyne*, *Grayia—Lycium*. NTS: Local along Forty-Mile Cyn wash (w. Shoshone Mtn); in or below w. Spotted Range (Red Mtn, Mercury Ridge), Ranger Mtns, CP Hills, and Halfpint Range (incl. Banded Mtn and Plutonium Valley). NYE CO.: N. Ash Meadows, n. Pahrump Valley, cent. Sarcobatus Flat; below Bare Mtn. 2200—4400 ft. Perennial, woody at base. Mar—May, some yrs autumn.

L. lasiocarpum Nutt. Widely distributed and common in *Larrea*, *Coleogyne*, occasional on disturbed sites in *Artemisia* and *Arte-*

misia—Pinyon—Juniper, and uncommon in *Grayia—Lycium—Atriplex*. NTS: Ne. Amargosa Valley, Mercury Valley, Rock Valley, n. Jackass Flats, Frenchman Flat, Mid Valley, Yucca Flat. NYE CO.: Ash Meadows, n. Amargosa Valley; below Stonewall Mtn, s. Kawich Range (Cedar Pass area), Reveille Range (Fang Ridge), cent. Belted Range. 2200—6800 ft. Winter annual. Mar—June.

L. montanum Nutt. ssp. **canescens** (Thell.) C. L. Hitchc. Abundant in usually local areas of basin floors of n. part of the region, in *Atriplex*, *A. canescens*, *Atriplex—Ceratoides*, or *Sarcobatus*. NTS: Pahute Mesa; s. Gold Flat, s. Groom Lake, w. Emigrant Valley. NYE CO.: Cactus Flat, Kawich Valley, cent. Groom Lake, sw. Penoyer Valley, cent. Reveille Valley, Railroad Valley. 4500—6000 ft. Perennial. May—Sept.

ssp. **cinereum** (C. L. Hitchc.) C. L. Hitchc. NYE CO.: Common in many areas of Ash Meadows, in *Atriplex*. 2200—2300 ft. Perennial. Apr—May.

L. perfoliatum L. Disturbed moist soils, usually in *Atriplex*. NTS: S. Frenchman Flat, cent. Jackass Flats. NYE CO.: N. Sarcobatus Flat, s. Hot Creek Valley. 3200—5400 ft. Introduced annual. Apr—May.

Lesquerella

L. kingii S. Wats. ssp. **kingii.** Locally common, *Artemisia*—Pinyon—Juniper, *Artemisia—Cercocarpus*, on talus slopes, flatrock areas and washes of usually volcanic mountains and mesas. NTS: Shoshone Mtn, Eleana Range, s. Belted Range (Rainier Mesa, Gold Meadows), Pahute Mesa, Halfpint Range (Cockeyed Ridge). NYE CO.: Grapevine Mtns, (head of Phinney Cyn), n. Kawich Range (upper Eden Creek cyn). 5800—8800 ft. Perennial. May—July.

ssp. **latifolia** (Nels.) Roll. & Shaw (*L. latifolia* A. Nels.). Locally common, *Artemisia*—Pinyon—Juniper, Yellow Pine—Fir, Yellow Pine—Pinyon, in mountain ranges to the east and south of those with ssp. *kingii*. CLARK CO.: Nw. Spring Mtns (upper Clark Cyn, Trough Spg, Wheeler Pass area). LINCOLN CO.: Cent. Groom Range (Bald Mtn). NYE CO.: White Blotch Spg W of n. Groom Range. 6100—9000 ft. Perennial. Late May—July.

L. ludoviciana (Nutt.) S. Wats. Local populations on steep sandy slopes and rock ledges, sands derived from light-colored tuffs; *Atriplex canescens*, *Artemisia*, *Artemisia*—Pinyon—Juniper. NTS: Sugar Loaves area of Eleana Range (e. Forty-Mile Cyn drainage), local on S face of Rainier Mesa (s. Belted Range). 5500—6800 ft. Perennial. May—June.

Malcolmia

M. africana (L.) R. Br. NTS: Local on disturbed irrigated site, Mercury townsite. 3700 ft. Introduced annual. Apr—June.

Phoenicaulis

P. cheiranthoides Nutt. in Torr. & Gray. NYE CO.: Common, on volcanic talus in *Artemisia*, summit of Sawtooth Mtn, Bullfrog Hills. 6000 ft. Perennial. Mar—Apr.

Physaria

P. chambersii Roll. Locally common, esp. on burned or other disturbed sites and washes, in *Artemisia*, *Artemisia*—Pinyon—Juniper, Yellow Pine—Pinyon, Pinyon—Fir. NTS: Shoshone Mtn, s. Belted Range (Gold Meadows), nw. Pahute Mesa. CLARK CO.: Nw. Spring Mtns (Wheeler Wash to upper Clark Cyn, Willow Creek—Cold Creek area, Trough Spg). LINCOLN CO.: N. and cent. Groom Range (incl. Bald Mtn). NYE CO.: Cent. Belted Range (Cliff Spg), s. Kawich Range (Cedar Pass). 5500—8800 ft. Perennial. May—June.

Raphanus

R. raphanistrum L. (CHARLOCK). NYE CO.: Irrigated field, Oasis Valley. 3400 ft. Introduced perennial. May—June.

Rorippa

R. nasturtium-aquaticum (L.) Schinz & Thell. [*Nasturtium officinale* R. Br.] (WATER-CRESS). Frequent along streams and in shallow pools; Ash—Screwbean, *Artemisia*—Pinyon—Juniper. CLARK CO.: Nw. Spring Mtns (Willow Creek), Indian Spgs. LINCOLN CO.: S. Groom Range. NYE CO.: Ash Meadows; nw. Spring Mtns (Crystal Spg cyn), Bullfrog Hills, n. Kawich Range (Longstreet and Eden Creek cyns). 2200—7100 ft. Introduced perennial. Apr—June.

Sibara

S. deserti (M. E. Jones) Roll. NTS: Occasional to locally common, banks of washes below S face of cent. Specter Range, in *Larrea*—*Atriplex*. 3000 ft. Annual, probably winter annual. Apr.

Sisymbrium

S. altissimum L. (TUMBLE-MUSTARD). Local populations on disturbed sites in *Larrea*, *Coleogyne*, *Atriplex*, *Grayia*—*Lycium*,

Artemisia, *Artemisia*—Pinyon—Juniper. NTS: Mercury Valley, e. Jackass Flats, nw. Yucca Flat; Timber Mtn (Cat Cyn), n. Pahute Mesa. CLARK CO.: Nw. Spring Mtns (common weed along length of Wheeler Wash cyn). NYE CO.: N. Pahrump Valley, n. Amargosa Valley, w. Crater Flat, n. Sarcobatus Flat; n. Kawich Range (Eden Creek cyn). 3200—7200 ft. Introduced biennial or perennial. Mar—July.

S. irio L. Locally common on disturbed sites in *Larrea—Ambrosia* or *Atriplex*. NTS: Mercury townsite and sw. Mercury Valley. CLARK CO.: Indian Springs townsite. NYE CO.: Ash Meadows; Beatty townsite. 2200—3800 ft. Introduced winter annual. Apr—May.

S. orientale L. NYE CO.: Bullfrog Hills (roadside W of Beatty), disturbed *Larrea—Ambrosia—Atriplex*. 3500 ft. Introduced annual or biennial. Mar—Apr.

Stanleya

S. elata M. E. Jones. Scattered plants, washes and canyon slopes of esp. limestone mountain ranges and bajadas below; *Atriplex*, *A. canescens*, *Coleogyne*, *Artemisia tridentata*. NTS: W. Spotted Range (Red Mtn, Mercury Ridge), CP Hills, Halfpint Range (French Peak area, Banded Mtn, Cockeyed Ridge); many areas of Forty-Mile Cyn drainage, s. Gold Flat. CLARK CO.: N-S axis of Spotted Range. NYE CO.: Cactus Flat. 4000—6000 ft. Perennial, woody at base. May—June.

S. pinnata (Pursh) Britt. var. **inyoensis** (Munz & Roos) Reveal. *Larrea*, *Atriplex*, Mesquite. CLARK CO.: Common on low dunes, Cactus Spgs. NYE CO.: Stewart Valley, common in n. and e. Ash Meadows, w. Pahrump Valley, ne. Amargosa Valley, cent. and ne. Stonewall Flat. 2200—4800 ft. Perennial, woody at base. Apr—June.

var. **pinnata.** Widely distributed, usually as scattered plants, mostly in washes in *Larrea*, *Atriplex*, *Atriplex—Ceratoides*, *Grayia—Lycium*, *Coleogyne*, *Artemisia*; uncommonly in Mesquite and *Artemisia*—Pinyon—Juniper. NTS: Mercury Valley, Rock Valley, Frenchman Flat (large population SE edge of playa), Yucca Flat, Forty-Mile Cyn, s. Groom Lake; Pahute Mesa; absent in Jackass Flats. CLARK CO.: Nw. Spring Mtns (Wheeler Wash). NYE CO.: Nw. Spring Mtns (Crystal Spg cyn), below n. Yucca Mtn, Bullfrog Hills (summit of Sawtooth Mtn), Reveille Range; occasional in Ash Meadows, Pahrump Valley and n. Amargosa Valley, where merging into var. *inyoensis*, and in s. and cent. Groom Lake. 2300—6000 ft. Perennial, woody at base. Apr—June.

Streptanthella

S. longirostris (S. Wats.) Rydb. Widely distributed, locally common in sandy soils; *Larrea* (esp. *Larrea—Ambrosia*), *Atriplex*, *Lycium pallidum—Grayia*, *Grayia—Lycium*, rarely *Artemisia tridentata.* NTS: N. Amargosa Valley, Mercury Valley, Rock Valley, Jackass Flats, Frenchman Flat, Yucca Flat, Forty-Mile Cyn. NYE CO.: N. Ash Meadows, Crater Flat. 2200—5800 ft. Winter annual. Mar—Apr.

Streptanthus

S. cordatus Nutt. var. **cordatus.** Common and widely distributed in *Artemisia*—Pinyon—Juniper, less frequent in *Artemisia* and Yellow Pine—Pinyon. NTS: Shoshone Mtn, Timber Mtn, Eleana Range, s. Belted Range (Rainier Mesa), Pahute Mesa, Halfpint Range (Cockeyed Ridge). CLARK CO.: Nw. Spring Mtns (Clark Cyn, Wheeler Wash area). LINCOLN CO.: Cent. Groom Range (Bald Mtn). NYE CO.: Nw. Spring Mtns (N slope below Mt. Stirling), Bullfrog Hills, Grapevine Mtns, Reveille Range, cent. Belted Range; white volcanic hills or outcrops in w. Emigrant Valley and s. Penoyer Valley. 5000—7800 ft. Perennial. Late Apr—June.

Thelypodium

T. integrifolium (Nutt.) Endl. ssp. **affine** (Greene) Al-Shehbaz. *Atriplex* and Mesquite. CLARK CO.: Low dunes, Cactus Spgs. NYE CO.: Common and conspicuous in many areas of Ash Meadows. 2200—3200 ft. Biennial or perennial. July—Oct.

T. laxiflorum Al-Shehbaz. Restricted to bases of volcanic boulder-cliffs, in *Artemisia*—Pinyon—Juniper. NTS: S. Belted Range (E and SW rims of Rainier Mesa), certain cyns tributary to South Silent Canyon of n. Pahute Mesa. NYE CO.: Cent. Belted Range (both slopes below Wheelbarrow Peak). 6000—7500 ft. Biennial. June—July.

Thysanocarpus

T. curvipes Hook. var. **eradiatus** Jeps. Locally common on talus slopes, ledges, and washes in *Coleogyne*, *Artemisia*, lower limits of *Artemisia*—Pinyon—Juniper, less commonly *Larrea.* NTS: Below Yucca Mtn (w. Jackass Flats), s. and e. Shoshone Mtn, Eleana Range, below S face of Pahute Mesa (n. Forty-Mile Cyn). NYE CO.: Below n. Bare Mtn, nw. Yucca Mtn, Bullfrog Hills, Grapevine Mtns. 3700—6000 ft. Annual. Apr—May.

T. laciniatus Nutt. ex Torr. & Gray var. **hitchcockii** Munz. NTS: *Atriplex*, hill outlier of Specter Range, e. Rock Valley. 3600 ft. Annual. Apr.

CACTACEAE. Cactus Family

Coryphantha

C. vivipara (Nutt.) Britt & Rose var. **deserti** (Engelm.) W. T. Marshall. Occasional in or below limestone hills or mountains, in *Atriplex* or *Coleogyne*. NTS: W. Spotted Range (Mercury Ridge), E of Shoshone Mtn (near saddle, N slope of Jackass Flats), below Eleana Range. 4400—4800 ft. Perennial. May.

var. **rosea** (Clokey) L. Benson. Occasional to locally frequent in *Artemisia nova* and *Artemisia*—Pinyon—Juniper. NTS: S. Belted Range (W slope of Rainier Mesa), Pahute Mesa. LINCOLN CO.: Cent. Groom Range (below W slope of Bald Mtn). 6100—6800 ft. Perennial. June—July.

Echinocactus

E. polycephalus Engelm. & Bigel. (BARREL CACTUS). Occasional, usually as small plants in many areas; common, plants often large, in *Larrea—Atriplex* or *Atriplex* of canyons and bajadas of limestone mountain ranges. NTS: W. Spotted Range (Red Mtn, Mercury Ridge); nw. Papoose Range, in *Grayia—Lycium/Artemisia*, where at its known n. limits. NYE CO.: N. Ash Meadows. 2300—4900 ft. Perennial. May.

Echinocereus

E. engelmannii (Parry) Lem. var. **armatus** L. Benson. NTS: *Artemisia tridentata* of nw. Pahute Mesa; *Atriplex* of Thirsty Cyn uplands. 5500—6000 ft. Perennial. May.

var. **chrysocentrus** (Engelm. & Bigel.) Engelm. ex Rümpler. Occasional on slopes, in washes, or bases of cliffs, in *Atriplex*, *A. canescens*, *Larrea—Atriplex*, *Larrea/Coleogyne*, *Coleogyne*, *Grayia—Lycium*, *Artemisia*, *Artemisia*—Pinyon—Juniper. NTS: In or below Ranger Mtns, Shoshone Mtn, Mine Mtn, nw. Papoose Range. LINCOLN CO.: Cent. Groom Range (below W slope of Bald Mtn). NYE CO.: Bullfrog Hills, Stonewall Mtn, cent. and n. Belted Range. 3300—6600 ft. Perennial. June.

var. **engelmannii** (HEDGEHOG CACTUS). Frequent, *Larrea—Atriplex*, *Atriplex*, *Coleogyne*, *Artemisia tridentata*. NTS: Specter Range, w. Spotted Range (Mercury Ridge), nw. Pahute Mesa.

CLARK CO.: Nw. Spring Mtns (Wheeler Wash area). NYE CO.: Mtn W side of Pahrump Valley. 2800—6000 ft. Perennial. Apr—May.

E. **triglochidiatus** Engelm. var. **melanacanthus** (Engelm.) L. Benson. Occasional, *Coleogyne/Artemisia*, *Artemisia*, *Artemisia*—Pinyon—Juniper. NTS: S. Belted Range (Oak Spring Butte). LINCOLN CO.: N. Groom Range. NYE CO.: Nw. Spring Mtns (N slope below Mt. Stirling), cent. and n. Belted Range. 5500—6200 ft. Perennial. June.

var. **mojavensis** (Engelm. & Bigel.) L. Benson (MOUND CACTUS). NYE CO.: Occasional, *Artemisia nova*, *Artemisia*—Pinyon—Juniper; nw. Spring Mtns (N slope below Mt. Stirling), s. and cent. Belted Range, cent. Reveille Range. 6200—6900 ft. Perennial. June.

Mammillaria

M. tetrancistra Engelm. Uncommon, limestone mountains, in *Atriplex* and *A. hymenelytra*. NTS: W. Spotted Range (Red Mtn). NYE CO.: Mtn W side of Pahrump Valley. 2800—4400 ft. Perennial. Apr.

Opuntia

O. basilaris Engelm. & Bigel. var. **basilaris** (BEAVERTAIL CACTUS). Widely distributed throughout the region, in *Larrea*, *Atriplex*, *Coleogyne*, *Artemisia*, rarely *Artemisia*—Pinyon—Juniper. NTS: In or below w. Spotted Range (Red Mtn, Mercury Ridge), Ranger Mtns, Shoshone Mtn. NYE CO.: N. Ash Meadows; abundant in overgrazed *Atriplex—Grayia* below ne. Stonewall Mtn. 2300—5500 ft. Perennial. May—June.

O. chlorotica Engelm. & Bigel. (PANCAKE PEAR). Locally common on slopes, esp. around rock outcrops, in *Coleogyne/Artemisia* and *Artemisia*—Pinyon—Juniper. CLARK CO.: Nw. Spring Mtns (lower Clark Cyn, lower Buck Spg cyn). NYE CO.: Nw. Spring Mtns (Crystal Spg cyn, near Rock Spg). 5500—6400 ft. Perennial. May—June.

O. echinocarpa Engelm. & Bigel. var. **echinocarpa** (GOLDEN CHOLLA). Widely distributed throughout the region, in *Larrea*, *Atriplex*, *Grayia—Lycium*, *Coleogyne*, *Artemisia*. NTS: Specter Range, w. Spotted Range (Red Mtn, Mercury Ridge), below Ranger Mtns. NYE CO.: N. Ash Meadows; below n. Bare Mtn, below cent. Belted Range and Groom Range (Groom Lake), and below s. Kawich Range (e. Cactus Flat). Golden-yellow spines over 3 cm long characterize the populations from below the Ranger Mtns, Specter

Range and Bare Mtn S to Ash Meadows. 2300—5800 ft. Perennial. May—June.

O. erinacea Engelm. & Bigel. var. **erinacea** (MOJAVE PRICKLY PEAR). The common cactus of *Artemisia*—Pinyon—Juniper throughout the region; less common in *Artemisia tridentata*, *Atriplex—Ceratoides*. NTS: Shoshone Mtn, s. Belted Range (Rainier Mesa), Pahute Mesa. LINCOLN CO.: Cent. Groom Range (W slope below Bald Mtn). NYE CO.: Nw. Spring Mtns (N slope below Mt. Stirling); cent. and n. Groom Lake, e. Cactus Flat. 5000—7500 ft. Perennial. June—July.

var. **ursina** (A. Weber) Parish (GRIZZLY BEAR CACTUS). Locally common, esp. in limestone or dolomite areas; *Atriplex*, *Coleogyne*, *Artemisia*, *Artemisia*—Pinyon—Juniper. NTS: In or below w. Spotted Range (incl. Red Mtn, Mercury Ridge), Ranger Mtns, Buried Hills, Halfpint Range (Scarp Cyn), Shoshone Mtn, Eleana Range, s. Belted Range, n. Pahute Mesa, nw. Papoose Range. CLARK CO.: N-S axis of Spotted Range. NYE CO.: Nw. Spring Mtns (N slope below Mt. Stirling), cent. Belted Range (Johnnies Water cyn). 3700—7200 ft. Perennial. May—June.

O. phaeacantha Engelm. var. **major** Engelm. CLARK CO.: Local populations in *Coleogyne/Artemisia*, nw. Spring Mtns (Willow Creek area). 6000 ft. Perennial. June.

O. polyacantha Haworth var. **rufispina** (Engelm.) L. Benson. NTS: Local, *Artemisia*—Pinyon—Juniper, S face of Pahute Mesa. 6800—7000 ft. Perennial. June.

O. pulchella Engelm. *Atriplex—Ceratoides*. NTS: Below N face of Pahute Mesa (s. Gold Flat). NYE CO.: Frequent in many areas of basin floors of Cactus Flat, Gold Flat, and Kawich Valley. 5300—5800 ft. Perennial. June.

O. ramosissima Engelm. (PENCIL CACTUS). Frequent in *Atriplex* or *Larrea—Atriplex* of limestone areas. NTS: Bajadas below w. Spotted Range (incl. Red Mtn, Mercury Ridge), e. Skull Mtn (below Mt. Salyer), and Ranger Mtns. CLARK CO.: N-S axis of Spotted Range. NYE CO.: N. Ash Meadows. 2200—4500 ft. Perennial. May.

O. stanlyi Engelm. var. **parishii** (Orcutt) L. Benson. NYE CO.: Local populations in *Grayia—Lycium*, below cent. Belted Range (w. Groom Lake). 5000 ft. Perennial. May.

Sclerocactus

S. polyancistrus (Engelm.) Britt. & Rose. Scattered local populations in *Atriplex—Ceratoides*, *Artemisia*—Pinyon—Juniper. NTS: Nw. Pahute Mesa and below N face (s. Gold Flat). NYE CO.: Bullfrog Hills (mouth of Box Cyn). 3900—6300 ft. Perennial. May.

CAMPANULACEAE. Bellflower Family

Nemacladus

N. glanduliferus Jeps. var. **orientalis** McVaugh. Usually in *Larrea—Atriplex*, *Coleogyne*, or *Artemisia nova* of bajadas below limestone mountains or hills, often in mixed populations with *N. sigmoideus.* NTS: Ne. Amargosa Valley, Mercury Valley, Rock Valley, s. Jackass Flats, most areas of Frenchman Flat, nw. and sw. Yucca Flat. NYE CO.: Stewart Valley; Beatty Mtn, n. Bare Mtn. 2500—4400 ft. Winter annual. Apr—May.

N. rubescens Greene. Locally common, usually in loose sands, below limestone or usually volcanic mountains and hills; *Larrea—Ambrosia*, *Larrea—Lycium—Grayia*, *Lycium pallidum—Grayia*, *Atriplex.* NTS: E. Rock Valley, w. Jackass Flats, most areas of Frenchman Flat. NYE CO.: N. Amargosa Valley (below N end of Spring Mtns). 3100—4000 ft. Winter annual. Apr—May.

N. sigmoideus Robbins. Usually in mixed populations with *N. glanduliferus* in limestone areas; *Larrea—Atriplex*, *Larrea—Grayia—Lycium*, *Coleogyne.* NTS: W. Mercury Valley, e. Rock Valley, nw. and e. Jackass Flats, sw. Frenchman Flat, sw., nw., and ne. Yucca Flat. NYE CO.: Bare Mtn. 3700—4200 ft. Winter annual. Apr—June.

CAPPARACEAE. Caper Family

Cleome

C. lutea Hook. Scattered populations in washes or disturbed, moist, sandy sites throughout the region; *Larrea—Ambrosia*, *Atriplex canescens*, *Artemisia tridentata*, *Artemisia*—Pinyon—Juniper. NTS: Cane Spg (w. Frenchman Flat), Tippipah Reservoir (se. Forty-Mile Cyn), esp. upper and lower Forty-Mile Cyn wash, cent.—w. Yucca Flat; s. Pahute Mesa and below N face (s. Gold Flat), s. Belted Range (sw. Groom Lake). NYE CO.: N. and s. Kawich Range (Longstreet cyn and below Cedar Spg). 2600—7000 ft. Annual. Apr—Oct.

C. serrulata Pursh. NYE CO.: Occasional plants in *Atriplex*, s. Penoyer Valley. 5000—5200 ft. Annual. Aug.

C. sparsifolia S. Wats. NYE CO.: With Mesquite and *Baccharis* on low dunes, former Carson Slough area of nw. Ash Meadows. 2200 ft. Annual. July.

Cleomella

C. brevipes S. Wats. NYE CO.: Locally common, *Distichlis* meadows in spring areas of n. and e. Ash Meadows. 2100—2300 ft. Annual. Apr—Sept.

C. hillmanii A. Nels. ESMERALDA CO.: Occasional, mining area near Goldfield townsite. NYE CO.: Occasional to common, Goldfield Hills, in *Atriplex*, *Sarcobatus*, *Artemisia nova*. 5600—5800 ft. Annual. Apr—June.

C. obtusifolia Torr. & Frém. NYE CO.: Populations frequent on locally moist soils, *Larrea—Atriplex*, *Atriplex*, in the Amargosa Valley drainage, incl. Ash Meadows, length of Oasis Valley, and area of Big Dune. 2200—4000 ft. Annual. June—Oct.

Oxystylis

O. lutea Torr. & Frém. NYE CO.: Occasional or locally common, in *Atriplex*; uncommon in Ash Meadows, common near playa N of Ash Meadows; ne. Stonewall Flat, where reaches its n. limits. 2200—4800 ft. Annual. June—Sept.

Polanisia

P. trachysperma Torr. & Gray. Local, along washes in *Artemisia*—Pinyon—Juniper. CLARK CO.: Cold Creek area of nw. Spring Mtns. NYE CO.: Common locally, base of cent. Belted Range (e. Kawich Valley). 6100—6400 ft. Annual. July.

Wislizenia

W. refracta Engelm. LINCOLN CO.: Occasional, *Grayia—Lycium*, e. Penoyer Valley. 4500 ft. Annual. Aug—Sept.

CAPRIFOLIACEAE. Honeysuckle Family

Lonicera

L. japonica Thunb. NYE CO.: Occasional escapes from cultivation, Beatty townsite. 3000 ft. Perennial. June.

Sambucus

S. caerulea Raf. Local, *Artemisia*—Pinyon—Juniper, *Artemisia—Cercocarpus*, Yellow Pine and Fir—Pinyon. CLARK CO.: Nw. Spring Mtns (upper Clark Cyn, Wheeler Pass, and Cold Creek—Willow Creek area). LINCOLN CO.: Cent. Groom Range (Bald Mtn). NYE CO.: Stonewall Mtn (grotto at Stonewall Spg), n. Kawich Range (upper Eden Creek cyn), cent. Belted Range (Cliff Spg). 5600—9100 ft. Shrub. June—July.

Symphoricarpos

S. longiflorus A. Gray. The common *Symphoricarpos* of foothills and dissected upper bajadas of the region, esp. at bases of cliffs, rock ledges, and along washes; *Coleogyne*, *Artemisia*, *Artemisia*—Pinyon—Juniper. NTS: In or below Specter Range, Skull Mtn, w. Spotted Range (Red Mtn), Buried Hills, Halfpint Range (French Peak mtn, Banded Mtn; w. Emigrant Valley), Shoshone Mtn, Yucca Mtn, CP Hills, Mine Mtn, Eleana Range, Pahute Mesa, Thirsty Cyn uplands. CLARK CO.: N-S axis of Spotted Range. NYE CO.: N. Bare Mtn, Tolicha Peak area, Stonewall Mtn, n. Kawich Range (Eden Creek cyn), Reveille Range (Fang Ridge), cent. Belted Range, White Blotch Spg W of n. Groom Range. 3800—7400 ft. Shrub. May—June.

S. parishii Rydb. Bases of cliffs, or in ravines, *Artemisia*—Pinyon—Juniper, Fir—Pinyon. NTS: S. Belted Range (Rainier Mesa), s. Pahute Mesa. CLARK CO.: Abundant, esp. on burned slopes, S face of nw. Spring Mtns (incl. upper Clark Cyn). LINCOLN CO.: Cent. Groom Range (Bald Mtn). NYE CO.: Stonewall Mtn. 6500—8500 ft. Shrub. June—July.

S. vaccinioides Rydb. LINCOLN CO.: Cent. Groom Range (with Limber Pine and Fir, Bald Mtn). NYE CO.: *Artemisia*—Pinyon—Juniper and *Artemisia*—*Cercocarpus*, n. Kawich Range (upper Eden Creek cyn). 7600—9100 ft. Shrub. June—July.

CARYOPHYLLACEAE. Pink Family

Arenaria

A. congesta Nutt. ex Torr. & Gray var. **charlestonensis** Maguire. Locally common, *Artemisia*—Pinyon—Juniper. CLARK CO.: Nw. Spring Mtns (near Wheeler Pass). NYE CO.: Nw. Spring Mtns (N slope below Mt. Stirling). 6200—7700 ft. Perennial. Apr—June.

var. **subcongesta** (S. Wats.) S. Wats. Widely distributed and common in *Artemisia*—Pinyon—Juniper of volcanic mountain ranges; occasional in *Artemisia nova*, and around rock outcrops in *Coleogyne*. NTS: Yucca Mtn, Shoshone Mtn, Timber Mtn, Eleana Range, s. Belted Range (Rainier Mesa, Gold Meadows, Oak Spring Butte), Pahute Mesa, Halfpint Range (Cockeyed Ridge, Raysonde Buttes area). LINCOLN CO.: Cent. Groom Range (Bald Mtn). NYE CO.: Nw. Spring Mtns (Rock Spg of N slope), Bullfrog Hills, n. and s. Kawich Range, n. and cent. Belted Range; populations abundant in *Artemisia*—*Cercocarpus* of treeless slopes and high ridges of n. Kawich Range (upper Eden Creek cyn), at 7600—9100 ft, appear to be worthy of taxonomic distinction. 4500—9100 ft. Perennial. Apr—July.

A. **kingii** (S. Wats.) M. E. Jones var. **glabrescens** (S. Wats.) Maguire. Local, around bases of cliffs, or on loose volcanic talus; *Artemisia nova*, *Artemisia*—Pinyon—Juniper. NTS: S rim of Pahute Mesa. ESMERALDA CO.: Goldfield Hills. 5600—7300 ft. Perennial. Apr—July.

A. **macradenia** S. Wats. ssp. **ferrisiae** Abrams. NTS: Occasional, *Larrea/Coleogyne*, *Artemisia nova*—Juniper; Shoshone Mtn, Eleana Range (near Capt. Jack Spg, occurring with *A. congesta* var. *subcongesta*). 4000—6000 ft. Perennial. May—June.

var. **parishiorum** Rob. Common, restricted to limestone mountains or calcareous outcrops elsewhere; known from essentially all limestone areas of the region; *Larrea*, *Atriplex*, *Grayia—Lycium*, *Coleogyne*, *Artemisia*, lower *Artemisia*—Pinyon—Juniper. NTS: Striped Hills, Specter Range, w. Spotted Range (Red Mtn, Mercury Ridge), Ranger Mtns, Buried Hills and limestone butte to W, Halfpint Range (French Peak mtn, Scarp Cyn, Banded Mtn), CP Hills, Mine Mtn, s. Belted Range (Oak Spg Butte), nw. Papoose Range. CLARK CO.: N-S Axis of Spotted Range. LINCOLN CO.: N end of Groom Range. NYE CO.: Foothills N end of Spring Mtns, e. Goldfield Hills, n. Stonewall Mtn. 3400—6200 ft. Perennial. Apr—May.

Paronychia

P. **jamesii** Torr. & Gray. LINCOLN CO.: Local populations in *Artemisia nova*, cent. Groom Range (common from N saddle to summit of Bald Mtn). 8000—9300 ft. Perennial. July.

Sagina

S. **saginoides** (L.) Karst. var. **hesperia** Fern. NYE CO.: Locally common on wet soils in *Artemisia*—Pinyon—Juniper, n. Kawich Range (Eden Creek cyn). 7100 ft. Perennial. June.

Scopulophila

S. **rixfordii** (Bdg.) Munz & Jtn. Limestone crevices and cliff faces, predictable in all limestone mountains and foothills, *Atriplex*, *Larrea—Atriplex*. NTS: Specter Range, w. Spotted Range (Red Mtn), Ranger Mtns, Buried Hills. CLARK CO.: N-S axis of Spotted Range. NYE CO.: Mtns N of Ash Meadows; base of nw. Stonewall Mtn. 2700—5600 ft. Perennial. May—Oct.

Silene

S. verecunda S. Wats. ssp. **andersonii** (Clokey) Hitchc. & Maguire. Occasional, sometimes locally common, below cliffs, around boulders and volcanic rock outcrops, in *Artemisia*—Pinyon—Juniper, *Artemisia—Cercocarpus.* NTS: Shoshone Mtn, Eleana Range, s. Belted Range (Rainier Mesa), Pahute Mesa. LINCOLN CO.: Cent. Groom Range (Bald Mtn). NYE CO.: Stonewall Mtn, n. Kawich Range (Eden Creek cyn). 6000—8500 ft. Perennial. June—July.

Stellaria

S. jamesiana Torr. NYE CO.: N. Kawich Range, abundant on steep treeless slopes in *Artemisia—Cercocarpus* and adjacent *Artemisia*—Pinyon—Juniper in upper Eden Creek cyn. 7600—9100 ft. Perennial. June—July.

CELASTRACEAE. Staff-Tree Family

Forsellesia

F. nevadensis (A. Gray) Greene. Uncommon, *Artemisia*—Pinyon—Juniper, Yellow Pine—Pinyon, around cliffs and boulders. NTS: S. Belted Range (local on e. Rainier Mesa), n. Pahute Mesa. CLARK CO.: Nw. Spring Mtns (upper Clark Cyn). NYE CO.: Nw. Spring Mtns (N slope below Mt. Stirling); ravine, s. Monitor Range. 5600—7800 ft. Shrub. May—June.

Mortonia

M. utahensis (Cov.) A. Nels. NTS: Rare, but common locally, *Atriplex*, limestone crevices and talus slopes, cent. Specter Range. 3800 ft. Shrub. May.

CHENOPODIACEAE. Goosefoot Family

Atriplex

A. argentea Nutt. ssp. **expansa** (S. Wats.) Hall & Clem. Local on disturbed sites in *Atriplex*, *Artemisia tridentata.* NTS: Below w. Spotted Range (s. Frenchman Flat), Pahute Mesa. ESMERALDA CO.: Roadside near Tonopah. NYE CO.: Nw. Stonewall Mtn. 3100—6000 ft. Annual. June—Sept.

A. canescens (Pursh) Nutt. var. **canescens** (FOUR-WINGED SALTBUSH). Occasional to common, usually in sandy soils, occurring to some extent in most vegetation types (occasional on disturbed sites in *Artemisia*—Pinyon—Juniper); a polymorphic species in this region. NTS: Dominant shrub in local areas in s. Frenchman Flat lowlands, with *A. tridentata* on sands of e. Forty-Mile Cyn; Jackass Flats, sandy areas of Frenchman Flat (incl. Cane Spg), Mid Valley, esp. *Grayia—Lycium* in Yucca Flat and s. Groom Lake, s. Gold Flat. NYE CO.: Sands in nw. Ash Meadows (in Screwbean—*Atriplex*), Oasis Valley (*Sarcobatus—Atriplex*), cent. and s. Kawich Valley, cent. and w. Groom Lake. 2200—7500 ft. Shrub. Late Apr—June.

A. confertifolia (Torr. & Frém.) S. Wats. (SHADSCALE). Probably the most abundant shrub of the region; in pure stands, or with *Ceratoides*, over extensive areas extending out from the playas of closed drainage basins, in the more northerly basins the communities covering most of the basin floors; usually the shrub matrix species of the limestone hills, mountains or outcrops throughout the region, commonly associated with *Larrea*, less frequently *Coleogyne*. There are large variations in vegetative and fruiting characters, and populations of some areas may be worthy of taxonomic distinction (esp. in Ash Meadows). NTS: N. Amargosa Valley, Mercury Valley, Rock Valley, Frenchman Flat, Yucca Flat, Thirsty Cyn, s. Groom Lake; uncommon in Jackass Flats, Topopah Valley, Forty-Mile Cyn. NYE CO.: Ash Meadows, Crater Flat, Cactus Flat, Gold Flat, Kawich Valley, Groom Lake. 2200—6200 ft. Shrub. Apr—May.

A. elegans (Moq.) D. Dietr. ssp. **fasciculata** (S. Wats.) Hall & Clem. Local populations on disturbed sites in *Larrea*. NTS: Below S end of Yucca Mtn. CLARK CO.: Below SE end of Spotted Range. 2600—3400 ft. Annual. Sept—Oct.

A. hymenelytra (Torr.) S. Wats. (DESERT HOLLY). Associated with soils derived from certain limestone or calcareous volcanic mountain ranges or parts of those ranges, as dominant, codominant, or associated species with *Atriplex* or *Larrea*. NTS: Striped Hills, common in a major wash and on disturbed sites below S face of Specter Range, scattered roadside plants below W end of Skull Mtn (s. Jackass Flats and e. Rock Valley); essentially the only shrub of cyns and slopes in buff-colored tuff below S face of French Peak mtn (n. Frenchman Flat), where at its known n. limits. CLARK CO.: Below se. Spotted Range. NYE CO.: Dominant species in s. Stewart Valley; mtns N of Ash Meadows, N end and W side of Pahrump Valley; basin floor of Pahrump Valley, Ash Meadows; local below nw. Yucca Mtn (Beatty Wash). 2200—4400 ft. Shrub. Dec—Mar.

A. lentiformis (Torr.) S. Wats. var. **lentiformis**. NTS: Occasional, washes in *Larrea* below w. Spotted Range (se. Mercury Valley). 3500 ft. Shrub. Aug—Sept.

A. parryi S. Wats. NYE CO.: Occurring with *Atriplex*, *A. hymenelytra*, or *Sarcobatus*; common in many areas of Ash Meadows; also cent. Sarcobatus Flat, and below N end of Reveille Range (Twin Spgs area). 2200—5200 ft. Shrub. July.

A. phyllostegia (Torr.) S. Wats. NYE CO.: Local populations, esp. on disturbed sites or near springs, in *Atriplex*, *A. hymenelytra*, Mesquite, *Juncus*, in a number of areas of Ash Meadows, ne. Amargosa Valley (near playa), Oasis Valley, and e. Bullfrog Hills. 2200—4000 ft. Annual. May—June.

A. polycarpa (Torr.) S. Wats. Usually in washes (where it may be the dominant species), in *Atriplex* or *Larrea*. NTS: Major wash below Specter Range, occasional plants in washes of s. Frenchman Flat below w. Spotted Range (Red Mtn), lower Forty-Mile Cyn. CLARK CO.: Low dunes, Cactus Spgs. NYE CO.: Locally dominant on slopes of e. Stewart Valley; sandy areas near playa N of Ash Meadows; n. Pahrump Valley, s. Oasis Valley, below N end of Yucca Mtn (Beatty Wash), and n. Amargosa Valley on calcareous outcrops below mtns between Bare Mtn and Yucca Mtn. 2300—3800 ft. Shrub. Aug—Sept.

A. rosea L. Weed of disturbed sites, esp. near springs, in *Atriplex—Ceratoides*, *Artemisia nova*. ESMERALDA CO.: E of Goldfield. NYE CO.: Cedar Pipeline Ranch (n. Kawich Valley), and spgs of Cactus Range. 5600—6300 ft. Introduced annual. Aug—Sept.

A. torreyi (S. Wats.) S. Wats. NYE CO.: Local to dominant shrub in many areas of Ash Meadows, Pahrump Valley, Oasis Valley, and elsewhere in the Amargosa Valley, with *Atriplex*, Mesquite, *Suaeda*, or sometimes *Scirpus*. 2200—3200 ft. Shrub. June—Aug.

Bassia

B. hyssopifolia (Pall.) Kuntze. Weed of disturbed sites, esp. on irrigated soils; *Atriplex*, *A. canescens*, *Artemisia*—Pinyon—Juniper. NTS: Mercury Valley, Rock Valley, s. Jackass Flats, Frenchman Flat; local on slopes of Rainier Mesa (s. Belted Range); roadside weed, with *Salsola*, on Pahute Mesa. ESMERALDA CO.: Common roadside weed W of Tonopah. NYE CO.: Esp. along irrigation channels in Ash Meadows, uncommon in Pahrump Valley, abundant weed of s. Oasis Valley; spgs of Cactus Range; foothills of s. Monitor Range. 2200—7300 ft. Introduced annual. July—Sept.

Ceratoides

Ceratoides lanata (Pursh) J. T. Howell [*Eurotia l.* (Pursh) Moq.] (WINTER FAT). Widely distributed as a dominant or associated species in *Larrea*, *Grayia—Lycium*, *Coleogyne—Grayia*, as a dominant in discrete areas in *Artemisia*, *Artemisia*—Pinyon—Juniper, or at higher elevations; dominant shrub, with *Atriplex*, over large areas of floors of closed drainage basins. Var. *subspinosa* (Rydb.) Kearn. & Peebl. is recognizable as the most common form of the lower elevations. NTS: Mercury Valley, Rock Valley, Jackass Flats, Frenchman Flat, Mid Valley, Yucca Flat, Forty-Mile Cyn, s. Groom Lake; Pahute Mesa. CLARK CO.: Nw. Spring Mtns (occasional, Wheeler Wash); N-S axis of Spotted Range. LINCOLN CO.: Cent. Groom Range (dominant on summit of Bald Mtn). NYE CO.: Cactus Flat, Gold Flat, Kawich Valley. 3100—9300 ft. Shrub. Apr—June, some yrs autumn.

Chenopodium

C. album L. (PIGWEED or LAMB'S QUARTERS). CLARK CO.: Weed of disturbed sites, Indian Springs townsite. NYE CO.: Ash Meadows; Beatty townsite. 2200—3300 ft. Introduced annual. Sept—Oct.

C. atrovirens Rydb. Occasional, sandy soils, usually *Artemisia*—Pinyon—Juniper, sometimes *Atriplex canescens* or *Artemisia tridentata*. NTS: Below N end of Shoshone Mtn, s. Belted Range (Rainier Mesa), Pahute Mesa. CLARK CO.: Nw. Spring Mtns (upper Clark Cyn). NYE CO.: N. Kawich Range (Eden Creek cyn). 4500—8200 ft. Annual. July—Sept.

C. berlandieri Moq. var. **sinuatum** (Murr) H. A. Wahl. Disturbed sites in *Artemisia*—Pinyon—Juniper. NTS: S. Pahute Mesa. NYE CO.: S. Monitor Range, s. Kawich Range. 6000—7200 ft. Annual. July—Sept.

var. **zschackei** (Murr) Murr. The common variety of the region, usually on disturbed sites in *Artemisia tridentata* or *Artemisia*—Pinyon—Juniper. NTS: Shoshone Mtn, s. Belted Range (Rainier Mesa), esp. Pahute Mesa; occasional plants in *Larrea* below w. Spotted Range (Mercury Valley), and disturbed *Coleogyne*—*Grayia* below Shoshone Mtn (Mid Valley). LINCOLN CO.: Cent. Groom Range (base of Bald Mtn). (3800—) 6000—7500 ft. Annual. July—Oct.

C. desiccatum A. Nels. var. **leptophylloides** (Murr) H. A. Wahl. Locally abundant in loose sands, *Atriplex* and *A. canescens*, uncommonly *Larrea*—*Ambrosia*. NTS: N. Amargosa Valley and nw. Jackass Flats (Forty-Mile Cyn wash), below n. Shoshone Mtn and Timber Mtn (washes of lower Cat Cyn and Forty-Mile Cyn); area of Sugar Loaves of e. Forty-Mile Cyn drainage. NYE CO.: Sw. Penoyer Valley. 2600—5200 ft. Annual. June—July.

C. fremontii S. Wats. Common to abundant on disturbed sites of *Artemisia* and *Artemisia*—Pinyon—Juniper, but belonging to undisturbed vegetation of canyons, at bases of cliffs, and around boulders. NTS: Shoshone Mtn, Timber Mtn, Eleana Range, s. Belted Range (Rainier Mesa), Pahute Mesa; Cane Spg (base of Skull Mtn, w. Frenchman Flat). CLARK CO.: Nw. Spring Mtns (Clark Cyn). NYE CO.: Stonewall Mtn, s. Kawich Range, cent. Reveille Range, cent. Belted Range, n. Groom Range. 4000—7500 ft. Annual. Aug—Oct.

C. gigantospermum Aellen. NTS: Local populations on disturbed sites in *Artemisia*—Pinyon—Juniper, n. and e. Pahute Mesa. 6300—6800 ft. Annual. June—July.

C. incanum (S. Wats.) Heller. Widely distributed, esp. at middle elevations, as scattered plants or locally common on disturbed sites; *Larrea*, *Atriplex*, *A. canescens*, *Grayia—Lycium*, *Coleogyne*, *Artemisia*, rarely *Artemisia*—Pinyon—Juniper. NTS: Bajadas of Rock Valley, Jackass Flats, Frenchman Flat, Yucca Flat, Forty-Mile Cyn (esp. washes), Thirsty Cyn; below s. Belted Range and nw. Papoose Range (s. Groom Lake), Pahute Mesa. LINCOLN CO.: Foothills of Groom Range. NYE CO.: Below n. Yucca Mtn (Beatty Wash), Goldfield Hills, Tolicha Peak area, s. Kawich Range, foothills of n. Belted Range. 2600—6800 ft. Winter annual, probably also germinating in spring. May—Aug.

C. leptophyllum Nutt. ex Moq. Locally common on disturbed sites, but also in undisturbed *Artemisia* and *Artemisia*—Pinyon—Juniper; occasional, disturbed moist sites at lower elevations. NTS: Shoshone Mtn, Eleana Range, s. Belted Range (Rainier Mesa), Pahute Mesa. CLARK CO.: Nw. Spring Mtns (Clark Cyn area). NYE CO.: Reveille Range. 5000—7000 ft. Annual. June—Oct.

C. missouriense Aellen. NTS: Local population on disturbed site in *Artemisia*—Pinyon—Juniper, Eleana Range below s. Belted Range (S face of Rainier Mesa), 6600 ft. Introduced annual. Sept—Oct.

C. nevadense (S. Wats.) Heller. Local populations in *Atriplex* and *Atriplex—Ceratoides*. NTS: Below N face of Pahute Mesa (s. Gold Flat). NYE CO.: Cent. Kawich Valley. 5200—5400 ft. Annual. June—July.

C. strictum Roth var. **glaucophyllum** (Aellen) H. A. Wahl. NTS: Disturbed moist soils, *Larrea* or *Artemisia*—Pinyon—Juniper; Mercury townsite, and Eleana Range. 3700—6000 ft. Annual. Sept.

Grayia

G. spinosa (Hook.) Moq. (HOP-SAGE). One of the most common and widely distributed shrubs of the region, esp. at middle elevations;

dominant or associated species in nearly all vegetation types except *Artemisia*—Pinyon—Juniper, codominant with *Lycium andersonii*, *L. pallidum*, *Larrea*, or *Coleogyne* over extensive areas below 5000 ft, esp. in *Grayia*—*Lycium* and *Lycium pallidum*—*Grayia* of floors of closed drainage basins; sometimes occurring locally in pure stand, or absent from large discrete areas. NTS: Mercury Valley, Rock Valley, Jackass Flats (absent on the long, low-gradient bajada below Shoshone Mtn), Frenchman Flat, Mid Valley, Yucca Flat, Forty-Mile Cyn (codominant with *Artemisia tridentata* over large areas). NYE CO.: Bullfrog Hills, Reveille Range. 3000—6500 ft. Shrub. Apr—May.

Halogeton

H. glomeratus (Bieb.) C. A. Mey. in Led. Abundant weed of disturbed soils in certain areas; much less common in the region than *Salsola*, and the two commonly not present on same sites; disturbed *Larrea*, *Atriplex*, *A. canescens*, *Grayia*—*Lycium*, *Artemisia*. NTS: Basin floors of Frenchman and Yucca Flats and Groom Lake near playas; Mercury Valley and Forty-Mile Cyn; nw. Pahute Mesa. NYE CO.: Dominant ground cover over many square miles of overgrazed *Atriplex* in cent. and n. Kawich Valley (becoming scarce in *Atriplex*—*Ceratoides*), many disturbed sites in Cactus Flat and Stonewall Flat; s. Quinn Cyn Range. 3100—6800 ft. Introduced annual. Aug—Sept.

Kochia

K. americana S. Wats. var. **vestita** S. Wats. (GREEN MOLLY). Dominant shrub, with *Atriplex*, near all playas of region, either over extensive areas or locally, and in local areas elsewhere. NTS: Frenchman Flat, Yucca Flat, se. Groom Lake; nw. Pahute Mesa, uplands of Thirsty Cyn. NYE CO.: Bullfrog Hills (summit of Sawtooth Mtn); Sarcobatus Flat, Stonewall Flat, Ralston Valley, Stone Cabin Valley, Kawich Valley, Hot Creek Valley, Groom Lake, w. Emigrant Valley. 3100—6200 ft. Shrub. June—July.

K. californica S. Wats. NYE CO.: Local in *Atriplex hymenelytra*, Stewart Valley. 2500 ft. Shrub. May—Sept.

K. iranica Bornm. (*K. scoparia* of Amer. auths.) Locally common, often occurring with *Salsola* as a weed of disturbed sites in *Artemisia*, *Artemisia*—Pinyon—Juniper and *Atriplex*. NTS: Pahute Mesa. NYE CO.: Amargosa River floodplain S of Beatty; Tonopah townsite; with *Bassia* at Warm Spgs. 3200—7300 ft. Introduced annual. Late Aug—Oct.

Monolepis

M. nuttalliana (Schult.) Greene. NYE CO.: Locally common, disturbed moist soils in *Atriplex* and *Artemisia nova*; Ash Meadows; spgs of Stonewall Mtn and Cactus Range. 2200—6300 ft. Annual. May—Sept.

M. spathulata A. Gray. Locally common, sandy washes in *Artemisia*—Pinyon—Juniper. NTS: Pahute Mesa. NYE CO.: N. and s. Kawich Range (Eden Creek cyn and Cedar Pass). 6800—7200 ft. Annual. June.

Nitrophila

N. mohavensis Munz & Roos. Known only from the type locality in sw. Ash Meadows, Inyo Co., Calif., near Nevada line, 3 mi E of Death Valley Junction at S end of Carson Slough; in *Atriplex—Distichlis.* Has not been found beyond the nearby Calif./Nevada boundary, but may be expected in Nye Co. in remote parts of Carson Slough N of the boundary. 2050 ft. Perennial. May—July.

N. occidentalis (Nutt.) Moq. NYE CO.: Locally common in seasonally wet soils of scattered localities, usually with *Juncus* and *Distichlis*; spg areas of Ash Meadows and Bullfrog Hills, and s. Oasis Valley. 2100—4000 ft. Perennial. May—Sept.

Salsola

S. iberica Sennen & Pau. [*S. pestifer* A. Nels.; *S. kali* L. var. *tenuifolia* Tausch] (RUSSIAN-THISTLE). (Beatley, J. C., *J. Range Manage.*, 26: 225. 1973). The common *Salsola* of disturbed sites of the higher elevations (above 6000 ft) in *Artemisia* and *Artemisia*—Pinyon—Juniper, esp. on soils derived from volcanic rocks. From 4000—6000 ft, frequently in mixed populations with *S. paulsenii*, with which it commonly hybridizes. NTS: Local, Mercury Valley, Rock Valley, Jackass Flats (Forty-Mile Cyn wash), common on basin floors, esp. near playas, of Frenchman Flat and Yucca Flat; Forty-Mile Cyn, s. Gold Flat; common on Rainier Mesa (s. Belted Range). NYE CO.: N. Pahrump Valley. 2600—7500 ft. Introduced annual. Late Aug—Oct.

S. paulsenii Litv. (RUSSIAN-THISTLE). (Beatley, J. C., *J. Range Manage.*, 26: 225. 1973). The common *Salsola* of disturbed sites of lower elevations (below 4000 ft), esp. on soils derived from limestone mountain ranges. From 4000—6000 ft, frequently in mixed populations with *S. iberica*, with which it commonly hybridizes. NTS: Mercury Valley, Rock Valley, Jackass Flats, Topopah Valley, Frenchman Flat, Mid Valley, Yucca Flat, e.

Forty-Mile Cyn. NYE CO.: Ash Meadows, Pahrump Valley, w. Indian Springs Valley, Oasis Valley (Beatty Wash), Stonewall Flat, Cactus Flat. 2200—6000 ft. Introduced annual. May—Oct.

Sarcobatus

S. vermiculatus (Hook.) Torr. (GREASEWOOD). NTS: Locally common on or near old playa at 4500 ft, se. Groom Lake. NYE CO.: Species primarily of the Great Basin Desert areas of the region, occurring extensively as a dominant with *Atriplex* or *Kochia* in the lowlands of Sarcobatus Flat, Ralston Valley, Railroad Valley, Penoyer Valley, s. Hot Creek Valley, near playa of Emigrant Valley, and on slopes of e. Goldfield Hills; small scattered populations in spring or seepage areas of Bullfrog Hills, foothills of Monitor Range, and Kawich Range; all usually above 5000 ft; also locally common in parts of Oasis Valley at 3300—4000 ft, and a small population below former Fairbanks Spg (Carson Slough) of n. Ash Meadows, at 2200 ft. 2200—6600 ft. Shrub. May—July.

Suaeda

S. intermedia S. Wats. NYE CO.: Common, usually with *Atriplex* in Ash Meadows and near playa to the N, less common in Pahrump Valley, in *Atriplex—Sarcobatus* in Sarcobatus Flat; in *Atriplex hymenelytra* in Stewart Valley. 2200—4000 ft. Perennial, woody at base. Apr—July.

S. occidentalis S. Wats. *Atriplex*, extreme sw. Ash Meadows, Inyo Co., Calif., near Nevada line; to be expected elsewhere in Ash Meadows, in Nye Co., Nevada. 2100 ft. Annual. June.

S. torreyana S. Wats. var. **ramosissima** (Standl.) Munz. CLARK CO.: Common, dunes at Indian Spgs and Cactus Spgs. NYE CO.: In *Atriplex* and *Larrea—Atriplex* on slopes of e. Stewart Valley; esp. with Mesquite or *Atriplex* on low dunes or deep sands of Ash Meadows and Pahrump Valley. 2200—3200 ft. Perennial, woody at base. Aug—Oct.

var. **torreyana**. In *Atriplex*, *A. torreyi*, *A. canescens*. NTS: Locally common on seepage slope at Cane Spg (w. Frenchman Flat); near playa, Emigrant Valley. NYE CO.: Common in Oasis Valley. 3200—4000 ft. Perennial, woody at base. May—July.

CONVOLVULACEAE. Morning-glory Family

Convolvulus

C. arvensis L. (BINDWEED). Occasional weed of disturbed moist soils and near springs, in *Atriplex*. CLARK CO.: Indian Springs townsite. NYE CO.: Ash Meadows; e. Goldfield Hills; s. Sarcobatus Flat. 2200—5800 ft. Introduced perennial. May—Sept.

Cressa

C. truxillensis HBK. var. **minima** (Heller) Munz. NYE CO.: Ground cover on disturbed (plowed) soils in one locality of Ash Meadows, previously *Atriplex.* 2200 ft. Perennial. June.

Ipomoea

I. purpurea (L.) Roth. CLARK CO.: Local on disturbed moist soils, Indian Springs townsite. 3100 ft. Introduced annual. July—Oct.

CORNACEAE. Dogwood Family

Cornus

C. stolonifera Michx. (AMERICAN DOGWOOD). NYE CO.: Along a stream bank in cent. Hot Creek Range, *Artemisia*—Pinyon—Juniper; to be expected in Kawich Range to S. 8700 ft. Shrub. May—July.

CROSSOSOMATACEAE. Crossosoma Family

Crossosoma

C. bigelovii S. Wats. CLARK CO.: Locally common on limestone ledges and cliff faces, in *Larrea/Coleogyne*, N-S axis of Spotted Range; plants less than 25 cm high, spreading near the rock surfaces. 5000—5500 ft. Shrub. Apr—May.

CUCURBITACEAE. Gourd Family

Citrullus

C. lanatus (Thunb.) var. **citroides** (Bailey) Mansf. NTS: Irrigated disturbed soils, Mercury townsite. 3700 ft. Introduced annual. Oct.

Cucurbita

C. foetidissima HBK. CLARK CO.: Common weed of disturbed sites, Indian Springs townsite. 3100 ft. Perennial. June—July.

C. pepo L. (PUMPKIN). NYE CO.: Occasional on disturbed sites, escaped from cultivation; Pahrump Valley and ne. Amargosa Valley. 2700 ft. Introduced annual. May—June.

CUSCUTACEAE. Dodder Family

Cuscuta

C. californica Hook. & Arn. NYE CO.: Local, on *Hymenoclea*, disturbed site in n. Pahrump Valley. 3200 ft. Perennial. May.

C. denticulata Engelm. Common in some areas, some years, on *Larrea;* occasional on *Lycium andersonii*, *Acamptopappus*, *Artemisia tridentata*, *Chrysothamnus viscidiflorus* ssp. *viscidiflorus*, and hybrids of *Salsola* spp.; in *Larrea* and *Artemisia tridentata* vegetation. NTS: Cent. Jackass Flats, sw. Frenchman Flat, w. Yucca Flat, e. and sw. Forty-Mile Cyn. NYE CO.: N. Pahrump Valley; esp. below Bare Mtn (Amargosa Valley). 2500—5000 ft. Perennial. June—July.

C. indecora Choisy. NYE CO.: Abundant on alfalfa and other nearby plants, abandoned field, cent. Pahrump Valley. 2600 ft. Perennial. Aug—Oct.

C. nevadensis Jtn. The most common Dodder of the region; grows on *Atriplex*, *A. hymenelytra*, *Ambrosia dumosa*, *Lycium andersonii*, *L. pallidum*, *L. shockleyi*, *Psorothamnus fremontii*, occasionally on *Grayia* and *Convolvulus;* in *Larrea* and *Atriplex* vegetation. NTS: N. Amargosa Valley, Mercury Valley, Rock Valley, cent. and e. Jackass Flats, sw. Frenchman Flat. NYE CO.: Ash Meadows, s. Stewart Valley. 2200—4000 ft. Perennial. Apr—June.

C. salina Engelm. On *Ambrosia acanthicarpa*, *Cleomella obtusifolia*, *Convolvulus*, and *Lactuca serriola*, usually in *Atriplex* vegetation. CLARK CO.: Indian Springs townsite. NYE CO.: N. Amargosa Valley, Ralston Valley. 2200—5600 ft. Known host plants mostly annuals. June—Sept.

ELAEAGNACEAE. Oleaster Family

Elaeagnus

E. angustifolia L. (RUSSIAN-OLIVE). Commonly planted as windbreak and for shade around townsites and ranch areas of the region, occasional escapes. CLARK CO.: Indian Springs townsite. NYE CO.: Planted at Beatty townsite and in n. Pahrump Valley; escaped at Warm Springs (below s. Hot Creek Range) 2600—5400 ft. Introduced small tree. Apr—May.

ERICACEAE. Heath Family

Arctostaphylos

A. pungens HBK. CLARK CO.: *Artemisia*—Pinyon—Juniper and Yellow Pine—Fir, nw. Spring Mtns; often the dominant plant on extensive burned sites of Wheeler Pass and Clark Cyn areas. 6500—8000 ft. Shrub. May—June.

EUPHORBIACEAE. Spurge Family

Euphorbia

E. albomarginata Torr. & Gray. Widely distributed and locally common, esp. in sandy soils and in washes of middle elevations; *Larrea*, *Coleogyne*, *Grayia—Lycium*, *Atriplex*, *A. canescens*, *Artemisia tridentata*, rarely *Artemisia*—Pinyon—Juniper. NTS: Mercury Valley, Rock Valley, Jackass Flats, Topopah Valley, Frenchman Flat, Yucca Flat, Forty-Mile Cyn, s. Gold Flat. LINCOLN CO.: Nw. Desert Valley (below n. Groom Range). NYE CO.: Crater Flat, n. Amargosa Valley (below Bullfrog Hills), w. Penoyer Valley. 3200—6500 ft. Perennial. Apr—Nov.

E. fendleri Torr. & Gray. The common perennial *Euphorbia* of the higher elevations; *Artemisia*, *Artemisia*—Pinyon—Juniper, Yellow Pine—Pinyon—Juniper. NTS: Associated esp. with white rock outcrops of Eleana Range, s. Belted Range (Rainier Mesa), w. Emigrant Valley; washes of Forty-Mile Cyn. CLARK CO.: Nw. Spring Mtns (Wheeler Wash, Clark Cyn); s. Spotted Range. LINCOLN CO.: S. Groom Range. NYE CO.: Cent. and n. Belted Range, n. Kawich Range, s. Monitor Range. (3400—) 5000—7400 ft. Perennial. May—Aug.

E. maculata L. NTS: Disturbed irrigated soils, Mercury townsite. 3800 ft. Introduced annual. May—Sept.

E. micromera Boiss. Widely distributed and often locally abundant following summer rains, esp. on disturbed sites; *Larrea*, *Lycium pallidum—Grayia*, *Grayia—Lycium*, *Atriplex*, *A. canescens*, *Atriplex—Ceratoides*, *Coleogyne*, *Artemisia tridentata*, Mesquite. NTS: N. Amargosa Valley, Mercury Valley, Rock Valley, Jackass Flats, Topopah Valley, Frenchman Flat, Mid Valley, Yucca Flat, s. Groom Lake. CLARK CO.: Cactus Spgs. LINCOLN CO.: N. Groom Lake. NYE CO.; Ash Meadows, Pahrump Valley, Crater Flat, cent. Gold Flat. 2300—5400 ft. Summer annual. Aug—Oct, rarely May—June.

E. parishii Greene. *Larrea—Ambrosia* of deep sands. NTS: N. Amargosa Valley (below red cinder cone near s. Yucca Mtn). NYE CO.: Ash Meadows. 2200—2700 ft. Perennial. June—Sept.

E. robusta (Engelm.) Small [*E. palmeri* Engelm.] CLARK CO.: Occasional small populations in Yellow Pine—Fir, nw. Spring Mtns (upper Clark Cyn). 8000—9200 ft. Perennial. June—July.

E. serpyllifolia Pers. Locally abundant, esp. in *Coleogyne*, *Grayia—Lycium*, *Artemisia*, lower limits of *Artemisia*—Pinyon—

Juniper, less common in *Larrea/Coleogyne*, *Larrea—Lycium—Grayia*, *Atriplex canescens*, and *Atriplex—Ceratoides*. NTS: N. Jackass Flats, Topopah Valley, Frenchman Flat, Mid Valley, most parts of Yucca Flat, Forty-Mile Cyn, uplands of Thirsty Cyn, Emigrant Valley; s. Belted Range (below W face of Rainier Mesa), Pahute Mesa. NYE CO.: Ash Meadows, Cactus Flat, cent. Gold Flat, nw. Kawich Valley, nw. Reveille Valley, s. Hot Creek Valley. 2200—7000 ft. Summer annual. July—Oct, most often Sept.

E. **setiloba** Engelm. Locally common, sandy soils, *Larrea*, *Grayia—Lycium*, *Coleogyne*, *Artemisia tridentata*, washes in *Atriplex*. NTS: Mercury Valley, Rock Valley, n. Jackass Flats, Topopah Valley, Frenchman Flat, Thirsty Cyn. NYE CO. Nw. Crater Flat, Cactus Flat, cent. Gold Flat, w. Railroad Valley, s. Hot Creek Valley. 3500—5800 ft. Summer annual. July—Oct, most often Sept.

Stillingia

S. **spinulosa** Torr. in Emory. NTS: Locally common, *Larrea—Ambrosia* of sands below S end of Striped Hills (n. Amargosa Valley). 2800—3300 ft. Perennial. Apr—Oct.

FABACEAE (LEGUMINOSAE). Pea Family

Astragalus

A. **acutirostris** S. Wats. The most widely distributed annual *Astragalus* of volcanic soils; common in *Larrea—Lycium—Grayia*, *Grayia—Lycium*, *Coleogyne*, *Artemisia nova*. NTS: Rock Valley (below Skull Mtn), most parts of Jackass Flats, Topopah Valley, w. Frenchman Flat, Mid Valley, most parts of Yucca Flat. NYE CO.: Crater Flat, Oasis Valley; Bullfrog Hills. 3400—5800 ft. Winter annual. Apr—June.

A. **aequalis** Clokey. CLARK CO.: Locally common and rather widely distributed in nw. Spring Mtns (areas of Wheeler Cyn, Trough Spg, Clark Cyn, Willow Creek), in *Artemisia*—Pinyon—Juniper. 6100—6800 ft. Perennial. May—June.

A. **beatleyae** Barneby (Barneby, R. C., *Aliso*, **7**: 161. 1970). NTS: Scattered plants in *Artemisia nova*, around volcanic flatrock areas or other shallow soils, n. and e. Pahute Mesa; type locality. 6100—6800 ft. Perennial. May.

A. **beckwithii** Torr. & Gray var. **purpureus** M. E. Jones. Local populations in *Artemisia tridentata*, *Artemisia*—Pinyon—Juniper. CLARK CO.: Nw. Spring Mtns (slopes of Wheeler Cyn, Willow Creek—Cold Creek area). NYE CO.: Nw. Spring Mtns (N slope below

Mt. Stirling), foothills of n. Belted Range. 5000—7000 ft. Perennial. Apr—early June.

A. calycosus Torr. var. **calycosus.** Common, esp. on shallow soils of talus slopes and disturbed sites, *Artemisia nova*, *Artemisia*—Pinyon—Juniper, *Artemisia—Cercocarpus.* NTS: Shoshone Mtn, s. Belted Range (Rainier Mesa and uplands to E), Pahute Mesa. NYE CO.: N. Belted Range, n. Kawich Range (upper Eden Creek cyn). 6100—9100 ft. Perennial. May—June.

A. casei A. Gray. *Atriplex*, *Grayia—Lycium*, *Coleogyne*, *Artemisia*, *Artemisia*—Pinyon—Juniper, uncommon except on burned sites, NTS: S and E faces of Shoshone Mtn, Timber Mtn, Eleana Range, foothills of s. Belted Range (sw. Groom Lake). ESMERALDA CO.: Near Goldfield. NYE CO.: Bullfrog Hills, Grapevine Mtns, Goldfield Hills. 4500—6000 ft. Perennial. Apr—May.

A. didymocarpus Hook. & Arn. var. **didymocarpus.** NYE CO.: Local in *Artemisia*, Bullfrog Hills. 3600—5200 ft. Winter annual. Apr—May.

var. **dispermus** (A. Gray) Jeps. NTS: Locally abundant, sandy soils, *Larrea* (esp. *Larrea—Ambrosia*); Rock Valley (below Skull Mtn), w. Jackass Flats (below foothills of Shoshone Mtn). 3100—3500 ft. Winter annual. Apr.

A. funereus M. E. Jones. Rock crevices and talus in *Atriplex*. NTS: Cyns of French Peak area (n. Frenchman Flat). NYE CO.: Below n. Yucca Mtn (Beatty Wash). 3700—5000 ft. Perennial. Apr—May.

A. inyoensis Sheld. LINCOLN CO.: Occasional, *Artemisia nova*—Pinyon—Juniper below n. Groom Range. 6500 ft. Perennial. May.

A. layneae Greene. *Larrea—Ambrosia—Atriplex*, *Atriplex*, *Coleogyne*, or *Coleogyne—Atriplex*. NTS: Common, slopes below Yucca Mtn (nw. Jackass Flats). NYE CO.: The common perennial *Astragalus* of n. Bare Mtn (Fluorspar Cyn), areas of n. Yucca Mtn (Beatty Wash, Tates Wash), Bullfrog Hills, and Tolicha Peak area; s. Crater Flat and s. Sarcobatus Flat. 3000—4800 ft. Perennial. Apr—May.

A. lentiginosus Dougl. var. **fremontii** (A. Gray) S. Wats. Widely distributed and more or less common, sometimes abundant over discrete areas in certain years (Beatley, J. C., *Madroño*, **20**: 326. 1970); *Larrea—Ambrosia*, *Larrea—Lycium—Grayia*, *Lycium pallidum—Grayia*, *Grayia—Lycium*, *Atriplex*, *Atriplex—Ceratoides*, *Coleogyne*, *Artemisia*, *Artemisia*—Pinyon—Juniper; uncommon in the limestone areas where *A. tidestromii* is the perennial *Astragalus*, and usually absent in areas or on sites where other perennial *Astragali* occur. NTS: Mercury Valley, Rock Valley, Jackass Flats (absent in w. Jackass Flats where var. *micans* occurs), Topopah Valley,

Frenchman Flat, Mid Valley, Yucca Flat, s. and e. Forty-Mile Cyn, s. Gold Flat, Thirsty Cyn, s. Groom Lake; s. Belted Range (Rainier Mesa), Pahute Mesa. LINCOLN CO.: Foothills of n. Groom Range. NYE CO.: Nw. Spring Mtns (N slope below Mt. Stirling); local below n. Yucca Mtn and parts of Crater Flat where *A. layneae* is the common perennial *Astragalus*; Oasis Valley, Sarcobatus Flat, Stonewall Flat, Cactus Flat, Kawich Valley, Railroad Valley, sw. Penoyer Valley. (3500—) 4000—7500 ft. Winter annual and biennial at lower elevs., perennial at higher elevs. Apr—May; also following summer rains, individuals any month of yr.

var. **micans** Barneby. The perennial *Astragalus* of *Larrea—Ambrosia* of a restricted sandy area of the n. Amargosa Valley drainage. NTS: Sands below Striped Hills, extending N on the bajada of w. Jackass Flats (below foothills of s. Shoshone Mtn). NYE CO.: Plants reaching their best development in the Big Dune area, with a northward extension of the population into areas of Striped Hills and w. Jackass Flats. 2400—3600 ft. Biennial or perennial. Apr—May, some yrs following summer rains.

A. minthorniae (Rydb.) Jeps. var. **villosus** Barneby. *Coleogyne*, *Artemisia*, *Artemisia*—Pinyon—Juniper. NTS: Locally common in *Artemisia tridentata*, bajada below ne. Shoshone Mtn. CLARK CO.: Nw. Spring Mtns (the common *Astragalus* of Wheeler Wash—Clark Cyn area, occasional in Willow Creek—Cold Creek area). NYE CO.: Nw. Spring Mtns (slopes and washes of N slope below Mt. Stirling). 5000—6500 ft. Perennial. Apr—May.

A. mohavensis S. Wats. var. **mohavensis.** NTS: Occasional, limestone crevices or loose talus, in *Larrea—Ambrosia—Atriplex*, *Atriplex*, *Coleogyne*; Specter Range, w. Spotted Range, Ranger Mtns, Buried Hills. 3400—4400 ft. Perennial. Apr—May.

A. newberryi A. Gray var. **newberryi.** Occasional to locally common, *Artemisia tridentata*, *Artemisia*—Pinyon—Juniper. NTS: CP Hills, Eleana Range, n. Pahute Mesa. CLARK CO.: Nw. Spring Mtns (slopes above and S of Willow Creek—Cold Creek). LINCOLN CO.: Nw. Spring Mtns (N slope below Mt. Stirling), area of Tolicha Peak, s. Monitor Range, n. Reveille Range (slopes above Reveille), cent. and n. Belted Range; common over S-cent. Kawich Valley in *Ceratoides—Chrysothamnus greenei.* 5100—7600 ft. Perennial. Mar—early June.

A. nuttallianus DC. var. **imperfectus** (Rydb.) Barneby. NYE CO.: *Artemisia*—Pinyon—Juniper, nw. Spring Mtns (N slope below Mt. Stirling). 6000 ft. Annual. Apr.

A. nyensis Barneby (Barneby, R. C., *Leafl. West. Bot.*, **7**: 195. 1954). NTS: Type coll. from Spotted Range toward Frenchman Flat;

rare and local, below 4500 ft, on calcareous gravel knolls (Barneby, R. C., *Contribs. Flora of Nev.*, **38.** 1956). Annual. Apr.

A. oophorus S. Wats. var. **clokeyanus** Barneby. CLARK CO.: Local, *Artemisia*—Pinyon—Juniper, nw. Spring Mtns (S slope below Wheeler Pass). 6800 ft. Perennial. May—June.

var. **oophorus.** Local populations, *Artemisia*—Pinyon—Juniper of volcanic mountain ranges. NTS: Cyns of Pahute Mesa and below S face, Halfpint Range (Cockeyed Ridge). LINCOLN CO.: Cent. Groom Range (Bald Mtn). NYE CO.: Frequent, cyns of Grapevine Mtns (Phinney Cyn), Stonewall Mtn (cyn with "Ruins"), n. to s. Kawich Range (Eden Creek cyn, Stinking Spg, Cedar Pass), Belted Range (Cliff Spg, Indian Spg, and Johnnies Water cyns). 5800—7500 ft. Perennial. Apr—May.

A. phoenix Barneby (Barneby, R. C., *Madroño*, **20:** 395. 1970). NYE CO.: Endemic to Ash Meadows, known from 6 sites in spg areas of n. and e. Ash Meadows, on light-colored clay soils with *Atriplex* and usually *Haplopappus acradenius;* type locality, e. Ash Meadows. 2200—2300 ft. Perennial. Apr—early May.

A. platytropis A. Gray. NYE CO.: Occasional plants on steep mtn slopes and ridges, in *Artemisia—Cercocarpus*, n. Kawich Range (upper Eden Creek cyn). 8600—9100 ft. Perennial. Aug.

A. preussii A. Gray var. **preussii.** Locally common, *Atriplex*, esp. sandy soils. CLARK CO.: N-S axis of Spotted Range; Indian Springs townsite. NYE CO.: Ash Meadows and Amargosa Valley to N and NW, Pahrump Valley. 2300—4400 ft. Perennial. Apr—May.

A. purshii Dougl. var. **purshii.** NYE CO.: Occasional, *Grayia—Lycium*, *Artemisia*, *Artemisia*—Pinyon—Juniper; Bullfrog Hills, s. Monitor Range, s. Kawich Range (Rose Spg cyn, Cedar Pass), n. Reveille Range (slopes above Reveille). 4600—7000 ft. Perennial. Apr—early June.

var. **tinctus** M. E. Jones. The most widely distributed and common perennial *Astragalus* of the higher elevations; *Artemisia nova* and *Artemisia*—Pinyon—Juniper of volcanic mountains and mesas. NTS: Shoshone Mtn, Eleana Range, s. Belted Range (Rainier Mesa and Twin Peaks area below W face), Pahute Mesa, Halfpint Range (Cockeyed Ridge). ESMERALDA CO.: W. Goldfield Hills. LINCOLN CO.: Cent. Groom Range (Bald Mtn). NYE CO.: Nw. Spring Mtns (N slope below Mt. Stirling), Bullfrog Hills (summit of Sawtooth Mtn), Grapevine Mtns, Stonewall Mtn, s. Kawich Range (Cedar Pass), White Blotch Spg W of n. Groom Range. 5500—8000 ft. Perennial. Apr—May, some yrs summer months.

A. tidestromii (Rydb.) Clokey. The perennial *Astragalus* of bajadas and canyons of limestone mountain areas of lower elevations; *Larrea* and *Coleogyne*, esp. with *Atriplex*. NTS: Specter Range, w.

Spotted Range (incl. Red Mtn, Mercury Ridge), Ranger Mtns, dolomite hills of Eleana Range. CLARK CO.: N-S axis of Spotted Range. NYE CO.: S. Stewart Valley (in *Atriplex hymenelytra*). 2500—4800 ft. Perennial. Apr—May, some yrs Sept—Oct.

Dalea

D. mollissima (Rydb.) Munz. Occasional in sandy soils, *Larrea—Ambrosia, Atriplex canescens.* NTS: Sands below S end of Striped Hills and red cinder cone of s. Yucca Mtn area (n. Amargosa Valley); below e. Specter Range and w. Spotted Range (s. Mercury Valley, n. Amargosa Valley); sw. Frenchman Flat (below e. Skull Mtn). NYE CO.: W. Ralston Valley (below hills NW of Mud Lake). 2500—5600 ft. Perennial. Apr—May, some yrs Sept—Oct.

D. searlsiae (A. Gray) Barneby [*Petalostemum s.* A. Gray]. Uncommon except locally, sandy soils of the middle elevations; *Atriplex canescens, Artemisia tridentata*, uncommonly *Coleogyne* and *Artemisia*—Pinyon—Juniper. NTS: S. and e. Forty-Mile Cyn (below n. Shoshone Mtn, and Sugar Loaves of Eleana Range), local in nw. Yucca Flat and s. Groom Lake (below s. Belted Range). CLARK CO.: Nw. Spring Mtns (occasional to common in Wheeler Cyn area). LINCOLN CO.: Foothills of n. Groom Range, and s. Groom Range. NYE CO.: Nw. Spring Mtns (common in Crystal Spg cyn), ne. Kawich Valley, Stone Cabin Valley. 4600—6800 ft. Perennial. May—July.

Glycyrrhiza

G. lepidota Pursh var. **glutinosa** (Nutt.) S. Wats. NYE CO.: Local populations, n. and se. Ash Meadows. 2200—2300 ft. Perennial. May—June.

Lathyrus

L. hitchcockianus Barneby & Reveal (Barneby, R. C. and J. L. Reveal, *Aliso*, 7: 361. 1971). Local along washes in *Artemisia tridentata* or lower *Artemisia*—Pinyon—Juniper. NTS: Yucca Mtn (nw. Jackass Flats). NYE CO.: Cent. Bullfrog Hills (type locality). 4500—5200 ft. Perennial. Apr—May.

Lotus

L. corniculatus L. NYE CO.: Irrigated site, s. Oasis Valley. 3400 ft. Introduced perennial. May.

L. humistratus Greene. Uncommon except locally; esp. in *Coleogyne*, less frequent in *Larrea—Ambrosia, Larrea—Atriplex*, and *Artemisia tridentata.* NTS: Bajadas below s. Shoshone Mtn (n.

Topopah Valley), Skull Mtn (sw. Frenchman Flat, cent. Rock Valley), Eleana Range (nw. Yucca Flat). NYE CO.: Nw. Spring Mtns (Johnnie Mine area), below n. Yucca Mtn (w. Crater Flat), Bullfrog Hills. 3500—5800 ft. Winter annual. Apr—May.

L. oblongifolius (Benth.) Greene. CLARK CO.: Local, wet soils in *Artemisia*—Pinyon—Juniper, nw. Spring Mtns (margin of Cold Creek). 6200 ft. Perennial. June—July.

L. rigidus (Benth.) Greene. NYE CO.: Common to scattered in washes of S slope of nw. Spring Mtns (Johnnie Mine area, Crystal Spg cyn), in *Larrea*, *Coleogyne*, and Pinyon—Juniper. 3500—5000 ft. Perennial. Apr—May.

Lupinus

L. alpestris A. Nels. CLARK CO.: Locally common, *Artemisia*—Pinyon—Juniper, nw. Spring Mtns (below Wheeler Wells). 6400 ft. Perennial. May—June.

L. argenteus Pursh var. **tenellus** (Dougl. ex D. Don) Dunn. Apparently the most common perennial Lupine of volcanic areas of region; *Artemisia*—Pinyon—Juniper, *Artemisia—Cercocarpus.* NTS: Common in s. Belted Range (Rainier Mesa) and on Pahute Mesa. NYE CO.: N. Kawich Range (Eden Creek and Longstreet cyns). 6500—7500 ft. Perennial. May—July.

L. aridus Dougl. ex Lindl. NTS: Local, sandy wash in *Artemisia tridentata*, s. Kawich Valley below ne. Pahute Mesa. 6200 ft. Perennial. May—June.

L. brevicaulis S. Wats. Locally common, most frequent on stony soils of *Coleogyne* and *Artemisia nova*, sometimes on sandy soils in *Atriplex canescens* and *Lycium pallidum—Grayia*, and in *Atriplex—Ceratoides* of basin floors. NTS: S. Frenchman Flat; below Shoshone Mtn (Topopah Valley, Mid Valley, s. Forty-Mile Cyn), Eleana Range (nw. Yucca Flat), s. Belted Range (Twin Peaks below W face of Rainier Mesa; s. Groom Lake); Pahute Mesa, below N rim (s. Gold Flat) and S rim (n. Forty-Mile Cyn). LINCOLN CO.: Below n. Groom Range. NYE CO.: Nw. Spring Mtns (N slope below Mt. Stirling), White Blotch Spg W of n. Groom Range, n. Belted Range, n. Reveille Range, s. Kawich Range. (3100—) 4500—6500 ft. Winter annual. May—early June.

L. caudatus Kell. *Artemisia* and *Artemisia*—Pinyon—Juniper. NTS: Small roadside population below Eleana Range (e. Forty-Mile Cyn). LINCOLN CO.: Below n. Groom Range (s. Penoyer Valley), and cent. Groom Range (Bald Mtn). NYE CO.: Washes below cent. Belted Range, n. Reveille Range, s. Kawich Range, s. Monitor Range. 5400—7500 ft. Perennial. May—July.

L. concinnus Agardh var. **orcuttii** (S. Wats.) C. P. Sm. Restricted to sandy washes, usually in *Larrea.* NTS: Nw. Jackass Flats (below Yucca Mtn), sw. Frenchman Flat (below Skull Mtn). NYE CO.: Nw. Spring Mtns (Crystal Spg cyn, Johnnie Mine area), below n. Yucca Mtn (Beatty Wash), Bullfrog Hills. 3000—4700 ft. Winter annual. Apr—May.

L. excubitus M. E. Jones var. **excubitus.** NYE CO.: Local, *Artemisia*—Pinyon—Juniper, Grapevine Mtns (head of Phinney Cyn). 7600 ft. Shrub. Late Apr—June.

L. flavoculatus Heller. The most common annual Lupine of the region, in *Larrea, Grayia—Lycium, Coleogyne, Artemisia.* NTS: Mercury Valley, cent. Rock Valley, n. and e. Jackass Flats, n. and w. Frenchman Flat, most parts of Yucca Flat, n. Forty-Mile Cyn; nw. Pahute Mesa and below N face (s. Gold Flat). NYE CO.: Below nw. Spring Mtns (N slope below Mt. Stirling), Bullfrog Hills, below n. Bare Mtn. Tolicha Peak—Obsidian Butte area; s. Sarcobatus Flat, cent. Groom Lake. 3400—6000 ft. Winter annual. Late Apr—June.

L. holmgrenanus C. P. Sm. Common locally along washes, *Artemisia tridentata, Artemisia*—Pinyon—Juniper. ESMERALDA CO. (W of area): Near Lida summit. NYE CO.: Bullfrog Hills, Grapevine Mtns, Tolicha Peak area (all colls. in Sarcobatus Flat drainage). 4500—7500 ft. Perennial. Apr—May.

L. palmeri S. Wats. *Artemisia*—Pinyon—Juniper. NTS: Common on S slope of Shoshone Mtn (esp. on the large burned area) and below E slope. LINCOLN CO.: Cent. Groom Range (Bald Mtn). 5600—7400 ft. Perennial. May—July.

L. pusillus Pursh var. **intermontanus** (Heller) C. P. Sm. NYE CO.: Locally abundant on loose sands, in *Atriplex canescens,* sw. Penoyer Valley. 5100 ft. Annual. May—June.

L. shockleyi S. Wats. Locally common on sandy soils, *Larrea* (usually *Larrea—Ambrosia*), *Atriplex, A. canescens.* NTS: N. Amargosa Valley (red cinder cone near S end of Yucca Mtn); the annual Lupine of w. Jackass Flats (below foothills of s. Shoshone Mtn), n. Rock Valley (below Skull Mtn), sands of ne. Frenchman Flat below Buried Hills and mtns of s. Halfpint Range. NYE CO.: Oasis Valley; Big Dune of n. Amargosa Valley. 2500—3500 ft. Winter annual. Apr—May.

L. uncialis S. Wats. NTS: Rare, *Artemisia nova,* nw. Pahute Mesa. 5600 ft. Annual. May.

Medicago

M. lupulina L. Commonly present to some extent in disturbed wet soils near springs of the region, *Larrea* or *Artemisia*—Pinyon—Juniper. CLARK CO.: Indian Springs townsite; nw. Spring

Mtns (Cold Creek—Willow Creek). NYE CO.: Sedge meadow, n. Kawich Range; Stonewall Mtn Spg. 3100—7200 ft. Introduced annual. May—June.

M. sativa L. (ALFALFA). Common escape from cultivation, occasional individuals throughout region. NTS: Roadsides of n. Amargosa Valley, Mercury Valley, s. Frenchman Flat. NYE CO.: Ash Meadows, Oasis Valley, Stone Cabin Valley. 2300—6000 ft. Introduced perennial. May—Sept.

Melilotus

M. albus Desr. (SWEET-CLOVER). Occasional on disturbed moist sites. NTS: S. Belted Range (Rainier Mesa area). CLARK CO.: Nw. Spring Mtns (abundant in Cold Creek area). NYE CO.: Occasional populations in Ash Meadows and Pahrump Valley, abundant in s. Oasis Valley. 2300—6500 ft. Introduced annual. June—Sept.

M. indicus (L.) All. Seepage sites or washes below springs, in *Coleogyne*, *Artemisia*—Pinyon—Juniper. NTS: Whiterock Spg (nw. Yucca Flat). NYE CO.: Nw. Spring Mtns (abundant along Crystal Spg cyn), n. Kawich Range. 4600—5400 ft. Introduced annual. June—Sept.

M. officinalis (L.) Lam. Occasional populations, moist soils. NTS: Nw. Pahute Mesa. CLARK CO.: Nw. Spring Mtns (common along washes from Wheeler Pass to Willow Spg). NYE CO.: Occasional in Ash Meadows, common in Pahrump Valley. 2300—6300 ft. Introduced annual. May—July.

Peteria

P. thompsonae S. Wats. NTS: Local small population in *Lycium pallidum—Grayia* of s. Frenchman Flat. NYE CO.: In *Sarcobatus* of gravel uplands S of Wildhorse Spg (e. Goldfield Hills). 3200—5800 ft. Perennial. May—June.

Prosopis

P. glandulosa Torr. var. **torreyana** (L. Benson) M. C. Jtn. [*P. juliflora* (Sw.) DC. var. *t.* L. Benson] (MESQUITE). The *Prosopis* of sandy areas, esp. on low dunes (where plants may be shrub-like), reaching its n. limits in the region. NTS: Three small trees, apparently planted at Cane Spg (w. Frenchman Flat) near base of Skull Mtn (at 4000 ft), are not known to flower. CLARK CO.: Indian Spgs, Cactus Spgs. NYE CO.: Stewart Valley, common in Ash Meadows and Amargosa Valley to N, small plants only (to 4800 ft) along Crystal Spg cyn (S slope of nw. Spring Mtns). 2100—3100 ft as reproducing small trees; to 4800 ft as small, non-reproducing plants. Apr—May.

P. pubescens Benth. (SCREWBEAN). NYE CO.: Common as small trees, usually growing with Ash and *Baccharis* in scattered groves in spg areas of n. and e. Ash Meadows; thicket of reproducing plants in s. Oasis Valley (Beatty townsite) perhaps had its origin in a cultivated plant. Not known elsewhere in region. 2100—3100 ft. Small tree. May—June.

Psoralea

P. lanceolata Pursh ssp. **scabra** (Nutt.) Piper. NYE CO.: Locally common, *Atriplex*, near playa of cent. Kawich Valley (below W slope of cent. Belted Range). Perennial. June.

Psorothamnus

P. fremontii (Torr.) Barneby var. **fremontii** (*Dalea f.* Torr.) (INDIGO-BUSH). Widely distributed in *Larrea* communities, esp. *Larrea—Atriplex* and *Larrea—Ambrosia*, sometimes as a codominant, and locally in *Grayia—Lycium* and *Coleogyne—Atriplex*, bajadas below limestone mountain ranges. NTS: N. Amargosa Valley (sands below Striped Hills), Mercury Valley, Rock Valley, s. and e. Jackass Flats, most parts of Frenchman Flat; local in sw. and e. Yucca Flat (incl. Plutonium Valley), where at its n. limits. NYE CO.: N. Ash Meadows, nw. Crater Flat. 2300—4400 ft. Shrub. Apr—May.

P. polydenius (Torr.) Rydb. (*Dalea polyadenia* Torr. ex S. Wats.). Usually on loose sands, *Atriplex canescens*, *Lycium pallidum—Grayia*, *Artemisia tridentata*. NTS: Nw. Jackass Flats (Forty-Mile Cyn wash); s. and w. Frenchman Flat, and in *Atriplex* of talus slopes below French Peak mtn; se. and e. Forty-Mile Cyn; local in ne. Rock Valley, nw. and ne. Yucca Flat, Thirsty Cyn, s. Groom Lake. NYE CO.: Tuff-conglomerate outcrops of n. Oasis Valley. 3500—5500 ft. Shrub. June—July, some yrs Sept.

Trifolium

T. andersonii A. Gray ssp. **beatleyae** Gillett (Gillett, J. M., *Can. J. Bot.*, **50**: 1975. 1972). Locally common to abundant around volcanic outcrops, esp. flatrock areas and along washes, in *Artemisia nova* and *Artemisia*—Pinyon—Juniper. NTS: S and E slopes of Shoshone Mtn (type locality of variety, Topopah Spg, base of S slope), Eleana Range (near Capt. Jack Spg), and SE rim of Pahute Mesa where occurs over many acres. NYE CO.: Wash below White Blotch Spg W of n. Groom Range. 5800—7300 ft. Perennial. Late Apr—June.

T. monanthum A. Gray var. **monanthum.** NYE CO.: Locally abundant on wet soils in *Artemisia*—Pinyon—Juniper, n. Kawich Range (Eden Creek cyn). 6600—7200 ft. Perennial. June—July.

T. repens L. (WHITE-CLOVER). Disturbed irrigated or other moist sites of region. CLARK CO.: Indian Springs townsite; near Cold Creek below nw. Spring Mtns. NYE CO.: Sedge meadows in *Artemisia*—Pinyon—Juniper, n. Kawich Range. 3100—7100 ft. Introduced perennial. May—July.

T. wormskioldii Lehm. Locally common to abundant, in standing water of swamps or other wet soils; with *Scirpus*, *Juncus*, *Carex*, or along streams in *Artemisia*—Pinyon—Juniper. NYE CO.: Oasis Valley; cent. and n. Kawich Range (Stinking Spg, Eden Creek and Longstreet cyns). 3600—7200 ft. Perennial. June—Sept.

FAGACEAE. Beech Family

Quercus

Q. gambelii Nutt. Occurring as thickets, or sometimes in stands of trees with closed canopies, along washes or near springs in *Artemisia*, or usually as thickets and understory in *Artemisia*—Pinyon—Juniper. NTS: Common on Shoshone Mtn, Eleana Range, s. Belted Range (Rainier Mesa, Gold Meadows), Pahute Mesa. CLARK CO.: Nw. Spring Mtns (esp. in burned areas of Wheeler Cyn and Clark Cyn). NYE CO.: Cent. Belted Range. 5400—7500 ft. Small tree (to 7 m tall, 30 cm diam.). May—June.

FUMARIACEAE. Fumatory Family

Corydalis

C. aurea Willd. var. **aurea.** Common locally in Pinyon—Fir, Yellow Pine—Fir. CLARK CO.: Nw. Spring Mtns (Clark Cyn, Trough Spg). LINCOLN CO.: Cent. Groom Range (Bald Mtn). 8000—9000 ft. Probably biennial. June—July.

var. **occidentalis** Engelm. NYE CO.: Local under shrubs, in *Grayia*—*Tetradymia*; cent. Reveille Range (E end of Fang Ridge). 6000 ft. Probably winter annual. May.

GARRYACEAE. Silk-Tassel Family

Garrya

G. flavescens S. Wats. var. **flavescens.** CLARK CO.: Common locally in *Artemisia*—Pinyon—Juniper and burned Yellow Pine—Fir, both slopes of nw. Spring Mtns (Clark Cyn, Wheeler Cyn, Cold Creek area). 6200—7400 ft. Shrub. May—June.

GENTIANACEAE. Gentian Family

Centaurium

C. namophilum Reveal, Broome & Beatley (Reveal, J. L., C. R. Broome, and J. C. Beatley, *Bull. Torr. Bot. Club*, **101**: 353. 1973). NYE CO.: Common locally in *Atriplex*, *Distichlis*, and Ash—Screwbean, in clay soils of spring areas of n. and e. Ash Meadows; type locality, S of Devil's Hole. Known also from Tecopa Spgs, Calif. 2200—2300 ft. Annual. July—Sept.

Frasera

F. albomarginata S. Wats. Local, sometimes common, *Artemisia*—Pinyon—Juniper. CLARK CO.: Nw. Spring Mtns (common in Clark Cyn area). LINCOLN CO.: S. Groom Range. NYE CO.: Nw. Spring Mtns (N slope below Mt. Stirling). 6000—7800 ft. Perennial. May—July.

F. pahutensis Reveal (Reveal, J. L., *Bull. Torr. Bot. Club*, **98**: 107. 1971). NTS: Large population near SE rim of Pahute Mesa, in *Artemisia*—Pinyon—Juniper; type locality. 7200—7300 ft. Perennial. May—June.

GERANIACEAE. Geranium Family

Erodium

E. cicutarium (L.) L'Her. (STORKSBILL). Widely distributed on disturbed sites and in washes, esp. in *Coleogyne* and *Larrea—Atriplex* and at middle elevations of region. NTS: Mercury Valley, Jackass Flats, Topopah Valley, Frenchman Flat, Mid Valley, Yucca Flat. CLARK CO.: Nw. Spring Mtns (Cold Creek area, in *Artemisia*). NYE CO.: Ash Meadows; Bullfrog Hills. 2200—6000 ft. Introduced winter annual. Mar—June, individuals any month of yr.

HYDROPHYLLACEAE. Waterleaf Family

Emmenanthe

E. penduliflora Benth. Around bases of cliffs or volcanic rock outcrops, in *Coleogyne* or *Artemisia*. CLARK CO.: Nw. Spring Mtns (Rock Spg of N slope). NYE CO.: N. Bare Mtn (head of Fluorspar

Cyn), foothills of cent. Grapevine Mtns. 4400—5800 ft. Annual. Apr—May.

Eriodictyon

E. angustifolium Nutt. Common associated species of *Artemisia*, *Artemisia*—Pinyon—Juniper, and Yellow Pine of both slopes of nw. Spring Mtns. CLARK CO.: Common esp. in burned areas of Clark Cyn and from Wheeler Pass to Willow Spg. NYE CO.: Crystal Spg cyn. 4500—7000 ft. Shrub. June—July.

Eucrypta

E. chrysanthemifolia (Benth.) Greene var. **bipinnatifida** (Torr.) Const. NYE CO.: Base of volcanic cliffs below n. Yucca Mtn (Beatty Wash), in *Larrea—Ambrosia.* 3600 ft. Winter annual. Apr.

E. micrantha (Torr.) Heller. The usual *Eucrypta* of the region, common to abundant at bases of ledges, crevices, and cliffs of limestone mountain ranges and local on bajadas below; *Atriplex*, *Larrea*, *Coleogyne.* NTS: In or below Specter Range, w. Skull Mtn, w. Spotted Range (incl. Red Mtn), Ranger Mtns, limestone butte W of Buried Hills, w. Shoshone Mtn (Forty-Mile Cyn wash area), dolomite hill of Eleana Range. CLARK CO.: N-S axis of Spotted Range. NYE CO.: Bare Mtn, Beatty Mtn. 3200—7400 ft. Winter annual. Apr—May.

Nama

N. aretioides (Hook. & Arn.) Brand. The *Nama* of loose sands and along sandy washes, in *Larrea—Ambrosia*, *Atriplex canescens*, *Artemisia tridentata.* NTS: Common in sands below n. Shoshone Mtn and W of Eleana Range (s. and e. Forty-Mile Cyn); occasional in sands of nw. Jackass Flats, sw. Frenchman Flat, cent. Yucca Flat, s. Groom Lake (below s. Belted Range), s. Gold Flat (below N face of Pahute Mesa). NYE CO.: S. Penoyer Valley, cent. Groom Lake. 3300—6000 ft. Winter annual. May—June.

N. demissum A. Gray var. **demissum.** The common *Nama* of lower and middle elevations; *Larrea* (esp. *Larrea—Lycium—Grayia*), *Atriplex*, *Atriplex—Ceratoides*, *Lycium pallidum—Grayia*, *Grayia—Lycium*, *Coleogyne*, uncommon in *Artemisia*, and rarely disturbed sites in *Artemisia*—Pinyon—Juniper. NTS: N. Amargosa Valley (incl. red cinder cone near S end of Yucca Mtn), Mercury valley, Rock Valley, e. half of Jackass Flats, Frenchman Flat, most parts of Yucca Flat; s. Belted Range (Rainier Mesa area), nw. Pahute Mesa. NYE CO.: Below n. Yucca Mtn, Bullfrog Hills, Tolicha Peak area. 2600—5600 (—7200) ft. Winter annual. Apr—May (—July).

N. densum Lemmon. NTS: Uncommon and local, *Artemisia* and *Artemisia*—Pinyon—Juniper. NTS: S. Belted Range (Rainier Mesa), nw. Pahute Mesa. 5600—7500 ft. Annual. May—June.

N. depressum Lemmon ex Gray. NTS: Local, *Larrea—Ambrosia* N of hill outlier, w. Jackass Flats. 3400 ft. Winter annual. Apr.

N. pusillum Lemmon ex Gray. Known from a number of populations in several areas in the spring of one year (1969, when winter annual germination was triggered by a late January rain), and over a 10-yr period not known previously or afterwards on the same or other sites; *Larrea—Ambrosia*, *Larrea—Atriplex*, *Larrea/Coleogyne*, *Atriplex*; esp. on sites with well-developed soil pavement. NTS: Mercury Valley, Rock Valley, sw. and n. Frenchman Flat. CLARK CO.: Below N-S axis of Spotted Range (extreme e. Frenchman Flat). NYE CO.: Ash Meadows. 2200—4500 ft. Winter annual. Early to late May.

Phacelia

P. affinis A. Gray. Local populations, bases of cliffs, washes or other disturbed sites in *Artemisia*—Pinyon—Juniper. NTS: Several sites of n. Pahute Mesa. NYE CO.: N. and s. Kawich Range (Eden Creek cyn and Stinking Spg). 6400—7000 ft. Annual. June—July.

P. beatleyae Reveal & Constance (Reveal, J. L., and L. Constance, *Brittonia*, 24: 199. 1972). NTS: Locally abundant along certain washes and nearby loose talus below French Peak (n. Frenchman Flat), restricted to light-brown tuff bedrock areas in *Atriplex hymenelytra* (type locality); with scattered *Coleogyne* and Juniper on white gravel talus of hill in w. Emigrant Valley. 4000—4500 (—5800) ft. Probably winter annual. Apr—May.

P. bicolor Torr. ex S. Wats. NTS: Locally common in sandy soils of middle elevations, *Atriplex canescens*, *Artemisia tridentata*, and lower *Artemisia*—Pinyon—Juniper, uncommonly *Larrea—Ambrosia*; in or below Eleana Range and Shoshone Mtn (esp. Sugar Loaves area), e. Pahute Mesa. (3600—) 4600—6800 ft. Winter annual. May—June.

P. calthifolia Brand. NYE CO.: Locally common on sandy soils, in *Atriplex*, s. Ash Meadows. 2200 ft. Probably winter annual. May.

P. crenulata Torr. var. **crenulata.** The common variety in *Larrea* communities, associated with soils derived from calcareous materials, esp. along washes. NTS: Disturbed sites of sw. Mercury Valley; n. Amargosa Valley, Rock Valley, cent. and s. Jackass Flats; washes of se., sw., and n. Frenchman Flat. 3200—4000 ft. Winter annual. Mar—May.

var. **funerea** J. Voss. The variety of middle and higher elevations, in washes or on talus, in *Coleogyne*, *Artemisia tridentata*, *Artemisia*—Pinyon—Juniper. NTS: In or near Shoshone Mtn, Timber Mtn, Mine Mtn, Eleana Range, s. Belted Range (Rainier Mesa), Pahute Mesa, nw. Papoose Range. LINCOLN CO.: Ne. Groom Range. NYE CO.: Grapevine Mtns, Tolicha Peak area, Cactus Range, Reveille Range. 4500—7000 ft. Annual. May—July.

P. curvipes Torr. ex S. Wats. Occasional populations along washes and on shallow soils, usually *Artemisia*—Pinyon—Juniper, uncommonly *Coleogyne*. NTS: S. Shoshone Mtn, Eleana Range (near Capt. Jack Spg); flatrock areas of s. Belted Range (Rainier Mesa and Twin Peaks below W face) and n. Forty-Mile Cyn. NYE CO.: Nw. Spring Mtns (Rock Spg of N slope), cent. Belted Range (Johnnies Water). 5800—7500 ft. Annual. May—June.

P. fremontii Torr. One of the most common and consistently present annual species of *Larrea* (conspicuously absent in *Larrea*—*Ambrosia* of some areas), *Grayia*—*Lycium*, *Coleogyne*, *Artemisia*, and disturbed sites of lower *Artemisia*—Pinyon—Juniper. NTS: Mercury Valley, Rock Valley, Jackass Flats, Topopah Valley, Frenchman Flat, Mid Valley, Yucca Flat, Forty-Mile Cyn, Emigrant Valley; s. Belted Range (Twin Peaks below W face of Rainier Mesa), Pahute Mesa. LINCOLN CO.: Below ne. Groom Range (Desert Valley). NYE CO.: Pahrump Valley, n. Amargosa Valley, Crater Flat, Sarcobatus Flat, Stonewall Flat, Railroad Valley; nw. Spring Mtns (N slope below Mt. Stirling), Bullfrog Hills, and n. Belted Range (Kawich Valley). 2600—6000 (—7200) ft. Winter annual. Apr—June.

P. gymnoclada Torr. ex Wats. NYE CO.: Locally common, slopes of e. Goldfield Hills, in *Sarcobatus*. 5800 ft. Probably winter annual. May—June.

P. hastata Dougl. ex Lehm. Common along washes, *Artemisia*—Pinyon—Juniper, *Artemisia*—*Cercocarpus*, Yellow Pine—Pinyon. CLARK CO.: Nw. Spring Mtns (Wheeler Wash; Wheeler Pass to Willow Spg, area of Trough Spg). NYE CO.: N. Kawich Range (Eden Creek and Longstreet cyns). 6000—8500 ft. Perennial. Late May—July.

P. lemmonii A. Gray. NYE CO.: Locally common in *Artemisia*—Pinyon—Juniper, s. Kawich Range (Cedar Pass). 6000 ft. Annual. June—July.

P. mustelina Cov. Local, but rather widely distributed, apparently restricted to calcareous (usually volcanic) cliff crevices, rock ledges, or gravel talus, in *Coleogyne*, *Artemisia*, *Artemisia*—Pinyon—Juniper, uncommonly in *Larrea* and *Atriplex*. NTS: Below e. Skull Mtn, Shoshone Mtn, Eleana Range (e. Forty-Mile Cyn),

Pahute Mesa, Thirsty Cyn, Halfpint Range (E of Banded Mtn). NYE CO.: N. Yucca Mtn (Beatty Wash). 3500—5500 (—6500) ft. Annual. June—Sept.

P. parishii A. Gray. NTS: Known only from a coll. by Ripley & Barneby, May 1941 (CAS), below w. Spotted Range (Mercury Ridge) near playa, s. Frenchman Flat. 3200 ft. Winter annual. May.

P. pedicellata A. Gray. Usually scattered plants only, along major washes near base of limestone mountain ranges of n. Amargosa Valley drainage; in *Larrea* and *Atriplex*. NTS: Below e. Specter Range. NYE CO.: Below cent. Bare Mtn. 3300—4000 ft. Annual. Apr—May.

P. peirsoniana J. T. Howell. Steep talus slopes, around boulders, cliff crevices and rock ledges of canyon slopes of volcanic mountains and mesas, in *Artemisia*—Pinyon—Juniper. NTS: S. Belted Range (Rainier Mesa), Pahute Mesa. NYE CO.: Cent. Belted Range (Cliff Spg, Johnnies Water). 6200—7300 ft. Annual. June—Sept.

P. rotundifolia Torr. ex S. Wats. The frequent *Phacelia* of limestone cliffs at lower elevations, in *Larrea—Atriplex* or *Larrea—Ambrosia*, *Coleogyne*, *Atriplex*. NTS: Specter Range, w. Spotted Range (incl. Red Mtn), below E end of Skull Mtn, Ranger Mtns, Buried Hills and limestone butte to W, Halfpint Range (Banded Mtn), limestone cliffs of w. Shoshone Mtn (Forty-Mile Cyn wash). CLARK CO.: N-S axis of Spotted Range. NYE CO.: Mtn E side of Stewart Valley, n. Yucca Mtn, Bare Mtn, 2600—5500 ft. Annual. Apr—June.

P. saxicola A. Gray. Uncommon except locally, rock crevices and usually shallow soils, esp. flatrock areas; *Coleogyne*, *Artemisia*, *Artemisia*—Pinyon—Juniper. NTS: S. Belted Range (Rainier Mesa), Pahute Mesa; local on sandy sites below n. Shoshone Mtn (in *Atriplex canescens*) and e. Skull Mtn (*Larrea—Ambrosia*). NYE CO.: Below n. Yucca Mtn. 3500—7400 ft. Annual. May—July.

P. tetramera J. T. Howell [*Miltitzia pusilla* (A. Gray) Brand]. Locally common, *Artemisia*—Pinyon—Juniper. NTS: Washes, e. Pahute Mesa. NYE CO.: N. Kawich Range (Eden Creek cyn). 6800—7000 ft. Annual. June.

P. vallis-mortae J. Voss var. **vallis-mortae.** Probably the most widely distributed *Phacelia* of the region, nearly always growing in shrubs in *Larrea*, *Grayia—Lycium*, *Coleogyne*, *Artemisia*, and either under trees or in open in *Artemisia*—Pinyon—Juniper. NTS: Mercury Valley, Rock Valley, Jackass Flats, Topopah Valley, Frenchman Flat, Mid Valley, Yucca Flat, Forty-Mile Cyn, Emigrant Valley; Eleana Range, s. Belted Range (Twin Peaks below W face of Rainier Mesa), Pahute Mesa. NYE CO.: Stewart Valley (with *Atriplex* *hymenelytra*), n. Amargosa Valley, Crater Flat, Sarcobatus Flat. 3500—7000 ft. Winter annual. Apr—June.

Tricardia

T. watsonii Torr. ex Wats. Occasional, usually around rock crevices or ledges, sometimes under shrubs on talus slopes, associated with limestone or dolomite outcrops in *Coleogyne*, *Artemisia nova*, *Atriplex*, rarely *Larrea*. NTS: Specter Range, summit of Skull Mtn, Yucca Mtn, w. Spotted Range (Red Mtn, Mercury Ridge), Ranger Mtns, below e. Shoshone Mtn, Mine Mtn, dolomite hill of Eleana Range. CLARK CO.: N-S axis of Spotted Range. 3300—6400 ft. Perennial. Apr—May.

JUGLANDACEAE. Walnut Family

Hicoria

H. pecan Britt. (PECAN). CLARK CO.: Common escape from cultivation, Indian Springs townsite. 3100 ft. Introduced small tree. Fr. July—Aug.

KRAMERIACEAE. Krameria Family

Krameria

K. parvifolia Benth. Common shrub species, usually occurring with *Psorothamnus fremontii* in *Larrea* and *Atriplex* communities, on bajadas below esp. limestone mountain ranges. NTS: Ne. Amargosa Valley, Mercury Valley, Rock Valley, Jackass Flats, Frenchman Flat, sw. Yucca Flat (where reaches its n. limits). NYE CO.: N. Ash Meadows. The glandular and non-glandular forms [var. *glandulosa* (Rose & Painter) Macbr. and var. *imparata* Macbr.] may occur together (e.g., *Beatley 1999*, n. Mercury Valley), and are therefore not here distinguished. 2300—4200 ft. Shrub. Apr—June.

LAMIACEAE (LABIATAE). Mint Family

Hedeoma

H. nanum (Torr.) Briq. ssp. **californicum** W. S. Stewart. NTS: Local, limestone ledges and crevices, in *Atriplex*; e. Specter Range, w. Spotted Range (Red Mtn). 3900—4400 ft. Perennial. Apr—May.

ssp. **nanum.** CLARK CO.: Yellow Pine—Fir, nw. Spring Mtns (base of slope, upper Clark Cyn). 8000 ft. Perennial. June—July.

Marrubium

M. vulgare L. (HOREHOUND). Disturbed moist sites in *Larrea* or *Artemisia*—Pinyon—Juniper, local as occasional individuals or large populations. NTS: Roadside, w. Mercury Valley. CLARK CO.: Nw. Spring Mtns (common along Wheeler Wash). LINCOLN CO.: Near spgs of n. and cent. Groom Range. NYE CO.: Nw. Spring Mtns (Crystal Spg cyn). 3600—7000 ft. Introduced perennial. May—Sept.

Mentha

M. arvensis L. var. **villosa** (Benth.) S. R. Stewart. Moist meadows or near springs, *Artemisia*—Pinyon—Juniper. CLARK CO.: Nw. Spring Mtns (Cold Creek Spg). NYE CO.: N. Kawich Range (Eden Creek and Longstreet cyns). 6000—7200 ft. Perennial, probably introduced. July—Sept.

M. piperita L. (PEPPERMINT). CLARK CO.: With *Juncus*; around lake margin of Indian Springs townsite, and nw. Spring Mtns (Cold Creek area). NYE CO.: Crop plant in Pahrump Valley. 2600—3100 ft. Introduced perennial. July.

Monardella

M. odoratissima Benth. Occasional to locally common, *Artemisia—Cercocarpus*, Yellow Pine—Fir, Fir—Pinyon. CLARK CO.: Nw. Spring Mtns (floor of upper Clark Cyn). LINCOLN CO.: Cent. Groom Range (Bald Mtn). NYE CO.: N. Kawich Range (upper Eden Creek cyn). 7500—9000 ft. Perennial. July—Sept.

Salazaria

S. mexicana Torr. (BLADDER-SAGE). Widely distributed and common in washes in *Larrea*, *Grayia—Lycium*, *Coleogyne*. NTS: Mercury Valley, Rock Valley, Jackass Flats, Topopah Valley, Frenchman Flat, Yucca Flat. NYE CO.: Crater Flat, n. Amargosa Valley (washes of Bullfrog Hills, Beatty Wash). 3500—4600 ft. Shrub. Apr—June.

Salvia

S. columbariae Benth. var. **columbariae.** Widely distributed but common only locally, steep talus slopes and washes, in *Coleogyne*, *Artemisia*, lower *Artemisia*—Pinyon—Juniper, less common in *Larrea* and *Atriplex*. NTS: Below Specter Range, Yucca Mtn, Skull Mtn, Halfpint Range (French Peak mtn), CP Hills, Shoshone Mtn, s. Belted

Range. NYE CO.: W end of Spring Mtns, Beatty Mtn, below n. Yucca Mtn, E slope of Grapevine Mtns. 2900—6000 ft. Annual. Apr—May.

S. dorrii (Kell.) Abrams ssp. **argentea** (Epl.) Munz. The common subspecies at higher elevations in areas of volcanic mountain ranges, on slopes or in washes, *Artemisia* or *Artemisia*—Pinyon—Juniper, rarely *Grayia—Lycium.* NTS: S. Belted Range (Rainier Mesa, Gold Meadows), Pahute Mesa. NYE CO.: Washes below n. Yucca Mtn, Bullfrog Hills, cent. Grapevine Mtns, Tolicha Peak area, cent. Belted Range. 4000—7500 ft. Shrub. Apr—June.

ssp. **gilmanii** (Epl.) Abrams. Common at lower elevations in washes of limestone mountain ranges, in *Larrea—Ambrosia, Larrea—Atriplex*, and *Coleogyne.* NTS: Specter Range, w. Spotted Range (Red Mtn, Mercury Ridge), Ranger Mtns, Buried Hills and limestone butte to W, Halfpint Range (incl. Banded Mtn). CLARK CO.: Nw. Spring Mtns (washes, of or below both slopes), N-S axis of Spotted Range. NYE CO.: Nw. Spring Mtns (Crystal Spg cyn, Johnnie Mine area), Bare Mtn. 3200—4800 ft. Shrub. Apr—May.

S. funerea M. E. Jones. NYE CO.: Local or common along washes of limestone mtns E side of Stewart Valley and N end of Pahrump Valley, in *Larrea—Ambrosia* or *Atriplex hymenelytra—Ambrosia.* 2600—3500 ft. Shrub. May.

S. mohavensis Greene. NYE CO.: Wash of limestone mtn E side of Stewart Valley, in *Larrea—Ambrosia.* 2600 ft. Shrub. May.

Scutellaria

S. nana A. Gray NYE CO.: *Artemisia*—Pinyon—Juniper of cent. Hot Creek Range; to be expected in n. Kawich Range. 7400 ft. Perennial. June.

LINACEAE. Flax Family

Linum

L. lewisii Pursh [*L. perenne* L. ssp. *l.* Hult.] Widely distributed, usually as small local populations, at middle and higher elevations; *Artemisia*, *Artemisia*—Pinyon—Juniper, occasionally *Coleogyne* or *Atriplex canescens*, along washes or around boulders and rock outcrops on flatter terrain, and cliff bases and washes in canyons, throughout volcanic mountain and mesa areas. NTS: In or below Shoshone Mtn, Eleana Range, s. Belted Range (Rainier Mesa, Oak Spring Butte), Pahute Mesa. CLARK CO.: Nw. Spring Mtns (Clark Cyn area). LINCOLN CO.: N. Groom Range. NYE CO.: Cent. and n. Belted Range. 4500—7500 ft. Perennial. May—July, individuals through autumn.

LOASACEAE. Stick-leaf Family

Eucnide

E. urens Parry. Cliff faces and rock ledges of canyons of limestone mountain ranges, rarely in washes or other disturbed sites of bajadas below; usually in *Atriplex*. NTS: Specter Range, w. Skull Mtn, w. Spotted Range (Red Mtn), Ranger Mtns, Buried Hills. NYE CO.: Mtns and hills of n. Pahrump Valley, Bare Mtn. 2600—4300 ft. Perennial. May—June.

Mentzelia

M. albicaulis (Hook.) Torr. & Gray. The widely distributed and common annual *Mentzelia* of middle elevations; *Larrea—Grayia—Lycium*, *Grayia—Lycium*, *Atriplex canescens*, *Coleogyne*, *Artemisia*, infrequently *Artemisia*—Pinyon—Juniper. NTS: N. and e. Jackass Flats, Topopah Valley, w. Frenchman Flat, Mid Valley, Yucca Flat, Forty-Mile Cyn; s. Belted Range (Gold Meadows), nw. Pahute Mesa. NYE CO.: Bullfrog Hills, Tolicha Peak area; n. Crater Flat, sw. Penoyer Valley. 3600—6800 ft. Winter annual. Apr—May.

M. congesta Torr. & Gray. Uncommon except locally, *Artemisia*—Pinyon—Juniper. NTS: Shoshone Mtn, Eleana Range, s. Belted Range (Rainier Mesa incl. slopes). LINCOLN CO.: Cent. Groom Range (Bald Mtn). NYE CO.: Cent. Belted Range (Cliff Spg, Johnnies Water cyn). 5800—7500 ft. Annual. Late May—June.

M. laevicaulis (Dougl. ex Hook.) Torr. & Gray. CLARK CO.: Along major washes in *Artemisia*—Pinyon—Juniper, both slopes of nw. Spring Mtns. NYE CO.: Local in wash in *Atriplex*, Ralston Valley. 5200—6400 ft. Perennial. June—Sept.

M. leucophylla Bdg. NYE CO.: Restricted to the type locality in n. Ash Meadows; in clay soils of spg areas, esp. along cyn washes, in *Atriplex*. 2200—2300 ft. Perennial. June—Sept.

M. montana (A. Davids.) A. Davids. Apparently uncommon in the region; *Coleogyne*, *Artemisia tridentata*, *Artemisia*—Pinyon—Juniper. NTS: S. Belted Range (below W face of Rainier Mesa, Gold Meadows), nw. Papoose Range. NYE CO.: Nw. Spring Mtns (N slope below Mt. Stirling). 5000—6700 ft. Annual. May—June.

M. nitens Greene. Replacing *M. obscura* in some areas as the common *Mentzelia* of lower elevations, in *Larrea—Ambrosia*, *Larrea—Lycium—Grayia*, and *Atriplex canescens*, commonly, but not exclusively, in loose sands. NTS: Local areas of nw. Mercury Valley and e. Rock Valley (below Specter Range), n. Frenchman Flat (Scarp Cyn), sand dune of e. Forty-Mile Cyn (W of Sugar Loaves). NYE

CO.: The common *Mentzelia* of s. Bullfrog Hills; Beatty Wash (below n. Yucca Mtn). 3500—5500 ft. Winter annual. Apr—May.

M. obscura Thompson & Roberts. The common and widely distributed annual *Mentzelia* of lower elevations; most *Larrea* types, *Atriplex*, *A. hymenelytra*, *Atriplex—Ceratoides*, *Lycium pallidum—Grayia*, *Grayia—Lycium*. NTS: N. Amargosa Valley, Mercury Valley, Rock Valley, Jackass Flats, Frenchman Flat, sw. Yucca Flat. NYE CO.: Stewart Valley, Ash Meadows, s. Crater Flat; ne. Bullfrog Hills. 2500—4200 ft. Winter annual. Late Mar—May.

M. oreophila Darl. The common perennial *Mentzelia* of limestone mountain ranges; *Atriplex*, *Larrea—Ambrosia*, *Larrea/Coleogyne*. NTS: Specter Range, w. Spotted Range (incl. Red Mtn, Mercury Ridge), Ranger Mtns, limestone butte W of Buried Hills. CLARK CO.: N-S axis of Spotted Range. NYE CO.: Mtns E side of Stewart Valley; ne. Ash Meadows. 2300—5000 ft. Perennial. Late Apr—Aug.

M. reflexa Cov. Locally common, banks and ledges of major washes of certain parts of limestone mountain ranges in n. Amargosa Valley drainage; *Larrea—Ambrosia*, *Atriplex*. NTS: S slope of Specter Range. NYE CO.: W slope of Bare Mtn. 3000—3800 ft. Annual. Apr—May.

M. torreyi A. Gray NYE CO.: Local in *Atriplex* near playa, cent. Kawich Valley. 5300 ft. Perennial. Aug.

M. veatchiana Kell. Locally abundant on certain disturbed sites, occasional in undisturbed vegetation, *Coleogyne*, *Grayia—Lycium*, *Artemisia*, lower *Artemisia*—Pinyon—Juniper. NTS: Below S and E faces of Shoshone Mtn (esp. on burned sites), below Eleana Range, Halfpint Range (Plutonium Valley and white volcanic hill of w. Emigrant Valley). NYE CO.: Nw. Spring Mtns (N slope below Mt. Stirling), Bullfrog Hills, Grapevine Mtns, Reveille Range. 4100—6200 ft. Winter annual. Apr—June.

Petalonyx

P. nitidus S. Wats. Sporadic populations along washes in *Larrea*, *Atriplex*, *Coleogyne*, hills or bajadas below limestone mountain ranges and other areas of calcareous outcrops. NTS: Below Specter Range, sw. and e. Skull Mtn, Ranger Mtns, Halfpint Range (incl. French Peak mtn, Banded Mtn), Shoshone Mtn. CLARK CO.: Nw Spring Mtns (Wheeler Wash area). NYE CO.: Nw. Spring Mtns (both slopes), Bare Mtn, Obsidian Butte area. 3000—4500 (—5800) ft. Shrub. Late May—July.

P. thurberi A. Gray var. **thurberi.** NTS: Local in *Larrea—Ambrosia* below Striped Hills. NYE CO.: Locally common on sandy sites, in *Atriplex*, or with *Larrea* or Screwbean, in Ash Meadows and scattered other sites in the n. Amargosa Valley, incl. esp. Big Dune. 2100—2500 ft. Shrub. June—July.

LOGANIACEAE. Logania Family

Buddleja

B. utahensis Cov. Occasional to abundant in *Atriplex* or *Larrea—Ambrosia*, along cliffs, ledges, and rock crevices of limestone mountain ranges and hill outliers at lower elevations. NTS: Specter Range, w. Spotted Range (incl. Red Mtn), Ranger Mtns, Buried Hills and limestone butte to W, Halfpint Range (incl. French Peak mtn and Banded Mtn), CP Hills. CLARK CO.: N-S axis of Spotted Range, below N face of nw. Spring Mtns, Bare Mtn. 3400—5200 ft. Shrub. Late May—early July.

LYTHRACEAE. Loosestrife Family

Lythrum

L. californicum Torr. & Gray. NYE CO.: Occasional, moist soils of spg areas, in Ash—Screwbean, n. and e. Ash Meadows. 2200—2300 ft. Perennial, woody at base. June—July.

MALVACEAE. Mallow Family

Althaea

A. rosea (L.) Cav. (HOLLYHOCK). Escape from cultivation. CLARK CO.: Indian Springs townsite. NYE CO.: Beatty townsite. 3100—3300 ft. Introduced perennial. June—July.

Eremalche

E. exilis (A. Gray) Greene [*Malvastrum e.* A. Gray] Uncommon, *Larrea—Ambrosia*, *Larrea—Sarcobatus.* NTS: Cent. Jackass Flats, present in yrs when *E. rotundifolia* is absent on same site. NYE CO.: Oasis Valley, Beatty Wash; s. Bullfrog Hills. 3300—4200 ft. Winter annual. Apr—May.

E. rotundifolia (A. Gray) Greene [*Malvastrum r.* A. Gray]. Uncommon, *Larrea—Ambrosia*, *Larrea—Atriplex*, *Atriplex hymenelytra*; present only certain years, and in communities in which *Sphaeralcea* is not represented. NTS: Below w. and ne. Skull Mtn (cent. Jackass Flats and sw. Frenchman Flat), below French Peak (n. Frenchman Flat); red cinder cone near s. Yucca Mtn and below S slope of Specter Range (n. Amargosa Valley). NYE CO.: Mtn E side of Stewart Valley, below Bare Mtn. 2500—4000 ft. Winter annual. Apr—May.

Malva

M. parviflora L. NTS: Local on irrigated sites, Mercury townsite. CLARK CO.: Below nw. Spring Mtns (seepage area near Cold Creek). 3700—6000 ft. Introduced annual. July—Oct.

Sida

S. hederacea (Dougl. ex Hook.) Torr. ex Gray. NYE CO.: Common locally in moist to wet clay or sandy soils, in *Atriplex* or *Distichlis*; many areas of Ash Meadows, n. Pahrump Valley. 2200—2600 ft. Perennial. May—July.

Sphaeralcea

S. ambigua A. Gray ssp. **ambigua.** The common *Sphaeralcea* of bajadas and lower canyons of limestone mountain ranges or calcareous outcrops elsewhere; *Larrea*, *Larrea/Coleogyne*. NTS: E. Rock Valley (below e. Specter Range), s. Jackass Flats (w. Skull Mtn), n. and cent. Mercury Valley (Red Mtn and Mercury Ridge of w. Spotted Range), se. Frenchman Flat (Ranger Mtns), sw. Yucca Flat (CP Hills), ne. Yucca Flat (Banded Mtn), nw. Yucca Flat (Eleana Range). CLARK CO.: Below N-S axis of Spotted Range. Ours with conspicuous yellow pubescence. 3300—5000 ft. Perennial, often woody near base. Apr—May, individuals at other seasons, esp. autumn.

Hybrids with *S. emoryi* var. *variabilis* are common on disturbed sites between the ranges of the two species, or where the ranges are contiguous. NTS: Rock Valley, s. and w. Jackass Flats, sw. Yucca Flat.

ssp. **monticola** Kearney. The most common and widely distributed *Sphaeralcea* of the region; *Grayia—Lycium*, *Coleogyne*, *Artemisia*, *Artemisia*—Pinyon—Juniper, locally in *Larrea—Lycium—Grayia*, and esp. characteristic of *Atriplex* and *Atriplex—Ceratoides* on basin floors. NTS: W. and n. Mercury Valley, n. and e. Jackass Flats, Topopah Valley, w. Frenchman Flat, Mid Valley, most parts of Yucca Flat, Forty-Mile Cyn, s. Groom Lake; s. Belted Range (Rainier Mesa, Gold Meadows), Pahute Mesa. NYE CO.: N. Crater Flat, Cactus Flat, Gold Flat, Kawich Valley (abundant over grazed central part), Stone Cabin Valley, s. Hot Creek Valley. 3700—7000 ft. Perennial. May—June, some yrs through summer into autumn.

Commonly hybridizes with *S. emoryi* var. *variabilis* on disturbed sites where ranges of the two species overlap or are contiguous. NTS: W. Mercury Valley, Cane Spg area of w. Frenchman Flat, se. Yucca Flat, n. Forty-Mile Cyn; Rainier Mesa. 3600—7500 ft.

S. emoryi Torr. var. **arida** Kearney. NTS: Locally common, disturbed *Lycium pallidum—Grayia* of s. Frenchman Flat, and *Artemisia tridentata* of n. Forty-Mile Cyn. 3100 and 6000 ft. Perennial. May—July.

var. **emoryi.** CLARK CO.: With var. *nevadensis* and apparent hybrids, Indian Springs townsite. NYE CO.: Mesquite, n. Pahrump Valley. 2600—3100 ft. Perennial. May—July.

var. **nevadensis** Kearney. CLARK CO.: Common on disturbed sites, Indian Springs townsite. 3100 ft. Perennial. May—Oct.

var. **variabilis** (Ckll.) Kearney. Common in some areas, *Larrea* (esp. *Larrea—Ambrosia*), *Larrea/Coleogyne*, *Artemisia tridentata*, and *Artemisia*—Pinyon—Juniper. Hybridizes readily with both subspecies of *S. ambigua.* NTS: N. and w. Jackass Flats, nw. and sw. Frenchman Flat; s. Belted Range (below W face of Rainier Mesa), s. Pahute Mesa. NYE CO.: Ash Meadows. 2300—7400 ft. Perennial. Apr—June.

S. grossulariaefolia (Hook. & Arn.) Rydb. var. **pedata** (Torr.) Kearney. The *Sphaeralcea* of washes (common also on other disturbed sites, incl. burned areas) in *Atriplex*, *Coleogyne*, *Artemisia*, *Artemisia*—Pinyon—Juniper, esp. in limestone mountain ranges and bajadas below. NTS: Mercury Valley, Topopah Valley, Frenchman Flat, Mid Valley, Yucca Flat, Forty-Mile Cyn; s. Belted Range (Rainier Mesa). NYE CO.: Foothills of s. Quinn Canyon Range. 3100—7500 ft. Perennial. Apr—June.

S. parvifolia A. Nels. NTS: Common, *Atriplex*, *A. canescens*, and *Lycium pallidum—Grayia*, near playa of Frenchman Flat, and slopes below N end of Ranger Mtns. 3100—3400 ft. Perennial. Apr—May.

MORACEAE. Mulberry Family

Ficus

F. carica L. (COMMON FIG). NYE CO.: Persistent and apparent escapes from cultivation, nw. Spring Mtns (Crystal Spg cyn). 5300 ft. Introduced small tree. Fr. July.

Morus

M. nigra L. (MULBERRY). NYE CO.: Occasional escapes, Beatty townsite. 3300 ft. Introduced small tree. Fr. June.

NYCTAGINACEAE. Four O'Clock Family

Abronia

A. elliptica A. Nels. Widely distributed, common on sandy soils, esp. in washes, *Artemisia* or *Artemisia*—Pinyon—Juniper; probably all volcanic mountains and mesas. NTS: Shoshone Mtn, Timber Mtn (Cat Cyn), s. Belted Range (Rainier Mesa, Gold Meadows), Pahute Mesa. NYE CO.: Frequent in or below Kawich Range (cyns of both slopes), Reveille Range, cent. and n. Belted Range. 5000—8000 ft. Perennial. May—Sept.

A. turbinata Torr. Common locally, usually on deep loose volcanic sands; *Larrea—Ambrosia*, *Lycium pallidum—Grayia*, *Atriplex*, *A. canescens*, *Artemisia tridentata*. NTS: Sandy areas in region of Sugar Loaves (and throughout se. Forty-Mile Cyn drainage); sands of Forty-Mile Cyn wash (n. Amargosa Valley), below S end of Striped Hills, below s. Shoshone Mtn (w. Jackass Flats), E end of Skull Mtn (sw. Frenchman Flat), and below Buried Hills. NYE CO.: S. Penoyer Valley. 3000—5500 ft. Winter annual. Apr—June.

Allionia

A. incarnata L. Occasional to common, *Larrea—Atriplex*, below limestone mountain ranges. NTS: Bajadas below Specter Range, w. Spotted Range (Mercury Ridge), Ranger Mtns. CLARK CO.: Foothills of s. Spotted Range. 3000—4200 ft. Perennial. May—June, some yrs Aug—Oct.

Hermidium

H. alipes S. Wats. NYE CO.: Occasional plants or small populations along disturbed roadsides, in *Atriplex* or *Artemisia nova*; s. Stonewall Flat, Cactus Flat, Stone Cabin Valley; s. Monitor Range. 4800—6200 ft. Perennial. May—June.

Mirabilis

M. bigelovii A. Gray. Widely distributed as occasional plants, or locally common. Plants vary from densely glandular-pubescent throughout, to stems subglabrous below the inflorescence and with scattered retrorse hairs. The more densely pubescent populations, referable to var. *bigelovii*, usually occur above 4500 ft in *Coleogyne*/*Artemisia*, *Artemisia*, or *Artemisia*—Pinyon—Juniper. Those with pubescence characters of var. *retrorsa* (Heller) Munz occur esp. in limestone mountain ranges, in *Larrea—Ambrosia*, *Atriplex*,

A. hymenelytra, and usually below 5000 ft; the most densely pubescent plants known, however, occur in washes below the limestone Spotted Range (n. Mercury Valley) at 3600 ft, and some colls. from *Artemisia* are subglabrous. Recognition of varieties does not appear justified on the basis of pubescence characters in the populations of this region. NTS: In or on bajadas below Specter Range, w. Spotted Range (Red Mtn), e. Skull Mtn, Ranger Mtns, Buried Hills, Halfpint Range (French Peak mtn, Banded Mtn), Shoshone Mtn, Mine Mtn, Eleana Range, n. Pahute Mesa, nw. Papoose Range. CLARK CO.: S. Spotted Range. NYE CO.: S. Stewart Valley; Yucca Mtn (Beatty Wash), Bare Mtn, Bullfrog Hills, Tolicha Peak, Reveille Range. 2500—6500 ft. Perennial. Apr—June.

var. **aspera** (Greene) Munz. Local in *Larrea—Ambrosia*. NTS: N. Mercury Valley. NYE CO.: N. Pahrump Valley near divide into Amargosa Valley. 3000—3600 ft. Perennial. May.

M. froebelii (Behr) Greene. Occasional around boulders, ledges, and bases of cliffs in *Artemisia*—Pinyon—Juniper, less common in *Coleogyne* and *Artemisia*. NTS: Slopes of cyns of Shoshone Mtn, Timber Mtn, Eleana Range, s. Belted Range (Rainier Mesa), esp. Pahute Mesa. CLARK CO.: Nw. Spring Mtns (common in lower Wheeler Cyn). NYE CO.: Nw. Spring Mtns (N slope below Mt. Stirling), nw. Yucca Mtn, Bare Mtn, Tolicha Peak area, Stonewall Mtn, cent. and n. Belted Range. 3600—7000 ft. Perennial. May—July.

M. pudica Barneby. Species restricted to s. Nevada, in *Larrea, Atriplex, A. canescens, Atriplex—Ceratoides, Lycium pallidum—Grayia, Grayia—Lycium*, often occurring as a weed on disturbed sites. NTS: Local in n. Mercury Valley; common on bajadas of Frenchman Flat [where densely pubescent plants occur with the usual glabrous plants (Beatley, J. C., *Leafl. West. Bot.*, **10:** 294. 1966)], Yucca Flat (incl. Plutonium Valley), s. Groom Lake, s. Emigrant Valley. CLARK CO.: Below N-S axis of Spotted Range (e. Frenchman Flat). NYE CO.: Local, sw. Penoyer Valley. 3000—4800 (—5500) ft. Perennial. May—June.

Oxybaphus

O. comatus (Small) Weath. Widely distributed, esp. in volcanic mountains and foothills, as small local populations around low rock outcrops, bases of low cliffs, and washes and talus slopes of shallow canyons; *Coleogyne, Atriplex, Artemisia, Artemisia*—Pinyon—Juniper, Yellow Pine—Fir. Plants vary from densely glandular-pilose [as in *O. pumilus* (Standl.) Standl.] to glabrate, and flowers pale to deep pink, without geographic definition and in the same local population. Since all have leaves distinctly longer than broad, erect scarcely-branched stems, and usually large (8—10 mm) bright pink—lavender flowers, all colls. are referred to *O. comatus.* NTS: Timber Mtn, Eleana Range, Halfpint Range (head of Nye Cyn, Raysonde Buttes area), Pahute Mesa. CLARK CO.: Nw. Spring Mtns (lower Wheeler Wash, upper Clark Cyn). LINCOLN CO.: N. and s. Groom Range. NYE CO.: E. Goldfield Hills, foothills of Monitor Range, n. and cent. Belted Range. 4500—7800 ft. Perennial. May—Sept.

Selinocarpus

S. diffusus A. Gray. NTS: Occasional to common locally, *Larrea, Atriplex*; bajadas below w. Specter Range, w. Spotted Range (Red Mtn), Ranger Mtns. 2700—3400 ft. Perennial. May—June.

NYMPHAEACEAE. Water-Lily Family

Nymphaea

N. odorata Ait. NYE CO.: Occasionally established in spgs of Ash Meadows. 2300 ft. Introduced perennial. July.

OLEACEAE. Olive Family

Forestiera

F. neomexicana A. Gray (DESERT OLIVE). Uncommon and local, occurring as thickets in moist soils in *Larrea—Atriplex*, *Atriplex*, *A. torreyi*, and *Artemisia.* NTS: Cyn bottom near Tippipah Reservoir (se. Forty-Mile Cyn). NYE CO.: Near spgs in e. and s. Bullfrog Hills; w. Pahrump Valley. 2600—5300 ft. Shrub. Late Apr—May.

Fraxinus

F. anomala Torr. CLARK CO.: Cliff grotto in *Coleogyne*, N-S axis of Spotted Range. 4600 ft. Small tree. Apr—May.

F. velutina Torr. var. **coriacea** (S. Wats.) Rehder (VELVET ASH). NTS: Planted in the 1950s on Frenchman Flat playa, where it survived for several years as sprouts. NYE CO.: The common tree, usually with Screwbean and *Baccharis*, in scattered groves of spg areas of Ash Meadows; planted as a windbreak and shade tree in Pahrump Valley. 2200—2300 ft (where not planted). Small tree. Mar—Apr.

Menodora

M. spinescens A. Gray. Widely distributed associated shrub in *Larrea*, *Grayia—Lycium*, *Atriplex*, *Coleogyne*, occasional in

Artemisia. NTS: Becoming the dominant species of extensive areas of otherwise *Larrea—Ambrosia* and *Coleogyne* of N slope of Jackass Flats; other basins in certain areas only—Mercury Valley, Frenchman Flat, Yucca Flat, Thirsty Cyn, s. Groom Lake. NYE CO.: Crater Flat. 3000—6000 ft. Shrub. Late Mar—Apr.

ONAGRACEAE. Evening-Primrose Family

Calyophus

C. lavandulifolius (Torr. & Gray) Raven [*Oenothera l.* Torr. & Gray]. CLARK CO.: Common locally, nw. Spring Mtns (upper Clark Cyn) in Yellow Pine—Pinyon. 8000 ft. Perennial. July.

Camissonia

C. boothii (Dougl.) Raven ssp. **condensata** (Munz) Raven. Widely distributed and locally common, *Larrea, Atriplex, Grayia—Lycium, Coleogyne,* esp. characteristic of soils derived from limestone. NTS: Mercury Valley, Rock Valley, n. and e. Jackass Flats, Frenchman Flat, Mid Valley, s. half of Yucca Flat. CLARK CO.: E. Frenchman Flat (below N-S axis of Spotted Range). NYE CO.: Ash Meadows, n. Amargosa Valley (below red cinder cone near s. Yucca Mtn, Bare Mtn), Oasis Valley, Stonewall Flat; Bullfrog Hills. 2100—5000 ft. Winter annual. Late Mar—May.

ssp. **intermedia** (Munz) Raven. Widely distributed as small local populations, sandy washes or other disturbed sites, *Atriplex canescens, Atriplex—Ceratoides, Artemisia tridentata, Artemisia—*Pinyon—Juniper. NTS: Areas of Shoshone Mtn, Timber Mtn, Eleana Range, s. Belted Range (Rainier Mesa, Gold Meadows), esp. n. Pahute Mesa; local below w. Spotted Range (Red Mtn). LINCOLN CO.: Below n. Groom Range. NYE CO.: Tolicha Peak—Obsidian Butte area, San Antonio Mtns, Kawich Range, cent. and n. Belted Range, s. Quinn Cyn Range. (3300—) 4100—6800 ft. Annual. June—Sept.

C. brevipes (A. Gray) Raven ssp. **brevipes.** Locally common in certain areas of region; usually in *Larrea—Ambrosia,* but also in other *Larrea* types (incl. *Larrea/Coleogyne*), *Atriplex, A. hymenelytra.* NTS: N. Amargosa Valley, Mercury Valley, Rock Valley, n. and e. Jackass Flats, n. and w. Frenchman Flat, disturbed sites in w. Yucca Flat. NYE CO.: S. Stewart Valley, n. Amargosa Valley (esp. below red cinder cone near s. Yucca Mtn, and Big Dune), w. Crater Flat, Oasis Valley, Sarcobatus Flat (Tolicha Peak area); s. Bullfrog Hills. 2500—4800 ft. Winter annual. Late Mar—early May.

ssp. **pallidula** (Munz) Raven. The common subspecies of certain limestone areas in which ssp. *brevipes* does not occur; *Larrea—*

Ambrosia, *Larrea—Atriplex*, *Atriplex*. NTS: Ne. Amargosa Valley (below cent. Specter Range), se. Frenchman Flat (below Ranger Mtns), e. Mercury Valley (below w. Spotted Range); disturbed sites of e. Rock Valley. CLARK CO.: Indian Springs Valley (below nw. Spring Mtns and Spotted Range). NYE CO.: Ash Meadows, n. Amargosa Valley. 2100—4100 ft. Winter annual. Apr—early June.

C. chamaenerioides (A. Gray) Raven. Locally common, bases of cliffs or along washes in limestone mountain ranges; *Larrea—Ambrosia*, *Larrea—Atriplex*, *Atriplex*, *Coleogyne*. NTS: E. Specter Range, Ranger Mtns, limestone butte W of Buried Hills. CLARK CO.: N-S axis of Spotted Range. NYE CO.: Below nw. Spring Mtns (S slope), s. Beatty Mtn, Grapevine Mtns. 3300—6000 ft. Winter annual. Apr—May.

C. claviformis (Torr. & Frém.) ssp. **funerea** (Raven) Raven. NYE CO.: Replacing in Ash Meadows the common ssp. *integrior;* locally common on sandy soils, in *Atriplex—Haplopappus*, s. Ash Meadows. 2100—2200 ft. Winter annual. Apr—May.

ssp. **integrior** (Raven) Raven. The most widely distributed *Camissonia* of the region, usually in scattered small local populations in washes or on disturbed sites; *Larrea*, *Atriplex*, *A. canescens*, *Atriplex—Eurotia*, *Lycium pallidum—Grayia*, *Grayia—Lycium*, *Coleogyne*, *Artemisia*. Occasional hybrids with *C. munzii*. NTS: Cent. Rock Valley, nw. Jackass Flats, Frenchman Flat, Mid Valley, Yucca Flat (incl. Plutonium Valley), Forty-Mile Cyn, w. Emigrant Valley; foothills of s. Belted Range (Rainier Mesa) and Pahute Mesa. NYE CO.: N. Amargosa Valley (Beatty Wash, Bullfrog Hills, below Grapevine Mtns), Oasis Valley, Sarcobatus Flat, Stonewall Flat, Stone Cabin Valley, Railroad Valley, Kawich Valley, Gold Flat, Groom Lake. 3200—6200 ft. Winter annual. Apr—June.

C. heterochroma (S. Wats.) Raven. Widely distributed in the region as local populations, esp. in washes and on sandy volcanic talus below cliffs; *Larrea—Ambrosia*, *Atriplex*, *A. canescens*, *Artemisia tridentata*, *Artemisia*—Pinyon—Juniper. NTS: Rock Valley (below sw. Skull Mtn), Forty-Mile Cyn, Thirsty Cyn, Yucca Flat. NYE CO.: W. Ash Meadows, n. Amargosa Valley (below n. Yucca Mtn), Oasis Valley, Cactus Flat, Gold Flat (below Pahute Mesa), Kawich Valley. A polymorphic species, with some populations (incl. esp. w. Ash Meadows) apparently distinguishable from *C. megalantha* only on basis of corolla length. 2200—7000 ft. Annual. Late May—Oct.

C. kernensis (Munz) Raven ssp. **gilmanii** (Munz) Raven. NTS: Common in certain areas; *Larrea*, *Coleogyne*, occasionally *Atriplex* and *Artemisia*, esp. in washes. NTS: Common, E half of Jackass Flats and w. Frenchman Flat (below hills N of Cane Spg and E end of Skull Mtn); nw. Jackass Flats (below Yucca Mtn); occasional, cent. Rock Valley (below sw. Skull Mtn), n. Frenchman Flat (Nye Cyn area), w. Yucca Flat (below CP Hills and in or below Eleana Range), e. Yucca Flat (incl. Plutonium Valley). NYE CO.: Beatty Wash (below n. Yucca Mtn). 3300—5800 ft. Winter annual. Late Mar—May.

C. **megalantha** (Munz) Raven. NTS: Occasional to locally common (depending on the yr) on soils derived from light-colored volcanic rocks in two areas of w. and n. Frenchman Flat: Bare soil of seepage slope in *Atriplex canescens* near Cane Spg (base of N slope of Skull Mtn, type locality), and washes and talus slopes in *Atriplex* and *A. hymenelytra* of cyns of French Peak mtn. Cane Spg population has varied from a few dozen to around 4000 plants (1960—1973), and one yr was absent altogether. 4000—4400 ft. Spring-germinating annual (Apr or May). Sept—Oct, some small plants beginning in June.

C. **munzii** (Raven) Raven. NTS: The common *Camissonia* of certain areas, bajadas and lower canyons of limestone mountain ranges; *Larrea*, *Atriplex*, rarely *Coleogyne*; in or below Specter Range, w. Skull Mtn, w. Spotted Range (incl. Red Mtn, esp. Mercury Ridge), Ranger Mtns. 3000—4400 ft. Winter annual. Late Mar—May.

C. **parvula** (Nutt. ex Torr. & Gray) Raven. NTS: Occasional, *Artemisia*, *Artemisia*—Pinyon—Juniper; below S rim and nw. Pahute Mesa. 5600—6500 ft. Annual. May—June.

C. **pterosperma** (S. Wats.) Raven. Locally common and widely distributed some years; *Coleogyne*, *Artemisia nova*, less common in *Artemisia*—Pinyon—Juniper, rarely in washes at lower elevations. NTS: In or below W Spotted Range, e. Shoshone Mtn, dolomite hill of Eleana Range, s. Belted Range (Gold Meadows) and foothills, nw. Pahute Mesa and flatrock area below S rim. NYE CO.: Below nw. Spring Mtns (N slope below Mt. Stirling), Grapevine Mtns, s. Kawich Range (Cedar Pass), cent. Belted Range, White Blotch Spg W of n. Groom Range. (3700—) 5000—7000 ft. Annual. Apr—May.

C. **pusilla** Raven. Local in *Artemisia* and *Artemisia*—Pinyon—Juniper, esp. in flatrock areas. NTS: S. Belted Range (Rainier Mesa, Gold Meadows), nw. Pahute Mesa. NYE CO.: S. Kawich Range (Rose Spg cyn and Cedar Pass). 5600—7500 ft. Annual. Late May—June.

C. **refracta** (S. Wats.) Raven. Occasional, sometimes locally common, washes in *Larrea* and *Atriplex*, limestone mountain ranges. NTS: Specter Range, w. Spotted Range (incl. Mercury Ridge), Ranger Mtns. CLARK CO.: Below N face of Spring Mtns. NYE CO.: S. Stewart Valley; Bare Mtn, Beatty Mtn. 2500—4500 ft. Winter annual. Late Mar—May.

C. walkeri (A. Nels.) Raven ssp. **tortilis** (Jeps.) Raven. Occasional, *Larrea, Atriplex, Coleogyne—Atriplex*; crevices, ledges, and washes in canyons of limestone mountain ranges, and disturbed sites on bajadas below. NTS: Specter Range, w. Spotted Range (incl. Red Mtn, Mercury Ridge), Ranger Mtns, Buried Hills and limestone butte to W, CP Hills. CLARK CO.: Below N face of Spring Mtns. NYE CO.: Below S face of Spring Mtns (Johnnie Mine area and Crystal Spg cyn), n. Bare Mtn. 3000—4800 ft. Winter annual, sometimes perennial. Apr—May, some yrs summer months.

Epilobium

E. angustifolium L. ssp. **circumvagum** Mosquin. CLARK CO.: Locally common in moist soils near spg, in Yellow Pine—Fir, nw. Spring Mtns (upper Clark Cyn). 9000 ft. Perennial. July—Aug.

E. ciliatum Raf. [*E. adenocaulon* Hausskn. var. *adenocaulon* and var. *parishii* (Trel.) Munz]. Locally common in wet soils, as along drainage channels or seepage sites near springs, *Coleogyne* or *Artemisia*—Pinyon—Juniper. NTS: Whiterock Spg (nw. Yucca Flat). CLARK CO.: Base of nw. Spring Mtns (Cold Creek Spg). NYE CO.: Oasis Valley (*Scirpus* swamp). 3400—6200 ft. Perennial. May—Oct.

E. glandulosum Lehm. var. **tenue** (Trel.) C. L. Hitchc. Locally common in moist to wet soils, *Artemisia*—Pinyon—Juniper, Yellow Pine—Fir. CLARK CO.: Nw. Spring Mtns (upper Clark Cyn). NYE CO.: Stonewall Mtn (at Spg), n. Kawich Range (Eden Creek and Longstreet cyns). 5600—9000 ft. Perennial. June—Sept.

Gaura

G. coccinea Pursh. Occasional in washes in *Larrea—Ambrosia, Larrea—Atriplex, Larrea/Coleogyne, Atriplex*, below limestone mountain ranges. NTS: Below w. Spotted Range (Red Mtn, Mercury Ridge), Ranger Mtns. CLARK CO.: Below nw. Spring Mtns (Indian Springs Valley), N-S axis of Spotted Range. 3400—5000 ft. Perennial, sometimes woody at base. Apr—June, some yrs Sept—Nov.

G. parviflora Dougl. ex Hook. NYE CO.: Locally common along stream in Ash—Screwbean, n. Ash Meadows. 2300 ft. Biennial. May—June.

Gayophytum

G. decipiens Lewis & Szweyk. Uncommon except locally, *Artemisia*—Pinyon—Juniper. NTS: S. Belted Range (n. Rainier Mesa, Gold Meadows, and cyns of vicinity). LINCOLN CO.: Cent. Groom

Range (Bald Mtn). NYE CO.: Nw. Spring Mtns (N slope below Mt. Stirling). 6600—8000 ft. Annual. May—Sept.

G. diffusum Torr. & Gray ssp. **parviflorum** Lewis & Szweyk. Locally common, *Artemisia*—Pinyon—Juniper, esp. in flatrock areas. NTS: S. Belted Range (s. Rainier Mesa). NYE CO.: N. and s. Kawich Range (Eden Creek cyn and Cedar Pass). 6800—7400 ft. Annual. June—July.

G. racemosum Torr. & Gray. Locally common, *Artemisia nova*, *Artemisia*—Pinyon—Juniper, *Artemisia—Cercocarpus*, around boulder cliffs, flatrock areas, gravel washes, or steep mountain slopes. NTS: S. Belted Range (Rainier Mesa), esp. Pahute Mesa. LINCOLN CO.: Cent. Groom Range (Bald Mtn). NYE CO.: N. and s. Kawich Range (Eden Creek cyn and Cedar Pass area). 5600—9000 ft. Annual. June—July.

G. ramosissimum Torr. & Gray. Apparently the most common member of the genus, *Artemisia tridentata* and *Artemisia*—Pinyon—Juniper, esp. in washes or on other disturbed sites. NTS: In or below s. Belted Range (Rainier Mesa, Gold Meadows), Pahute Mesa. NYE CO.: Cent. Belted Range (Indian Spg, Cliff Spg), s. Kawich Range (Cedar Pass). 5600—7500 ft. Annual. Late May—July.

Oenothera

O. avita (W. Klein) W. Klein ssp. **avita.** Widely distributed, usually as small local populations in sandy soils, esp. washes and other disturbed sites; *Atriplex*, *A. canescens*, *Atriplex—Ceratoides*, *Grayia—Lycium*, esp. *Artemisia tridentata*, *Artemisia*—Pinyon—Juniper. NTS: E. Rock Valley (rare, disturbed *Larrea*), s. Frenchman Flat (*Lycium pallidum—Grayia*), s. and e. Forty-Mile Cyn, s. Groom Lake; in or below s. Belted Range (n. Yucca Flat; Gold Meadows), Pahute Mesa. CLARK CO.: Indian Springs townsite. LINCOLN CO.: Below n. Groom Range. NYE CO.: Sarcobatus Flat, Gold Flat, Kawich Valley, Groom Lake, s. Penoyer Valley, Reveille Valley, Railroad Valley, Ralston Valley. (3100—) 4200—6900 ft. Perennial. May—Sept.

O. brachycarpa A. Gray. CLARK CO.: Common on steep slopes in Yellow Pine—Pinyon—Juniper, nw. Spring Mtns (upper Clark Cyn). 7600—8000 ft. Perennial. June—July.

O. caespitosa Nutt. ssp. **crinita** (Rydb.) Munz. NYE CO.: Scattered populations in *Atriplex—Ceratoides*, w. and s. Cactus Flat below Cactus Range. 5500 ft. Perennial. May—July.

ssp. **marginata** (Nutt.) Munz. Widely distributed, but usually only local, often in washes; *Larrea—Atriplex*, *Atriplex*, *Coleogyne*, *Artemisia*, *Artemisia*—Pinyon—Juniper, mostly associated with limestone mountains or outcrops. NTS: In or below Specter Range, w.

Spotted Range (incl. Red Mtn, Mercury Ridge), Ranger Mtns, Buried Hills and limestone butte to W, Halfpint Range (Nye Cyn), CP Hills, Eleana Range, Yucca Mtn (nw. Jackass Flats), s. Shoshone Mtn, s. Belted Range (Rainier Mesa), Pahute Mesa. CLARK CO.: In or below nw. Spring Mtns (incl. Wheeler Wash, Crystal Spg cyn, Cold Creek—Willow Creek area), N-S axis of Spotted Range. NYE CO.: Nw. Spring Mtns (below W end), n. Yucca Mtn (Beatty Wash), e. Goldfield Hills, Stonewall Mtn, cent. and n. Belted Range. 3700—8200 ft. Perennial. Apr—Sept, mostly June.

O. deltoides Torr. & Frém. ssp. **deltoides.** NTS: Known only from sands of w. Jackass Flats and e. Frenchman Flat, where locally common in *Larrea—Ambrosia.* 3000—3400 ft. Winter annual. Apr—May.

O. hookeri Torr. & Gray ssp. **angustifolia** (Gates) Munz. Moist soils near springs or streams, in *Artemisia*—Pinyon—Juniper or *Juncus* meadows at lower elevations. CLARK CO.: Nw. Spring Mtns (common in Cold Creek—Willow Creek area). NYE CO.: N. Ash Meadows; n. Kawich Range (Eden Creek cyn). 2300—7000 ft. Biennial. June—Sept.

O. pallida Lindl. ssp. **pallida.** NTS: Known from two localities: Abundant on sand dune below Sugar Loaves (e. Forty-Mile Cyn) with *Atriplex canescens*, and local on disturbed sites in *Artemisia tridentata* of nw. Gold Meadows (below s. Belted Range). 5500—6500 ft. Perennial. May—Sept.

O. primiveris A. Gray. NTS: Uncommon, occurring as local populations, esp. in washes or on other disturbed sites, *Larrea—Ambrosia*, *Larrea—Atriplex*, *Lycium pallidum—Grayia*, *Coleogyne*; e. Rock Valley, ne. and e. Frenchman Flat and local in w. part, Mid Valley, s. Groom Lake. 3100—4800 ft. Winter annual. Apr—early May.

OROBANCHACEAE. Broom-Rape Family

Orobanche

O. cooperi (A. Gray) Heller. NTS: Occasional, *Larrea—Ambrosia*, *Atriplex*, shrub species to which attached not known; below Specter Range (e. Rock Valley), French Peak mtn (n. Frenchman Flat). 3500—4200 ft. Perennial. Apr—June.

O. corymbosa (Rydb.) Ferris. Occasional to locally common, *Artemisia tridentata* and *A. nova* (attached to both spp.), *Artemisia*—Pinyon—Juniper. NTS: In or below s. Belted Range (Rainier Mesa), Pahute Mesa; Thirsty Cyn. LINCOLN CO.: Below n. Groom Range.

NYE CO.: N. Belted Range. 6000—7400 ft. Perennial. Late May—July.

O. fasciculata Nutt. Occasional, *Artemisia*, *Artemisia*—Pinyon—Juniper, *Atriplex canescens;* shrub species to which attached variable, but commonly either *Artemisia tridentata* or *A. nova.* NTS: In or below Shoshone Mtn, Eleana Range, Pahute Mesa. LINCOLN CO.: Cent. Groom Range (Bald Mtn). NYE CO.: Nw. Spring Mtns (N slope below Mt. Stirling), s. Monitor Range. 5500—8000 ft. Perennial. Late May—June.

PAPAVERACEAE. Poppy Family

Arctomecon

A. merriamii Cov. Occasional populations in *Larrea—Ambrosia* and *Atriplex* of limestone mountain ranges, often on gravel terraces and sometimes as a weed of disturbed soils. NTS: Specter Range, w. Spotted Range (incl. Mercury Ridge), Ranger Mtns. NYE CO.: Mtns E side of Stewart Valley, n. Ash Meadows. 2200—4800 ft. Perennial. Apr—May.

Argemone

A. corymbosa Greene. Washes and disturbed sites of lower elevations; *Larrea, Lycium pallidum—Grayia.* NTS: Below w. Spotted Range (Red Mtn, Mercury Ridge) in s. Frenchman Flat. NYE CO.: Common on bare sands of Big Dune (Amargosa Valley). 2400—3200 ft. Perennial. Apr—May.

A. munita Dur. & Hilg. ssp. **rotundata** (Rydb.) Ownbey. Scattered small populations in washes and on other disturbed sites of middle and higher elevations, usually in sandy soils; *Atriplex canescens*, *Coleogyne*, *Artemisia*, most frequent in *Artemisia*—Pinyon—Juniper. NTS: In or below Shoshone Mtn, Timber Mtn, s. Belted Range (Rainier Mesa), Pahute Mesa and foothills below S rim. CLARK CO.: Nw. Spring Mtns (Wheeler Wash). LINCOLN CO.: Foothills below n. Groom Range. NYE CO.: N. Cactus Range, n. Kawich Range. 4500—7500 ft. Perennial. Late May—Aug.

Eschscholzia

E. covillei Greene [*E. minutiflora* var. *darwinensis* M. E. Jones]. Common in certain washes, usually in same locality as *E. minutiflora*, but not on same site; *Larrea—Ambrosia*, *Larrea—Atriplex*, *Coleogyne.* NTS: Below Shoshone Mtn (Forty-Mile Cyn wash), washes below e. Skull Mtn and Mine Mtn. NYE CO.: Nw. Spring

Mtns (Crystal Spg cyn), below n. Yucca Mtn (Beatty Wash). 3200—4800 ft. Winter annual. Apr—early June.

E. **glyptosperma** Greene. Widely distributed and locally common; *Larrea, Atriplex, Grayia—Lycium,* and *Coleogyne.* NTS: Bajadas below Shoshone Mtn, Skull Mtn, Specter Range, w. Spotted Range (Red Mtn), Ranger Mtns, Halfpint Range (French Peak mtn, Plutonium Valley), CP Hills. NYE CO.: S. Stewart Valley. 2500—4100 ft. Winter annual. Late May—early June.

E. **minutiflora** S. Wats. Local, usually in washes, *Atriplex, A. canescens, Coleogyne, Artemisia tridentata,* less frequently *Larrea.* NTS: Shoshone Mtn, Timber Mtn (Cat Cyn), below Yucca Mtn (Forty-Mile Cyn wash), CP Hills, Mine Mtn, below N rim of Pahute Mesa (s. Gold Flat). NYE CO.: Nw. Spring Mtns (washes of N slope below Mt. Stirling), Yucca Mtn (Beatty Wash and n. Crater Flat), Bullfrog Hills. 3500—6000 ft. Winter annual. Apr—May.

PLANTAGINACEAE. Plantain Family

Plantago

P. **insularis** Eastw. var. **fastigiata** (Morris) Jeps. Locally common to abundant, *Larrea—Ambrosia, Larrea—Atriplex, Atriplex, Coleogyne,* slopes of canyons and bajadas below limestone mountain ranges. NTS: In or below Specter Range, w. Spotted Range (Red Mtn), Ranger Mtns, Buried Hills and limestone butte to W. NYE CO.: Ash Meadows; below nw. Spring Mtns; Bare Mtn, Beatty Mtn, and other limestone hills or mtns of Bare Mtn—Yucca Mtn area (all in n. Amargosa Valley drainage). 2300—3800 ft. Winter annual. Late Mar—early May.

P. **lanceolata** L. (ENGLISH PLANTAIN). NYE CO.: Local on irrigated sites, Oasis Valley. 3400 ft. Introduced perennial. May—June.

P. **major** L. (COMMON PLANTAIN). Local in wet soils near streams or springs, at lower elevations usually in shade; in Ash—Screwbean, *Artemisia*—Pinyon—Juniper. CLARK CO.: Indian Springs townsite, nw. Spring Mtns (Willow Creek area). NYE CO.: Ash Meadows; nw. Spring Mtns (Crystal Spg cyn), Stonewall Mtn Spg, n. Kawich Range (Eden Creek cyn). 2200—7200 ft. Introduced perennial. May—Sept.

P. **purshii** Roem. & Schult. var. **oblonga** (Morris) Shinners. Uncommon except locally; *Coleogyne, Artemisia* (usually *A. nova*), *Artemisia*—Pinyon—Juniper, esp. around flatrock areas and other shallow soils. NTS: Below s. Shoshone Mtn (Topopah Valley), Eleana Range (nw. Yucca Flat), foothills of s. Belted Range, on and below

Pahute Mesa. LINCOLN CO.: Below n. Groom Range. NYE CO.: Reveille Range. 4700—6400 ft. Winter annual. May—June.

POLEMONIACEAE. Phlox Family

Collomia

C. tenella A. Gray. NTS: Rare, white gravel talus, with scattered *Coleogyne* and Juniper, w. Emigrant Valley. 5800 ft. Annual. May.

Eriastrum

E. eremicum (Jeps.) Mason. NTS: Common, usually sandy or disturbed soils, *Larrea* (esp. *Larrea—Ambrosia*), *Lycium pallidum—Grayia*, *Grayia—Lycium*, *Atriplex*, *A. canescens*, disturbed *Coleogyne*, rarely *Artemisia* or *Artemisia*—Pinyon—Juniper. Mercury Valley, Jackass Flats, Topopah Valley, Frenchman Flat, Mid Valley, Yucca Flat, s. Forty-Mile Cyn. 3000—4800 (—6400) ft. Winter annual. May—June.

E. sparsiflorum (Eastw.) Mason. Common locally, *Coleogyne/Artemisia*, *Artemisia*, and *Artemisia*—Pinyon—Juniper. NTS: E. Shoshone Mtn, s. Belted Range (Rainier Mesa, Gold Meadows, and foothills in sw. Groom Lake), n. Pahute Mesa. LINCOLN CO.: Below n. Groom Range. NYE CO.: Tolicha Peak area, below White Blotch Spg W of n. Groom Range. 4700—7500 ft. Annual. May—June.

E. wilcoxii (A. Nels.) Mason. Widely distributed and common at higher elevations; *Artemisia tridentata*, *Artemisia*-Pinyon-Juniper, rarely *Atriplex* and *A. canescens*. NTS: In or near Shoshone Mtn, Eleana Range, s. Belted Range (incl. Rainier Mesa), Pahute Mesa, e. Thirsty Cyn, and Forty-Mile Cyn (Buckboard Mesa). NYE CO.: Tolicha Peak, Stonewall Mtn, s. San Antonio Mtns, n. and s. Kawich Range (Eden Creek cyn, Cedar Pass area), cent. Belted Range. 5000—7200 ft. Annual. June—July.

Gilia

G. aliquanta A. & V. Grant ssp. **breviloba** A. & V. Grant. Uncommon except locally; *Coleogyne*, *Artemisia*, *Artemisia*—Pinyon—Juniper. NTS: Below s. and e. Shoshone Mtn (nw. Jackass Flats, w. Yucca Flat), Eleana Range, hills of se. Forty-Mile Cyn), s. Belted Range (below W face of Rainier Mesa and large volcanic flatrock area to W). 4400—6500 ft. Winter annual. Apr—May.

G. brecciarum M. E. Jones ssp. **brecciarum.** Widely distributed and common in *Coleogyne*, *Artemisia*, *Artemisia*—Pinyon—Juniper. NTS: Areas of Shoshone Mtn, Eleana Range, s. Belted Range (Rainier

Mesa and slopes; large volcanic flatrock area to W), s. Pahute Mesa. LINCOLN CO.: Cent. Groom Range (Bald Mtn). NYE CO.: Nw. Spring Mtns (uplands and washes of N slope below Mt. Stirling), below nw. Yucca Mtn (Beatty Wash), Grapevine Mtns, Stonewall Mtn. (3800—) 4800—8000 ft. Winter annual at lower elevs. Late Apr—June.

G. campanulata A. Gray. NTS: Locally common, usually in sandy soils, *Larrea* (esp. *Larrea—Ambrosia*), *Atriplex*, *A. canescens*, *Atriplex—Ceratoides*, *Lycium pallidum—Grayia*, *Grayia—Lycium*, rarely *Larrea/Coleogyne;* e. Rock Valley, w. and cent. Jackass Flats, sw. and ne. Frenchman Flat, s. and e. Yucca Flat, s. Forty-Mile Cyn, sw. Groom Lake. 3000—5000 ft. Winter annual. Apr—May.

G. cana (M. E. Jones) Heller ssp. **speciformis** A. & V. Grant. The common and conspicuous *Gilia* in certain areas of the region; *Larrea*, *Coleogyne*, *Grayia—Lycium*, sometimes *Atriplex*, rarely *Artemisia/Coleogyne.* NTS: Large populations below N slope and E end of Skull Mtn (s. and e. Jackass Flats, w. Frenchman Flat) and Yucca Mtn (w. Jackass Flats); scattered small populations, usually in washes or on disturbed sites, s. Mercury Valley, e. Rock Valley, s. Topopah Valley, s. Frenchman Flat, extreme s. Yucca Flat, ne. Mid Valley. NYE CO.: Nw. Spring Mtns (Johnnie Mine area, Mt. Stirling area of N slope), areas of Yucca Mtn and E face of Bare Mtn (incl. Beatty Wash, Tungsten Cyn, red cinder cone near S end of Yucca Mtn, and Beatty Mtn), Bullfrog Hills, Stonewall Mtn, Tolicha Peak; s. Sarcobatus Flat. 2600—4500 (—5400) ft. Winter annual. Apr—May.

ssp. **triceps** (Brand) A. & V. Grant. NTS: The most widely distributed *Gilia* of the Test Site, usually locally common; *Larrea*, *Atriplex*, *A. canescens*, *Atriplex—Ceratoides*, *Grayia—Lycium*, *Lycium pallidum—Grayia*, *Coleogyne.* Intermediates with ssp. *speciformis* where the two subspecies occur together (as at E end of Skull Mtn). Mercury Valley, Rock Valley, n. and e. Jackass Flats, Frenchman Flat (uncommon below Skull Mtn), Mid Valley, most parts of Yucca Flat (incl. Plutonium Valley), sw. Groom Lake, w. Emigrant Valley; esp. characteristic of and common below limestone mtn ranges (Specter Range, w. Spotted Range, Ranger Mtns, CP Hills, Mine Mtn, Banded Mtn). 3100—4500 (—5800) ft. Winter annual. Apr—May.

G. clokeyi Mason. Common some years, esp. in washes in *Larrea* (usually *Larrea—Atriplex*), *Atriplex*, *A. hymenelytra*, *Coleogyne*, restricted to areas of limestone mountain ranges. NTS: Bajadas below Specter Range, w. Spotted Range, Ranger Mtns, limestone butte W of Buried Hills, CP Hills. CLARK CO.: Below N-S axis of Spotted

Range. NYE CO.: Mtns E side of Stewart Valley; nw. Spring Mtns (N slope below Mt. Stirling). 2500—6000 ft. Winter annual. Apr—early May.

G. filiformis Parry ex Gray. NTS: Common locally in washes and around ledges of calcareous rock outcrops in *Larrea* or *Atriplex;* Specter Range, w. Spotted Range, Ranger Mtns, certain areas below E and W ends of Skull Mtn. 3100—4000 ft. Winter annual. Apr—May.

G. hutchinsifolia Rydb. Locally common and widely distributed in sandy soils, esp. along washes, or around low rock outcrops, in two population segments: (1) *Larrea—Ambrosia, Lycium pallidum—Grayia,* or *Atriplex* at elevations usually below 4500 ft; and (2) *Artemisia tridentata* and *Artemisia*—Pinyon—Juniper at elevations over 6000 ft. NTS: Cent. Rock Valley, sandy areas or washes of ne. and sw. Frenchman Flat, Forty-Mile Cyn, w. Emigrant Valley; Pahute Mesa. NYE CO.: Low dunes of Ash Meadows, n. Amargosa Valley (and between Bare Mtn and s. Yucca Mtn), s. and cent. Groom Lake, s. Kawich Valley; s. Kawich Range (Cedar Pass), cent. Belted Range. 2200—6800 ft. Annual, or winter annual at lower elevs. Late Apr—May at lower elevs., June—July at higher elevs.

G. latifolia S. Wats. Local populations in or near limestone mountain ranges; *Larrea—Ambrosia, Larrea—Atriplex, Larrea/Coleogyne, Coleogyne, Atriplex, A. hymenelytra.* NTS: Specter Range, Ranger Mtns, rimrock and steep slopes of East Thirsty Cyn. CLARK CO.: N-S axis of Spotted Range. NYE CO.: Se. Ash Meadows; mtns N of Ash Meadows, Bare Mtn, Beatty Mtn. 2200—5200 ft. Annual. Apr—Aug, summer-flowering plants (on disturbed soils) usually without fetid odor of spring-flowering plants.

G. leptomeria A. Gray. Locally common in sandy soils; characteristic of *Artemisia tridentata* and *Atriplex canescens,* but also in *Larrea—Ambrosia, Atriplex—Ceratoides, Sarcobatus, Coleogyne,* and *Artemisia*—Pinyon—Juniper. NTS: W. Jackass Flats, ne. Frenchman Flat, n. and sw. Yucca Flat; Pahute Mesa. LINCOLN CO.: Below n. Groom Range (nw. Desert Valley). NYE CO.: Below nw. Yucca Mtn (w. Crater Flat), e. Goldfield Hills, Tolicha Peak, n. and s. Kawich Range, cent. and n. Belted Range; sandy areas of Oasis Valley, s. Sarcobatus Flat, sw. Penoyer Valley. 3000—6800 ft. Winter annual at lower elevs. Apr—June.

G. malior A. Day & V. Grant. Local in *Coleogyne* and *Artemisia nova,* known with certainty only from certain areas, but probably widely distributed in the region. NTS: Below E and S faces of Shoshone Mtn (w. Mid Valley, n. Topopah Valley), in and below Eleana Range (nw. Yucca Flat), below Halfpint Range (ne. Yucca Flat). NYE CO.: Stonewall Mtn. 4800—6000 ft. Winter annual. Apr—early June.

G. modocensis Eastw. *Artemisia*, *Artemisia*—Pinyon—Juniper, probably widely distributed. NTS: Below S face of Shoshone Mtn (n. Topopah Valley), below Eleana Range (nw. Yucca Flat), s. Belted Range (Rainier Mesa and slopes), Pahute Mesa. NYE CO.: Bullfrog Hills (summit of Sawtooth Mtn), Tolicha Peak area. 5000—7500 ft. Annual. Late Apr—June.

G. nyensis Reveal (Reveal, J. L., *Bull. Torr. Bot. Club*, **96**: 476. 1969). The common and in some years abundant magenta-flowered *Gilia* of *Atriplex canescens* and *Artemisia tridentata* (sometimes *Artemisia*—Pinyon—Juniper or *Atriplex—Ceratoides*), on sands usually derived from white or buff-colored tuffs; known only from s. Nye County, Nevada. NTS: Common in s. and e. Forty-Mile Cyn (Sugar Loaves area and N of Shoshone Mtn; type locality, E of Timber Mtn); nw. Jackass Flats (Forty-Mile Cyn wash, as low as 3600 ft); local, s. Belted Range (S and E slopes of Rainier Mesa, Gold Meadows) and Pahute Mesa. NYE CO.: Scattered usually small populations, cent. and s. Kawich Valley, sw. Penoyer Valley, cent. Groom Lake; cent. Belted Range. (3600—) 4700—7900 ft. Winter annual. May—June, individuals through summer into autumn (Nov).

G. ophthalmoides Brand. Widely distributed and locally common *Coleogyne/Artemisia*, *Artemisia* (esp. *A. nova*), and *Artemisia*—Pinyon—Juniper. NTS: In or below Shoshone Mtn, Eleana Range, s. Belted Range (Rainier Mesa and slopes), Pahute Mesa. LINCOLN CO.: Cent. Groom Range (Bald Mtn). NYE CO.: Below nw. Yucca Mtn (Fluorspar Cyn), Tolicha Peak area, Stonewall Mtn, s. Monitor Range, n. and s. Kawich Range, n. Reveille Range. 4700—8000 ft. Winter annual at lower elevs. Apr—June.

G. ripleyi Barneby. Locally occasional to common in *Atriplex*, talus slopes, rock crevices, ledges, and cliffs of certain areas of limestone mountain ranges. NTS: Specter Range (incl. type locality), w. Spotted Range (incl. Red Mtn), sw. Ranger Mtns. CLARK CO.: N-S axis of Spotted Range. NYE CO.: Bare Mtn. 4000—4300 (—4800) ft. Perennial. May—Oct, peak in June—July.

G. scopulorum M. E. Jones. Locally common, *Larrea—Atriplex* and *Atriplex*, washes of canyons, around rock outcrops or talus slopes, limestone mountain ranges. NTS: Specter Range, w. Spotted Range (incl. Mercury Ridge), Ranger Mtns, below Halfpint Range (Nye Cyn wash), below E end of Skull Mtn (Mara Wash). CLARK CO.: N-S axis of Spotted Range. NYE CO.: Bare Mtn. 3000—4500 ft. Winter annual. Apr—May.

G. sinuata Dougl. ex Benth. Widely distributed, usually in sandy soils; uncommon in *Larrea*, *Atriplex*, *A. canescens*, *Lycium pallidum—Grayia*, more frequent in *Grayia—Lycium* and *Coleogyne*, and common and characteristic of *Artemisia* and *Artemisia*—Pinyon—

Juniper of volcanic mountain and mesa areas. NTS: N. Amargosa Valley (red cinder cone near s. Yucca Mtn), Mercury Valley, Rock Valley, Jackass Flats, Frenchman Flat, Mid Valley, Yucca Flat, e. Forty-Mile Cyn; s. Belted Range (Rainier Mesa, Gold Meadows, and foothills in sw. Groom Lake), Pahute Mesa. NYE CO.: Crater Flat; below nw. Yucca Mtn (Beatty Wash), s. Kawich Range, cent. Reveille Range. 2600—7000 ft. Winter annual. Apr—June.

G. stellata Heller. Uncommon except locally, present only in certain years; washes, rock ledges, or on bajadas, in *Larrea* (esp. *Larrea—Ambrosia* and *Larrea—Lycium—Grayia*) on calcareous soil materials. NTS: Sw. Mercury Valley (below e. Specter Range), cent. Rock Valley (below sw. Skull Mtn), ne. and sw. Frenchman Flat (Nye Cyn wash and adjacent uplands, and below e. Skull Mtn). NYE CO.; Below nw. Spring Mtns (Johnnie Mine area), E and W face of Bare Mtn, below nw. Yucca Mtn (Beatty Wash), Beatty Mtn, Bullfrog Hills. 2900—3900 ft. Winter annual. Apr—early May.

G. tenerrima A. Gray. NYE CO.: Locally common, moist soils in cyn bottom, *Artemisia*—Pinyon—Juniper, n. Kawich Range (Eden Creek cyn). 7200 ft. Annual. June—July.

G. transmontana (Mason & A. Grant) A. & V. Grant. Widely distributed and common, lower canyons and bajadas usually below limestone mountain ranges; *Larrea*, *Atriplex*, *Grayia—Lycium*, *Coleogyne.* NTS: Specter Range, Little Skull Mtn, w. Spotted Range (incl. Red Mtn, Mercury Ridge), Ranger Mtns, Halfpint Range (Scarp Cyn, Banded Mtn), CP Hills, and below se. Belted Range. CLARK CO.: N-S axis of Spotted Range. NYE CO.: Nw. Spring Mtns (N slope below Mt. Stirling, and foothills at W end), Bare Mtn, nw. Yucca Mtn, e. Bullfrog Hills. 2600—4500 (—5200) ft. Winter annual. Apr—May.

Ipomopsis

I. aggregata (Pursh) V. Grant ssp. **arizonica** (Greene) V. & A. Grant [*Gilia a.* Spreng var. *a.* Fosb.] CLARK CO.: Common on slopes of nw. Spring Mtns, *Artemisia*—Pinyon—Juniper and Yellow Pine—Fir. 6000—7600 ft. Perennial. May—Sept.

I. congesta (Hook.) V. Grant [*Gilia c.* Hook. var. *palmifrons* (Brand) Cronq.]. Common at higher elevations of volcanic mountain ranges; *Artemisia*—Pinyon—Juniper, *Artemisia—Cercocarpus*, or Fir, infrequent in *Artemisia.* NTS: Common in s. Belted Range (Rainier Mesa), and Pahute Mesa. LINCOLN CO.: Cent. Groom Range (Bald Mtn). NYE CO.: White Blotch Spg W of n. Groom Range, cent. Belted Range, n. Kawich Range (Eden Creek cyn). 5900—9100 ft. Perennial. May—Sept.

I. depressa (M. E. Jones) V. Grant [*Gilia d.* M. E. Jones]. Locally common in usually sandy soils, *Atriplex*, *A. canescens*, *Atriplex—Ceratoides;* uncommon in *Larrea—Lycium pallidum* and *Coleogyne.* NTS: Cent. Rock Valley, cent. Frenchman Flat, sw. Yucca Flat, sw. Forty-Mile Cyn, s. Groom Lake, s. Gold Flat. NYE CO.: Oasis Valley, n. Kawich Valley. 3100—5600 ft. Annual. Late May—Sept.

I. polycladon (Torr.) V. Grant [*Gilia p.* Torr.]. Widely distributed and often common locally; *Larrea*, *Atriplex*, *Atriplex—Ceratoides*, *Grayia—Lycium*, *Coleogyne*, *Artemisia*, rarely *Artemisia*—Pinyon—Juniper. NTS: Mercury Valley, Rock Valley, Jackass Flats, Topopah Valley, Frenchman Flat, Mid Valley, Yucca Flat, s. Gold Flat. NYE CO.: Ash Meadows, n. Amargosa Valley (below se. Bullfrog Hills), Stonewall Flat, e. Kawich Valley, sw. Penoyer Valley. 2200—4700 (—6600) ft. Winter annual. Apr—July.

Langloisia

L. punctata (Cov.) Goodd. NTS: Known from only one site in the region, where locally common in *Atriplex—Ceratoides*, s. Gold Flat (below N rim of Pahute Mesa). 5200 ft. Annual. June—July.

L. schottii (Torr.) Greene. The *Langloisia* of sandy soils, commonly below volcanic mountain ranges; *Larrea—Ambrosia*, *Atriplex*, *A. canescens*, *Sarcobatus*, *Grayia—Lycium*, *Coleogyne—Grayia*, uncommon in *Artemisia tridentata.* NTS: W. Mercury Valley, cent. and e. Rock Valley, w. and cent. Jackass Flats, sw. and ne. Frenchman Flat, n. and e. Yucca Flat, sw. Forty-Mile Cyn, sw. Groom Lake. NYE CO.: Ash Meadows, n. Amargosa Valley (Beatty Wash, below red cinder cone near S end of Yucca Mtn, ne. Bullfrog Hills), Oasis Valley, s. Sarcobatus Flat, n. Stonewall Flat, sw. Penoyer Valley. 2200—5800 ft. Winter annual. Apr—June.

L. setosissima (Torr. & Gray) Greene. The *Langloisia* of calcareous soils and bajadas below limestone mountain ranges; *Larrea—Ambrosia*, *Larrea—Lycium—Grayia*, *Grayia—Lycium*, *Atriplex*, *Atriplex—Ceratoides*, *Coleogyne*, rarely *Artemisia.* NTS: N. Amargosa Valley, n. Mercury Valley, e. Rock Valley, s. Jackass Flats, se. Frenchman Flat, cent. Mid Valley, sw. Yucca Flat, s. Groom Lake. LINCOLN CO.: Nw. Desert Valley. NYE CO.: E. Ash Meadows, scattered areas of n. Amargosa Valley, sw. Crater Flat. 2200—5600 ft. Winter annual. Apr—June.

Leptodactylon

L. caespitosum Nutt. NYE CO.: On volcanic boulders, *Artemisia*—Pinyon—Juniper, at Indian Spg on W slope of cent. Belted Range. 6800 ft. Shrub. July.

L. pungens (Torr.) Rydb. Widely distributed in the hills and mountains of the region. In this area two subspecies are doubtfully distinct: In general, plants with cream-colored corollas and short lateral leaf-segments (ssp. *hallii*) occur in limestone mountain areas and at lower elevations, and white-flowered plants with longer lateral leaf-segments (ssp. *pulchriflorum*) occur most frequently in volcanic mountain and mesa areas at higher elevations. There is, however, no clear morphologic, altitudinal, geographic, or ecologic separation, and in some areas within a single population individuals variously exhibit combinations of the leaf and corolla characters.

ssp. **hallii** (Parish) Mason (plants with cream-colored corollas). Frequent on ledges and in crevices of mountain canyons, nearly always this form in limestone mountain ranges, occurring sparingly also in volcanic mountain areas; *Larrea—Ambrosia, Atriplex, Coleogyne*, less frequent in *Artemisia, Artemisia*—Pinyon—Juniper. NTS: Striped Hills, Specter Range, Yucca Mtn (nw. Jackass Flats), Ranger Mtns, Buried Hills, Halfpint Range (limestone hills at head of Nye Cyn, Banded Mtn), Shoshone Mtn, nw. Papoose Range. NYE CO.: Bare Mtn, e. Bullfrog Hills, nw. Kawich Range (upper Longstreet cyn), n. Reveille Range (Reveille cyn). 3400—5000 (—7800) ft. Shrub. Apr—June.

ssp. **pulchriflorum** (Brand) Mason (plants with white corollas). Widely distributed at higher elevations throughout volcanic mountain and mesa areas; *Artemisia, Artemisia*—Pinyon—Juniper, Fir—Pinyon, Yellow Pine—Fir. NTS: Shoshone Mtn, Timber Mtn, Eleana Range, s. Belted Range (Rainier Mesa), Pahute Mesa, Halfpint Range (large white volcanic hill of w. Emigrant Valley). CLARK CO.: Nw. Spring Mtns (Trough Spg). LINCOLN CO.: Cent. Groom Range (Bald Mtn). NYE CO.: Nw. Spring Mtns (Gold Spg of N slope), nw. Yucca Mtn (Beatty Wash), Tolicha Peak—Obsidian Butte area, cent. Belted Range. (3800—) 6000—8200 ft. June—July.

Linanthus

L. arenicola (M. E. Jones) Jeps. & Bail. NTS: Uncommon, but inconspicuous and therefore probably infrequently seen; *Larrea—Ambrosia* of sandy soils, e. Rock Valley (below Skull Mtn), w. Jackass Flats (below Shoshone Mtn), cent. Frenchman Flat (below E end of Skull Mtn), n. Frenchman Flat (N of playa and below Halfpint Range). 3100—3500 ft. Winter annual. Apr—early May.

L. bigelovii (A. Gray) Greene. Apparently rare in region. NTS: Local, *Larrea—Ambrosia—Lycium*, e. Rock Valley and w. Mercury Valley below N slope of Specter Range. 3500 ft. Winter annual. Mar—Apr.

L. demissus (A. Gray) Greene. Rather widely distributed and locally common in *Larrea—Ambrosia*, *Larrea—Atriplex*, *Larrea—Lycium—Grayia*, occasional in *Coleogyne*, *Atriplex hymenelytra;* esp. common on bajadas below limestone mountain ranges. NTS: N. Amargosa Valley, Mercury Valley, Rock Valley, w. Jackass Flats, most parts of Frenchman Flat, Mid Valley, sw. Yucca Flat. NYE CO.: S. Stewart Valley. 2500—4500 ft. Winter annual. Apr—early May.

L. dichotomus Benth. Common to locally abundant in *Larrea/Coleogyne* and *Coleogyne*, less frequent in *Artemisia nova* and *Artemisia*—Pinyon—Juniper. NTS: Esp. common in Shoshone Mtn—Eleana Range region of n. Jackass Flats, Topopah Valley, Mid Valley, and w. Yucca Flat, coincident with large areas of *Coleogyne*; also se. Jackass Flats (below Skull Mtn), Forty-Mile Cyn (below Timber Mtn), ne. Yucca Flat (below Banded Mtn), and sw. Groom Lake (below se. Belted Range). NYE CO.: W. Crater Flat (nw. Yucca Mtn), s. Sarcobatus Flat (Bullfrog Hills). 3600—6200 ft. Winter annual. Apr—June.

L. jonesii (A. Gray) Greene. Locally common, *Larrea*, *Larrea/Coleogyne*, *Atriplex*, usually below limestone mountain ranges. NTS: Below Specter Range, w. Spotted Range (Red Mtn), e. Skull Mtn, Ranger Mtns, CP Hills; also lower n. slope of Jackass Flats and cent. Frenchman Flat. CLARK CO.: S. Spotted Range. NYE CO.: Below Bare Mtn (sw. Crater Flat). 3000—4100 ft. Winter annual. Late Mar—May.

L. nuttallii (A. Gray) Greene ex Mlkn. ssp. **nuttallii** [*Linanthastrum n.* Ewan]. Widely distributed in the mountains of the region; *Artemisia*—Pinyon—Juniper, Yellow Pine—Pinyon, less common in *Artemisia.* NTS: Frequent in s. Belted Range (Rainier Mesa, Gold Meadows) and Pahute Mesa. CLARK CO.: Nw. Spring Mtns (Wheeler Pass, upper Clark Cyn). 6000—7600 ft. Perennial, woody at base. May—July, some yrs autumn.

L. septentrionalis Mason. Local, *Artemisia*, *Artemisia*—Pinyon—Juniper. NTS: S. Belted Range (Rainier Mesa, Gold Meadows) and Pahute Mesa. NYE CO.: N. Kawich Range (Eden Creek cyn). 6600—7500 ft. Annual. June.

Microsteris

M. gracilis (Hook.) Greene ssp. **humilis** (Greene) V. Grant. Locally occasional to common, shallow soils, esp. along washes or disturbed sites, *Artemisia nova* or *Artemisia*—Pinyon—Juniper. NTS: S. Shoshone Mtn, Eleana Range, s. Belted Range (Rainier Mesa and flatrock areas to W), Pahute Mesa (S rim, at base of cliffs). NYE CO.:

Nw. Spring Mtns (N slope below Mt. Stirling), Bullfrog Hills, Grapevine Mtns, n. Kawich Range (Eden Creek cyn). 5800—7200 ft. Annual. Late Apr—June.

Navarretia

N. breweri (A. Gray) Greene. NTS: Locally common, *Artemisia*—Pinyon—Juniper, uncommon in *Artemisia;* esp. on flat-rock areas of s. Belted Range (Rainier Mesa) and Pahute Mesa. 6900—7500 ft. Annual. Late May—July.

Phlox

P. lanata Piper. Locally common, *Artemisia nova* or *Artemisia*—Pinyon—Juniper. NTS: S. Belted Range (nw. Gold Meadows). NYE CO.: N. and s. Kawich Range (Longstreet cyn and Cedar Pass). 6500—8000 ft. Perennial. May—June.

P. pulvinata (Wherry) Cronq. NYE CO.: Locally common on steep slope, *Artemisia—Cercocarpus*, n. Kawich Range (upper Eden Creek cyn). 9000 ft. Perennial. July.

P. stansburyi (Torr.) Heller. Widely distributed and common in esp. the volcanic mountain ranges; *Larrea/Coleogyne*, *Coleogyne*, *Grayia—Lycium*, *Artemisia*, *Artemisia*—Pinyon—Juniper. Corolla length and color highly variable between and within populations. NTS: In or below Yucca Mtn (nw. Jackass Flats), Skull Mtn, Shoshone Mtn, Eleana Range, s. Belted Range (Rainier Mesa, Gold Meadows), Pahute Mesa, Halfpint Range (French Peak area, Scarp Cyn, Plutonium Valley), nw. Papoose Range. CLARK CO.: Nw. Spring Mtns (Trough Spg, Wheeler Pass area). LINCOLN CO.: Cent. Groom Range (Bald Mtn). NYE CO.: Nw. Spring Mtns (N slope below Mt. Stirling), below n. Bare Mtn, nw. Yucca Mtn, Bullfrog Hills (summit of Sawtooth Mtn), Grapevine Mtns, Tolicha Peak area, n. Kawich Range (Eden Creek cyn), n. Belted Range. 4100—8000 ft. Perennial. Late Apr—June.

POLYGALACEAE. Milkwort Family

Polygala

P. acanthoclada A. Gray. NYE CO.: Known in the region only from Ash Meadows, where it is locally common, usually in sandy soils, esp. on low dunes with *Atriplex* and *Haplopappus*, or Ash—Screwbean. 2200—2300 ft. Shrub. May—July.

P. subspinosa S. Wats. var. **heterorhyncha** Barneby. Common locally, usually in sandy soils; *Larrea* (esp. *Larrea—Ambrosia*),

Atriplex, *A. canescens*, *Lycium pallidum—Grayia*, *Grayia—Lycium*, and local in *Coleogyne*. Doubtfully distinct from var. *subspinosa* in this region. NTS: E. Rock Valley (below Specter Range), se. to sw. Frenchman Flat (below w. Spotted Range) and N side of playa, S-cent. Jackass Flats (below Skull Mtn), cent.—e. Yucca Flat and Plutonium Valley (below Halfpint Range). NYE CO.: Tolicha Peak area. 3100—4600 ft. Perennial, slightly to strongly woody at base. Late Mar—June.

var. **subspinosa.** Common locally, *Artemisia nova*, Juniper, or *Artemisia*—Pinyon—Juniper. NTS: Flatrock outcrops and "Juniper Flats" of nw. Pahute Mesa. LINCOLN CO.: W slope of cent. Groom Range (below Bald Mtn). NYE CO.: N. Groom Lake, Penoyer Valley (below n. Belted Range and s. Quinn Cyn Range). 5200—6600 ft. Perennial, slightly to strongly woody at base. Late May—June.

POLYGONACEAE. Buckwheat Family

Chorizanthe

C. brevicornu Torr. var. **brevicornu.** Occasional to locally common, *Larrea*, *Larrea/Coleogyne*, *Atriplex*, *Atriplex—Ceratoides*, *Grayia—Lycium*, rarely *Coleogyne*. NTS: Mercury Valley, Rock Valley, E half of Jackass Flats, cent., s., and w. Frenchman Flat, n. Mid Valley, cent.—e. and sw. Yucca Flat. NYE CO.: Beatty Mtn, below n. Yucca Mtn (upper Beatty Wash) s. Bullfrog Hills. 3100—4400 ft. Winter annual. May.

var. **spathulata** (Small) C. L. Hitchc. Local, *Artemisia*, *Artemisia*—Pinyon—Juniper, *Atriplex*, often on disturbed soils. NTS: Sw. Forty-Mile Cyn; n. Pahute Mesa and below N rim (sw. Gold Flat). NYE CO.: S. Kawich Range (Cedar Pass area). 5500—7000 ft. Annual. June—July.

C. rigida (Torr.) Torr. & Gray. Occasional to abundant locally, calcareous soils, esp. below limestone mountain ranges; *Larrea—Atriplex*, *Larrea—Ambrosia*, *Atriplex*, uncommon in *Coleogyne* and *Grayia—Lycium*. NTS: Mercury Valley, Rock Valley, n. and s. Jackass Flats SW into Amargosa Valley, n. Topopah Valley, esp. s. and e. Frenchman Flat, cent. and s. Yucca Flat, s. Gold Flat. NYE CO.: Ash Meadows, n. Pahrump Valley, n. Amargosa Valley. 2100—4200 (—5200) ft. Winter annual. Apr—June.

C. thurberi (A. Gray) S. Wats. Common locally, *Larrea*, *Atriplex*, *Grayia—Lycium*, esp. *Coleogyne*, *Artemisia*, lower limits of *Artemisia*—Pinyon—Juniper. NTS: Rock Valley (below Skull Mtn), n. and s. Jackass Flats, Topopah Valley, w. Frenchman Flat, Mid Valley, most parts of Yucca Flat, e. Forty-Mile Cyn, s. Gold Flat, sw. Groom

Lake; Pahute Mesa. NYE CO.: W. Crater Flat, Oasis Valley, nw. Kawich Valley (s. Kawich Range), w. Groom Lake (below cent. Belted Range); Bullfrog Hills. 3400—6000 (—7000) ft. Winter annual. May—June.

C. **watsonii** Torr. & Gray. Occasional to common locally, characteristic of *Coleogyne* and *Artemisia nova*, but also in *Artemisia*—Pinyon—Juniper and *Grayia—Lycium—Atriplex*, rare in *Larrea*. NTS: N. Jackass Flats, n. Topopah Valley, Mid Valley, w. and ne. Yucca Flat, e. Forty-Mile Cyn; s. Belted Range (flatrock areas below W slope of Rainier Mesa and Gold Meadows), Pahute Mesa. NYE CO.: Bullfrog Hills; Stonewall Flat (below Stonewall Mtn) and nw. Kawich Valley (s. Kawich Range). 4200—7000 ft. Winter annual. Apr—May.

Eriogonum

E. baileyi S. Wats. var. **baileyi.** Frequent and locally common in sandy soils in *Artemisia*—Pinyon—Juniper of volcanic mountain areas. NTS: In or near Shoshone Mtn, Timber Mtn, s. Belted Range (Rainier Mesa, Gold Meadows), Pahute Mesa. LINCOLN CO.: Foothills of s. Groom Range. NYE CO.: Reveille Range, n. Kawich Range (Eden Creek cyn). 5800—7500 ft. Annual. July—Oct.

E. beatleyae Reveal (Reveal, J. L., *Aliso*, 7: 415. 1972). NYE CO.: Locally common in *Artemisia nova*, Silver Leaf Mine area of foothills of s. Monitor Range (type locality). 6200 ft. Perennial. May—June.

E. bifurcatum Reveal (Reveal, J. L., *Aliso*, 7: 357. 1971). NYE CO.: The common *Eriogonum* near Nevada/California line in Pahrump Valley and Stewart Valley (type locality); *Atriplex* and *A. canescens*. 2500 ft. Annual. June.

E. brachyanthum Cov. Common locally in sandy soils, *Artemisia tridentata*, *Atriplex*, *A. canescens*. NTS: E. and s. Forty-Mile Cyn (in or below Shoshone Mtn, Timber Mtn, Eleana Range, s. Belted Range). NYE CO.: E. Cactus Flat (below s. Kawich Range), s. Penoyer Valley. 4600—6100 ft. Annual. July—Oct.

E. brachypodum Torr. & Gray. Washes and other disturbed sites in areas of calcareous soils; *Larrea*, *Atriplex*, *Coleogyne*, *Coleogyne*/*Artemisia*. NTS: Lower cyns and bajadas below Striped Hills, Specter Range, sw. Skull Mtn, w. Spotted Range (incl. Red Mtn, Mercury Ridge), Ranger Mtns, Halfpint Range (incl. French Peak mtn), CP Hills, Mine Mtn, s. Belted Range (Oak Spring Butte), w. Shoshone Mtn; Thirsty Cyn. CLARK CO.: N-S axis of Spotted Range. NYE CO.: Ash Meadows, Pahrump Valley; Bare Mtn, Bullfrog Hills,

Tolicha Peak area, e. Cactus Range. 2200—6200 ft. Winter annual. Late May—Oct.

E. caespitosum Nutt. Widely distributed and common in *Artemisia nova* and *Artemisia*—Pinyon—Juniper, often the dominant ground cover species of esp. shallow soils. NTS: Yucca Mtn, Shoshone Mtn, Eleana Range, s. Belted Range (Rainier Mesa), Pahute Mesa. NYE CO.: Tolicha Peak, Stonewall Mtn, n. and s. Kawich Range (Longstreet cyn and Cedar Pass area). 5800—8000 ft. Perennial. May—early June.

E. cernuum Nutt. var. **cernuum.** Locally abundant, usually on disturbed sites, *Artemisia*—Pinyon—Juniper, Yellow Pine—Pinyon, Yellow Pine—Fir. NTS: S. Belted Range (Rainier Mesa), Pahute Mesa. CLARK CO.: Nw. Spring Mtns (upper Clark Cyn). NYE CO.: N. and s. Kawich Range (Eden Creek cyn, Cedar Pass). 6000—8800 ft. Annual. July—Sept.

var. **viminale** (S. Stokes) Reveal in Munz. *Atriplex*, *Artemisia*, *Artemisia*—Pinyon—Juniper, sometimes growing with var. *cernuum*. NTS: E. Pahute Mesa. NYE CO.: E. Cactus Flat, nw. Reveille Valley. 5800—6800 ft. Annual. Late July—Sept.

E. concinnum Reveal (Reveal, J. L., *Bull. Torr. Bot. Club*, **96:** 476. 1969). Sandy talus below low volcanic cliffs or around tuff outcrops; common locally in *Atriplex canescens*, *Artemisia tridentata*, *Artemisia*—Pinyon—Juniper. NTS: S. and e. Forty-Mile Cyn (type locality, base of volcanic outcrops E of Timber Mtn); n. Pahute Mesa; Thirsty Cyn, w. Emigrant Valley. NYE CO. Tolicha Peak. 4500—6700 ft. Annual. Late May—early Sept.

E. contiguum (Reveal) Reveal (Reveal, J. L., *Phytologia*, **23:** 163. 1972). NYE CO.: *Atriplex* and *A. hymenelytra;* common locally on lower bajadas, s. Stewart Valley, sw. Pahrump Valley, Ash Meadows and (the type locality) near playa to N in Amargosa Valley; known only from this region. 2200—2500 ft. Probably winter annual. Apr—June.

E. deflexum Torr. var. **baratum** (Elmer) Reveal. Small local populations in *Coleogyne*, *Artemisia*, *Artemisia*—Pinyon—Juniper. NTS: Below n. Shoshone Mtn, Timber Mtn (Cat Cyn, and incl. Dome Mtn), Eleana Range, s. Belted Range (Oak Spring Butte). NYE CO.: Bullfrog Hills, Grapevine Mtns, Tolicha Peak area. 4800—6600 ft. Annual. June—July (—Sept).

var. **deflexum.** Widely distributed and common as local populations below limestone mountain ranges, often with *E. brachypodum*, in *Larrea*—*Ambrosia*, *Larrea*—*Atriplex*. NTS: N. Amargosa Valley (below Specter Range), n. and cent. Mercury Valley (below w. Spotted Range). CLARK CO.: Below se. Spotted Range, and nw.

Spring Mtns (Wheeler Wash, in *Artemisia*—Pinyon—Juniper of cent. Clark Cyn). NYE CO.: N. and w. Pahrump Valley. 2600—6800 ft. Winter annual. June—Oct.

var. **nevadense** Reveal (Reveal, J. L., *Phytologia*, **25**: 169. 1973). The most common variety of the region, usually in areas of volcanic hills and mountains, esp. sandy soils and disturbed sites; *Larrea—Ambrosia*, *Larrea—Grayia—Lycium*, *Lycium pallidum—Grayia*, *Grayia—Lycium*, *Atriplex*, *A. canescens*, *Atriplex—Ceratoides*, *Coleogyne*, *Artemisia*, *Artemisia*—Pinyon—Juniper. NTS: W. Mercury Valley, e. Rock Valley, Jackass Flats, Frenchman Flat (esp. sw. part), Mid Valley, Yucca Flat, most parts of Forty-Mile Cyn, Thirsty Cyn, s. Gold Flat, s. Kawich Valley s. Groom Lake; Pahute Mesa. LINCOLN CO.: E. Groom Lake (Groom Range). NYE CO.: N. Amargosa Valley (below nw. Yucca Mtn), nw. Crater Flat (below Bare Mtn), s. Stonewall Flat, e. Cactus Flat, Ralston Valley, n. and e. Kawich Valley, s. Hot Creek Valley, s. Penoyer Valley. 3200—6600 ft. Winter annual, at least some yrs. Late May—early Oct.

E. esmeraldense S. Wats. var. **esmeraldense.** Common on certain sites in volcanic mountain ranges and mesas, in *Artemisia*—Pinyon—Juniper or Fir—Pinyon. NTS: Eleana Range, s. Belted Range (Rainier Mesa and slopes, Gold Meadows), Pahute Mesa. LINCOLN CO.: Cent. Groom Range (Bald Mtn). NYE CO.: Cent. Belted Range (cyns of both slopes). 5800—8500 ft. Annual. June—Aug.

E. fasciculatum Benth. var. **polifolium** (Benth. in DC.) Torr. & Gray. Characteristic of lower- and middle-elevation bajada or mountain washes throughout most of region, in *Larrea*, *Atriplex*, *Grayia—Lycium*, *Coleogyne*, infrequently *Artemisia*. NTS: Ne. Amargosa Valley, Rock Valley, Jackass Flats (esp. N slope), Topopah Valley, Frenchman Flat, Mid Valley, Yucca Flat, Forty-Mile Cyn (esp. along Forty-Mile Cyn wash), Thirsty Cyn. NYE CO.: N. Pahrump Valley, n. Amargosa Valley (Bare Mtn and Bullfrog Hills). 3200—5800 ft. Shrub. May—July.

E. glandulosum (Nutt.) Nutt. ex Benth. in DC. [*E. carneum* (J. T. Howell) Reveal in Munz]. Common locally in canyons of limestone mountain ranges; *Larrea—Atriplex*, *Atriplex*, *A. hymenelytra*. NTS: Specter Range, w. Spotted Range (incl. Red Mtn, Mercury Ridge), Ranger Mtns. CLARK CO.: S. Spotted Range. NYE CO.: Mtns of n. Pahrump Valley, Bare Mtn. 2900—5000 ft. Winter and summer annual. Late May—early Oct.

E. heermannii Dur. & Hilg. var. **argense** (M. E. Jones) Munz. Occasional to locally common, around boulders and along cliffs; *Artemisia*, *Artemisia*—Pinyon—Juniper. NTS: N. and w. Pahute Mesa, rimrock of Thirsty Cyn. NYE CO.: Bullfrog Hills (Sawtooth Mtn), Grapevine Mtns, Tolicha Peak area, nw. Stonewall Mtn, cent. Reveille Range, cent. Belted Range (cyns of both slopes), White Blotch Spg W of n. Groom Range. 5600—7800 ft. Shrub. Aug—Sept.

var. **heermannii.** Restricted to limestone canyon walls of certain areas; *Atriplex.* NTS: W. Spotted Range (Red Mtn, Mercury Ridge). CLARK CO.: Nw. Spring Mtns (steep talus slope near junction of Wallace Cyn and Wheeler Wash). 3900—4900 ft (at higher elevs. elsewhere in Spring Mtns). Shrub. Aug—Sept.

var. **humilius** (S. Stokes) Reveal. NYE CO.: Common along certain washes, in *Larrea—Ambrosia*, *Atriplex*, *A. canescens;* Beatty Wash (below nw. Yucca Mtn), Tolicha Peak—Obsidian Butte area. 3500—5100 ft. Shrub. Aug—Sept.

var. **sulcatum** (S. Wats.) Munz & Reveal [**E. s.** S. Wats.]. Restricted to limestone mountain ranges, around rock outcrops, crevices, and on cliff faces, in *Larrea*, *Atriplex*, or *Coleogyne*, sometimes as high as Yellow Pine—Fir. NTS: Specter Range, cyns below sw. Skull Mtn, w. Spotted Range (incl. Red Mtn, Mercury Ridge), Ranger Mtns, Buried Hills and limestone butte to W, Halfpint Range (upper Nye Cyn, Banded Mtn), CP Hills, Mine Mtn. CLARK CO.: N-S axis of Spotted Range, nw. Spring Mtns (Clark Cyn and along Wheeler Wash). NYE CO.: Nw. Spring Mtns (Crystal Spg cyn, and foothills of W end); mtns N end of Ash Meadows. 2700—8000 ft. Shrub. Sept—early Oct.

E. hookeri S. Wats. Scattered local populations in washes and esp. in foothills of volcanic ranges; *Atriplex*, *A. canescens*, *Artemisia tridentata.* NTS: Throughout most of Forty-Mile Cyn, Thirsty Cyn; Pahute Mesa. NYE CO.; Gold Flat, w. Cactus Flat, n. and e. Kawich Valley, w. Penoyer Valley. 4400—6500 ft. Annual. July—Sept.

E. howellianum Reveal (Reveal, J. L., *Phytologia*, **25**:169. 1973). NTS: Local on S slope of Mine Mtn, in *Coleogyne.* 5100—5300 ft. Annual. July.

E. inflatum Torr. & Frém. Widely distributed, but characteristic of calcareous soils in *Larrea*, *Coleogyne*, esp. *Atriplex*, rarely *Artemisia* and *Grayia—Lycium*, of bajadas below and lower cyns of limestone mountain ranges. NTS: N. Amargosa Valley, Mercury Valley, Rock Valley, Jackass Flats (esp. N slope), Frenchman Flat (esp. below Ranger Mtns), Yucca Flat (below CP Hills and Banded Mtn), uplands of Thirsty Cyn, s. Groom Lake (below nw. Papoose Range). CLARK CO.: Below nw. Spring Mtns. NYE CO.: Stewart Valley, Ash Meadows, Crater Flat (below Bare Mtn), Stonewall Flat (below Stonewall Mtn); Bullfrog Hills. 2200—5000 ft. Perennial. Apr—June.

E. insigne S. Wats. NTS: *Larrea—Ambrosia*, *Larrea—Lycium—Grayia;* locally the abundant *Eriogonum* of volcanic sands below Halfpint Range (ne. Frenchman Flat); local on disturbed sites near w. Skull Mtn (n. Rock Valley and s. Jackass Flats). 2800—3800 ft. Annual. Sept.

E. kearneyi Tidestr. var. **kearneyi.** Areas of volcanic sands in *Artemisia tridentata*, *Atriplex canescens*, *Atriplex—Ceratoides.* NTS: Upper Nye Cyn (below Halfpint Range) and on blowsand in Buried Hills; abundant in Forty-Mile Cyn (esp. Sugar Loaves area). NYE CO.: Local, Goldfield Hills; e. Kawich Valley (below cent. Belted Range), loose sands of sw. Penoyer Valley, major wash of cent. Gold Flat, sporadic in Ralston Valley (below San Antonio Mtns). 3700—6500 ft. Perennial, woody at base. Aug—Sept.

E. maculatum Heller [*E. angulosum* Benth. ssp. *m.* S. Stokes]. Widely distributed, but occurring usually as scattered plants in *Larrea—Ambrosia*, *Larrea—Lycium—Grayia*, *Lycium pallidum—Grayia*, *Grayia—Lycium*, *Atriplex*, *A. hymenelytra*, *Atriplex—Ceratoides;* locally common to abundant on disturbed sites. NTS: Mercury Valley, Rock Valley, Jackass Flats, Topopah Valley, Frenchman Flat, Mid Valley, Yucca Flat, Forty-Mile Cyn, Thirsty Cyn; Pahute Mesa. NYE CO.: Stewart Valley, n. Amargosa Valley, Crater Flat, Oasis Valley, Tolicha Peak area, se. Penoyer Valley, cent. Groom Lake; Bullfrog Hills. 2500—6600 ft. Winter annual. Apr—July.

E. microthecum Nutt. var. **foliosum** (Torr. & Gray) Reveal. The common and widely distributed variety in *Artemisia*, lower-elevation *Artemisia*—Pinyon—Juniper, rarely *Coleogyne*, *Atriplex*, in volcanic mountain and hill regions. NTS: W. Spotted Range (Mercury Ridge), hills N of Cane Spg, Shoshone Mtn, area of Buckboard Mesa, Mine Mtn, Eleana Range, s. Belted Range (Rainier Mesa, Oak Spring Butte), Pahute Mesa, Thirsty Cyn uplands, Halfpint Range. CLARK CO.: Nw. Spring Mtns (area of lower Clark Cyn). NYE CO.: Nw. Spring Mtns (N slope below Mt. Stirling), Bare Mtn, nw. Yucca Mtn, Grapevine Mtns, Stonewall Mtn, Cactus Range, n. and s. Kawich Range, cent. Belted Range, White Blotch Spg W of n. Groom Range. 3800—7500 ft. Shrub. July—Oct.

var. **lapidicola** Reveal (Reveal, J. L., *Brigham Young Univ. Sci. Bull., Biol. Ser.*, 13(1). 1971). Frequently the dominant ground cover of shallow soils, esp. in flatrock areas; *Coleogyne/Artemisia*, *Artemisia nova*, *Artemisia*—Pinyon—Juniper. NTS: Shoshone Mtn, n. Mine Mtn, Eleana Range, s. Belted Range (Rainier Mesa, slopes and foothills), Pahute Mesa. NYE CO.: Foothills of cent. Belted Range. 4500—7500 ft. Shrub. Aug—Sept.

var. **laxiflorum** Hook. The high-elevation variety of the region; *Artemisia tridentata*, *Artemisia*—Pinyon—Juniper, *Artemisia*—*Cercocarpus*. LINCOLN CO.: Cent. Groom Range (Bald Mtn). NYE CO.: N. Kawich Range (upper Eden Creek cyn). 7600—9100 ft. Shrub. July—Sept.

E. nidularium Cov. Widely distributed, usually as scattered plants in *Larrea*, *Lycium pallidum*—*Grayia*, *Grayia*—*Lycium*, *Coleogyne*, *Atriplex*, *Atriplex*—*Ceratoides*, *Artemisia*, lower *Artemisia*—Pinyon—Juniper; common to abundant on disturbed (esp. burned) sites. NTS: Mercury Valley, Rock Valley, Jackass Flats, Topopah Valley, Frenchman Flat, Mid Valley, Yucca Flat, Forty-Mile Cyn, Thirsty Cyn, s. Groom Lake, w. Emigrant Valley; Pahute Mesa. NYE CO.: Ne. Amargosa Valley (foothills of nw. Spring Mtns, below Bare Mtn, Beatty Wash), Stonewall Flat, w. Cactus Flat, s. Penoyer Valley. 3500—7000 ft. Winter annual. Apr—Sept.

E. nutans Torr. & Gray var. **nutans.** Occasional plants or small local populations in *Artemisia tridentata* and *Artemisia*—Pinyon—Juniper. NTS: Nw. Pahute Mesa. NYE CO.: E. Goldfield Hills, n. Kawich Range (Eden Creek cyn), W slope of n. Belted Range. 5800—6600 ft. Annual. Late May—July.

E. ovalifolium Nutt. var. **multiscapum** Gand. NYE CO.: Locally common, *Artemisia* and *Artemisia*—Pinyon—Juniper, Bullfrog Hills (summit of Sawtooth Mtn) and Grapevine Mtns (Phinney Cyn). 6000—6400 ft. Perennial. Apr—May.

var. **ovalifolium.** Occasional, usually in sandy soils, *Artemisia tridentata*, *Artemisia*—Pinyon—Juniper, infrequently *A. nova* and *Coleogyne*. NTS: In or below Yucca Mtn (nw. Jackass Flats), Shoshone Mtn, Timber Mtn, Eleana Range, s. Belted Range (Rainier Mesa, Oak Spring Butte), Pahute Mesa. NYE CO.: Bullfrog Hills, Tolicha Peak area, n. Kawich Range, n. Reveille Range, n. Belted Range, White Blotch Spg W of n. Groom Range. 5100—6800 ft. Perennial. May—June.

E. palmerianum Reveal in Munz. Common, sandy washes and other disturbed soils, *Coleogyne*/*Artemisia*, *Artemisia*, *Artemisia*—Pinyon—Juniper. NTS: In or near Shoshone Mtn, Timber Mtn, Pahute Mesa and s. foothills, local in s. Belted Range (Oak Spring Butte). CLARK CO.: Nw. Spring Mtns (Willow Creek area and Wheeler Wash). LINCOLN CO.: N. Groom Range. NYE CO.: Grapevine Mtns, Tolicha Peak area, s. Cactus Range, n. and s. Kawich Range, cent. Belted Range (cyns of both slopes), White Blotch Spg W of n. Groom Range. 4800—7200 ft. Annual. June—Oct.

E. panamintense Morton var. **panamintense.** Common locally, *Artemisia tridentata*, *Artemisia*—Pinyon—Juniper, Yellow Pine—

Pinyon. CLARK CO.: Nw. Spring Mtns (slopes near Trough Spg, Clark Cyn). NYE CO.: Bullfrog Hills (summit of Sawtooth Mtn), Grapevine Mtns (Phinney Cyn). 6000—8100 ft. Perennial. Aug.—Sept.

E. **pusillum** Torr. & Gray. Common locally, usually in sandy soils; *Larrea*, *Grayia—Lycium*, rarely *Artemisia*—Pinyon—Juniper. NTS: Bajadas and Forty-mile Cyn wash below Yucca Mtn and Shoshone Mtn (w. Jackass Flats); Mara Wash and Nye Cyn wash (sw. and n. Frenchman Flat); below Eleana Range (nw. Yucca Flat) and Halfpint Range (e. Yucca Flat); w. Emigrant Valley. LINCOLN CO.: Washes below Groom Range (e. Groom Lake and nw. Desert Valley). NYE CO.: Below nw. Yucca Mtn (Beatty Wash), Bullfrog Hills, Tolicha Peak area, Reveille Range, foothills of s. Quinn Canyon Range. 3200—6100 ft. Winter annual. Late Apr—early June.

E. **racemosum** Nutt. Occasional populations, talus slopes and below boulder-cliffs, *Artemisia*—Pinyon—Juniper, *Artemisia—Cercocarpus*, Fir—Pinyon; abundant on certain burned sites. NTS: Shoshone Mtn, Eleana Range, s. Belted Range (Rainier Mesa), Pahute Mesa. LINCOLN CO.: Cent. Groom Range (Bald Mtn). NYE CO.: Stonewall Mtn, n. Kawich Range (common in Eden Creek cyn). 5600—9200 ft. Perennial. July—Oct.

E. **reniforme** Torr. & Frém. One of the most common to abundant annuals of certain areas, in *Larrea—Ambrosia*, *Larrea—Grayia—Lycium*, *Grayia—Lycium*, *Lycium pallidum—Grayia*, *Atriplex*, *A. hymenelytra*. NTS: N. Amargosa Valley (red cinder cone S end of Yucca Mtn), e. Rock Valley, E half of Jackass Flats, n. and w. Frenchman Flat. NYE CO.: Ash Meadows, sw. Crater Flat, Oasis Valley, nw. Amargosa Valley (below Bare Mtn, se. Bullfrog Hills, Beatty Wash), Sarcobatus Flat. 2100—4400 ft. Winter annual. Apr—July.

E. **rixfordii** S. Stokes. NYE CO.: Locally common, rock crevices, washes, or disturbed sites, in *Atriplex*, *Larrea—Atriplex;* cent. and n. Bare Mtn, in or near Bullfrog Hills, s. Grapevine Mtns. 3000—5200 ft. Annual. Aug.—Oct.

E. **rupinum** Reveal (Reveal, J. L., *Aliso*, 7: 217. 1970). NYE CO.: Locally abundant in *Artemisia nova* in cent. Goldfield Hills, and *Artemisia*—Pinyon—Juniper in Rose Spg cyn of s. Kawich Range (type locality). 5600—7200 ft. Perennial. Aug—Sept.

E. **saxatile** S. Wats. Frequently in lower *Artemisia*—Pinyon—Juniper, occasional in *Artemisia tridentata* or *Coleogyne*, often in soils derived from white tuff. NTS: Yucca Mtn (nw. Jackass Flats), n. and s. Shoshone Mtn, Timber Mtn, Eleana Range, s. Belted Range (Oak Spring Butte W to E face of Rainier Mesa). NYE CO.: Nw.

Spring Mtns (Crystal Spg cyn, and N slope below Mt. Stirling), n. Yucca Mtn, Grapevine Mtns. 3800—6700 ft. Perennial. May—June.

E. shockleyi S. Wats. var. **shockleyi.** NYE CO.: Locally common in *Artemisia nova* and *Atriplex;* abundant on slopes of cent. and e. Goldfield Hills, local below n. Belted Range and foothills of s. Monitor Range. 5600—6400 ft. Perennial. May—June.

E. thomasii Torr. Common locally on ledges and talus, *Atriplex*, *Larrea—Ambrosia—Atriplex*, *Coleogyne;* restricted to areas of certain limestone mountain ranges. NTS: Specter Range. CLARK CO.: N-S axis of Spotted Range. NYE CO.: Ash Meadows; Bare Mtn, Beatty Mtn. 2200—5200 ft. Winter annual. Apr—early May.

E. trichopes Torr. Where present, usually the dominant annual; *Larrea—Atriplex*, *Larrea—Ambrosia*, *Atriplex*, *A. hymenelytra*, *Larrea/Coleogyne*, *Artemisia nova*, in or below usually limestone mountain ranges. NTS: Specter Range, w. Spotted Range (incl. Red Mtn, Mercury Ridge), Ranger Mtns, CP Hills. CLARK CO.: Below nw. Spring Mtns. NYE CO.: Stewart Valley, Ash Meadows, ne. Amargosa Valley, Pahrump Valley; Bare Mtn, Bullfrog Hills (summit of Sawtooth Mtn). 2100—3300 (—6000) ft. Winter annual. Late Apr—June.

E. umbellatum Torr. var. **dichrocephalum** Gand. NTS: Occasional to locally common in *Artemisia nova* and *Artemisia*—Pinyon—Juniper; Eleana Range (area of Capt. Jack Spg), n. and e. Pahute Mesa. 6200—7000 ft. Perennial, woody at base. June—July.

var. **subaridum** S. Stokes. Locally common, *Artemisia*— Pinyon—Juniper, *Artemisia—Cercocarpus*, Yellow Pine—Fir. NTS: S. Belted Range (Rainier Mesa and cyns below), S rim of Pahute Mesa. CLARK CO.: Nw. Spring Mtns (middle and upper Clark Cyn, slopes of Cold Creek area). LINCOLN CO.: Cent. Groom Range (saddle at base of Bald Mtn). NYE CO.: Cent. Belted Range, n. and s. Kawich Range (Eden Creek cyn and Cedar Pass area). 6200—9000 ft. Perennial, woody base. July—Sept.

var. **vernum** Reveal (Reveal, J. L., *Great Basin Natur.*, **28:** 157. 1968). Locally common, *Artemisia tridentata* and *Artemisia*—Pinyon—Juniper, sandy soils, esp. along washes. NTS: In or below Shoshone Mtn (type locality below NE end), Timber Mtn, Mine Mtn, Eleana Range. NYE CO.: Tolicha Peak area, Stonewall Mtn, s. Monitor Range, White Blotch Spg W of n. Groom Range. 4500—7000 ft. Perennial, woody at base. Late May—June (—Aug).

var. **versicolor** S. Stokes. Common locally, *Artemisia*—Pinyon—Juniper, Yellow Pine—Fir. NTS: Eleana Range (area of Capt. Jack Spg), n. Pahute Mesa. CLARK CO.: Nw. Spring Mtns (upper Clark Cyn). LINCOLN CO.: Cent. Groom Range (E slope of Bald Mtn). 6200—9000 ft. Perennial, woody at base. Late June—Aug.

Oxytheca

O. dendroidea Nutt. NYE CO.: Common locally in disturbed *Artemisia*, nw. Reveille Valley (below E face of Kawich Range). 6000 ft. Annual. July—Oct.

O. perfoliata Torr. & Gray. Widely distributed, occasional to common on bajadas of most basins at lower elevations; *Larrea*, *Grayia—Lycium*, *Coleogyne*, *Atriplex*, *Atriplex—Ceratoides*. NTS: Mercury Valley, Rock Valley, Jackass Flats (except w. part), Topopah Valley, Frenchman Flat (except cent. and e. parts), Mid Valley, Yucca Flat, s. Gold Flat, Thirsty Cyn. NYE CO.: N. Pahrump Valley (nw. Spring Mtns), w. Cactus Flat; Bullfrog Hills, Tolicha Peak area, Stonewall Mtn. 3000—5600 ft. Winter annual. Apr—June.

Polygonum

P. argyrocoleon Steud. ex Kunze. Weed of irrigated soils. NTS: Mercury townsite. CLARK CO.: Indian Springs townsite. NYE CO.: Common near irrigation channels, Ash Meadows, Pahrump Valley. 2200—3700 ft. Introduced annual. May—Oct.

P. aviculare L. Locally common weed of disturbed moist soils, scattered localities. NTS: Mercury townsite. NYE CO.: Ash Meadows; spgs in and near Stonewall Mtn, Kawich Range, s. Hot Creek Range. 2200—7000 ft. Introduced annual. July—Oct.

P. douglasii Greene var. **johnstonii** Munz. Local populations in *Artemisia*—Pinyon—Juniper, *Artemisia—Cercocarpus*, in washes, base of boulder-cliffs, flatrock areas, or steep talus slopes. NTS: Eleana Range (near Capt. Jack Spg), s. Belted Range (W rim of Rainier Mesa), S rim of Pahute Mesa. LINCOLN CO.; Cent. Groom Range (base of Bald Mtn). NYE CO.: N. Kawich Range (upper Eden Creek cyn). 6000—7600 ft. Annual. June—Sept.

P. pensylvanicum L. NTS: Disturbed moist soil near well, w. Jackass Flats. 3200 ft. Introduced annual. July—Aug.

Rumex

R. crispus L. Frequent and locally common, moist or wet soils near springs or water impoundments in *Artemisia*, *Artemisia*—Pinyon—Juniper, Atriplex, uncommonly *Larrea*. NTS: W. Jackass Flats; Topopah Spg (n. Topopah Valley), Cane Spg (w. Frenchman Flat). CLARK CO.: Cold Creek—Willow Creek (nw. Spring Mtns), Indian Springs townsite. LINCOLN CO.: Bald Mtn Spg (cent. Groom Range). NYE CO.: Oasis Valley; spgs of cent. and n. Belted Range and n. Kawich Range; White Blotch Spg W of n. Groom Range. 3600—6800 ft. Introduced perennial. May—July.

R. salicifolius Weinm. Frequent and locally common, moist or wet soils near springs in *Coleogyne*, *Artemisia*, *Artemisia*—Pinyon—Juniper. NTS: Topopah Spg (n. Topopah Valley), Cane Spg (w. Frenchman Flat), Whiterock Spg (nw. Yucca Flat), wash near Tippipah Reservoir (se. Forty-Mile Cyn). CLARK CO.: Cold Creek (nw. Spring Mtns). NYE CO.: Gold Spg (N slope of nw. Spring Mtns), spgs of Bullfrog Hills and Stonewall Mtn. 4000—6200 ft. Perennial. Late Apr—Sept.

PORTULACACEAE. Purslane Family

Calyptridium

C. monandrum Nutt. in Torr. & Gray. NTS: Occasional, mostly under shrubs, *Larrea* (with no other shrub spp); below red cinder cone of s. Yucca Mtn (n. Amargosa Valley). 2600 ft. Annual. Apr.

C. parryi A. Gray var. **nevadense** J. T. Howell. Occasional to common locally some years, *Artemisia*—Pinyon—Juniper (usually *A. nova*) shallow or disturbed soils, or washes. NTS: S. Belted Range (Rainier Mesa, Gold Meadows), s. and e. Pahute Mesa. NYE CO.: Cent. Belted Range (Johnnies Water cyn), n. and s. Kawich Range (Eden Creek cyn, Cedar Pass). 6500—7500 ft. Annual. June—Oct.

Lewisia

L. rediviva Pursh. In our populations, length of pedicels usually much exceeds 1 cm (commonly 1.5—2 cm), as in the typical variety, but perianth is usually less than 1.5 cm long, as in var. *minor;* two varieties may not be taxonomically distinguishable in this region.

var. **minor** (Rydb.) Munz. NTS: Uncommon except locally, *Artemisia nova*, *Artemisia*—Pinyon—Juniper, shallow soils, esp. around volcanic flatrock areas; s. Belted Range (Gold Meadows), s. Pahute Mesa and below S face. 6000—6700 ft. Perennial. May—June.

var. **rediviva.** NYE CO.: Local on gravel talus, *Artemisia*—Pinyon—Juniper, s. Kawich Range (Cedar Pass area). 7000 ft. Perennial. June.

Montia

M. chamissoi (Ledeb.) Dur. & Jacks. NYE CO.: Locally abundant in running water or on banks of mountain stream, in *Artemisia*—Pinyon—Juniper, n. Kawich Range (Eden Creek cyn). 7200 ft. Perennial. June.

M. perfoliata (Donn) Howell var. **utahensis** (Rydb.) Munz. NTS: Occasional, cliff crevices or along cyn bottom below cliffs, in

Coleogyne or *Artemisia*—Pinyon—Juniper; Yucca Mtn (nw. Jackass Flats), s. Shoshone Mtn (Topopah Spg). 5100—5700 ft. Annual. May.

Portulaca

P. oleracea L. (PURSLANE). Occasional weed of disturbed irrigated sites in *Larrea—Atriplex.* NTS: Mercury townsite. NYE CO.: N. Pahrump Valley. 2700—3800 ft. Introduced annual. June—Sept.

PRIMULACEAE. Primrose Family

Androsace

A. septentrionalis L. var. **puberulenta** (Rydb.) Knuth. NYE CO.: Local populations in *Artemisia—Cercocarpus*, near high ridges of n. Kawich Range (Kawich Peak area of upper Eden Creek cyn). 8800—9100 ft. Perennial. July.

Dodecatheon

D. pulchellum (Ref.) Merr. var. **pulchellum** [*D. radicatum* Greene] (Reveal, J. L., and E. L. Styer, *Southwest. Natur.*, **18**: 397. 1974). NYE CO.: Known in region only from n. Ash Meadows, common in cyn bottom of Ash—Screwbean area S of Devils Hole; collected by Coville & Funston from same site in March 1891 on Death Valley Expedition. 2300 ft. Perennial. Apr.

D. redolens (Hall) H. J. Thomps. CLARK CO.: Locally common at stream margin, in Yellow Pine—Fir, near head of Clark Cyn, nw. Spring Mtns. 9000 ft. Perennial. July.

Samolus

S. parviflorus Raf. NYE CO.: Known in the region only from spring areas of Ash Meadows where predictably present along stream banks, usually in running water and Ash—Screwbean communities. 2200—2300 ft. Perennial. Late Apr—July.

RANUNCULACEAE. Crowfoot Family

Anemone

A. tuberosa Rydb. Uncommon except locally, crevices along washes and canyons in limestone mountain ranges; *Larrea*, *Atriplex*, *Coleogyne.* NTS: E. Specter Range, w. Spotted Range (Red Mtn), Buried Hills. CLARK CO.: N-S axis of Spotted Range. 4000—5200 ft. Perennial. Mar—Apr.

Aquilegia

A. shockleyi Eastw. (Fig. 25). Bases of cliffs and crevices, near springs or along perennial streams, in *Artemisia*—Pinyon—Juniper, Yellow Pine—Pinyon, or Yellow Pine—Fir. NTS: N. Pahute Mesa (Columbine Cyn). CLARK CO.: Nw. Spring Mtns (Clark Cyn, Trough Spg, Cold Creek—Willow Creek area). NYE CO.: Stonewall Mtn at Spg), n. Kawich Range (Eden Creek and Longstreet cyns). 5600—8200 ft. Perennial. May—Aug.

Clematis

C. ligusticifolia Nutt. ex Torr. & Gray. Occasional to common, with *Salix* in *Artemisia*—Pinyon—Juniper. CLARK CO.: Nw. Spring Mtns (Cold Creek—Willow Creek area). NYE CO.: N. Kawich Range (Eden Creek cyn). 6000—7200 ft. Woody perennial (vine). July.

Delphinium

D. andersonii A. Gray. The widely distributed and locally common *Delphinium* of *Artemisia* (usually *A. nova*) and *Artemisia*—Pinyon—Juniper. NTS: In or below s. Belted Range (Rainier Mesa, Gold Meadows), Pahute Mesa. LINCOLN CO.: Cent. Groom Range (Bald Mtn). NYE CO.: S. Kawich Range (Cedar Pass). 5600—7200 ft. Perennial. May—June.

D. parishii A. Gray var. **parishii.** The widely distributed *Delphinium* of *Larrea*, *Grayia—Lycium*, *Coleogyne*, lower *Artemisia* (usually *A. tridentata*), uncommon in *Artemisia*—Pinyon—Juniper; often growing under shrubs. NTS: In or below Specter Range, Skull Mtn, w. Spotted Range (Red Mtn), Halfpint Range (French Peak mtn), CP Hills, Yucca Mtn, Shoshone Mtn, Timber Mtn, Mine Mtn, Eleana Range, s Belted Range (below W slope of Rainier Mesa, Gold Meadows), and n. Pahute Mesa. CLARK CO.: Nw. Spring Mtns (Wheeler Cyn, Trough Spg). NYE CO.: Nw. Spring Mtns (N slope below Mt. Stirling), below Yucca Mtn (Crater Flat), Bullfrog Hills, Reveille Range. 3000—6500 (—8000) ft. Perennial. Apr—May.

Ranunculus

R. andersonii A. Gray. Uncommon and local, around boulders and on talus slopes, *Artemisia*—Pinyon—Juniper, Pinyon—Fir. NTS: Shoshone Mtn, Eleana Range. LINCOLN CO.: Cent. Groom Range (Bald Mtn). NYE CO.: Nw. Spring Mtns (N slope below Mt. Stirling), Grapevine Mtns, Reveille Range. 6000—8000 ft. Perennial. Mar—Apr.

R. cymbalaria Pursh var. **saximontanus** Fern. Locally abundant, in running water of perennial streams or moist soils of spring areas, in

Artemisia, *Artemisia*—Pinyon—Juniper, or *Juncus* swamps at lower elevations. CLARK CO.: Nw. Spring Mtns (Willow Creek). LINCOLN CO.: Spgs of Groom Range, NYE CO.: Oasis Valley; spgs of Bullfrog Hills, Stonewall Mtn, Cactus Range, Goldfield Hills, and n. and cent. Kawich Range (Eden Creek and Longstreet cyns, Stinking Spg). 3800—8100 ft. Perennial. Apr.—Sept.

Thalictrum

T. fendleri Engelm. ex Gray. CLARK CO.: Locally common on lower slopes of cyns, esp. on old burned sites, in *Artemisia*—Pinyon—Juniper, Yellow Pine—Pinyon, Yellow Pine—Fir; nw. Spring Mtns (slopes above Cold Creek Spg, Trough Spg, and esp. upper Clark Cyn). 6200—9000 ft. Perennial. June—July.

RESEDACEAE. Mignonette Family

Oligomeris

O. linifolia (Vahl) Macbr. NYE CO.: Common in moist sandy soils, with *Atriplex* or Mesquite in Ash Meadows, and on dunes in Amargosa Valley N of Ash Meadows. 2200—2400 ft. Annual. Apr—May.

RHAMNACEAE. Buckthorn Family

Ceanothus

C. cordulatus Kell. (SNOW-BUSH). NYE CO.: Locally common in washes of *Artemisia*—Pinyon—Juniper, nw. Spring Mtns (N slope below Mt. Stirling). 6500—6800 ft. Shrub. June.

C. greggii A. Gray var. **vestitus** (Greene) McMinn. Local to common in *Artemisia*—Pinyon—Juniper and Yellow Pine—Pinyon—Juniper; also *Coleogyne/Artemisia* and *Artemisia nova.* NTS: Small local populations, Shoshone Mtn, Eleana Range (Capt. Jack Spg area and dolomite hill), s. Belted Range (E slope of Rainier Mesa). CLARK CO.; Nw. Spring Mtns (abundant on burned slopes of upper Clark Cyn and Wheeler Pass area). NYE CO.: Nw. Spring Mtns (common on N slope below Mt. Stirling). 5500—7800 ft. Shrub. May—June.

C. martinii M. E. Jones CLARK CO.: Nw. Spring Mtns (slopes of upper Clark Cyn) in Yellow Pine—Fir. LINCOLN CO.: Cent. Groom Range (common on NE slope of Bald Mtn) with Fir or Fir—Pinyon. 7800—8500 ft. Shrub. June—July.

ROSACEAE. Rose Family

Amelanchier

A. pallida Greene. NYE CO.: The *Amelanchier* near upper limits of *Artemisia*—Pinyon—Juniper, and *Artemisia—Cercocarpus* to the high ridges, n. Kawich Range (Kawich Peak area of upper Eden Creek cyn). 7200—9100 ft. Shrub. June.

A. utahensis Koehne (SERVICE-BERRY). Apparently the widely distributed and common *Amelanchier* of the region. NTS: Esp. common and conspicuous at and just below lower limits of *Artemisia*—Pinyon—Juniper, esp. along washes, in depressions, and around cliffs and rock outcrops; in or near Shoshone Mtn, Timber Mtn, s. Belted Range (Rainier Mesa and foothills to E, Gold Meadows), Pahute Mesa. CLARK CO.: *Artemisia*—Pinyon—Juniper, Yellow Pine—Pinyon, Yellow Pine—Fir, nw. Spring Mtns (upper Clark Cyn, Wheeler Pass area). LINCOLN CO.: Fir—Pinyon and upper *Artemisia*—Pinyon—Juniper, cent. Groom Range (Bald Mtn). NYE CO.: *Artemisia*—Pinyon—Juniper of nw. Spring Mtns (Crystal Spg cyn) and cent. Belted Range (both slopes); *Artemisia tridentata*, San Antonio Mtns, and White Blotch Spg W of n. Groom Range; common in *Artemisia*—Pinyon—Juniper of n. Kawich Range (Eden Creek and Longstreet cyns). 5000—9000 ft. Large shrub to small tree. Apr—May.

Cercocarpus

C. intricatus S. Wats. Uncommon in region, restricted to limestone hills and mountains; *Atriplex*, *Coleogyne*, *Artemisia*—Pinyon—Juniper, Yellow Pine—Fir. NTS: Local in e. Specter Range, w. Spotted Range (Red Mtn), Buried Hills. CLARK CO.: Nw. Spring Mtns (upper Clark Cyn), N-S axis of Spotted Range. 3900—8500 ft. Shrub. Apr—May.

C. ledifolius Nutt. var. **ledifolius** (MOUNTAIN MAHOGANY). *Artemisia*—Pinyon—Juniper, *Artemisia—Cercocarpus*, Yellow Pine—Pinyon, Fir—Pinyon, from rare and local in most mountain areas to common in certain canyons or on certain slopes. NTS: Small local populations in Eleana Range (Capt. Jack Spg area), s. Belted Range (Rainier Mesa). CLARK CO.: Nw. Spring Mtns (common in Clark Cyn and Cold Creek area). NYE CO.: Occasional small trees on Stonewall Mtn; n. Kawich Range (common, as small trees, on mtn slopes and high ridges of Kawich Peak area, upper Eden Creek cyn); cent. Belted Range (common in Johnnies Water cyn, where trees are to 7 m tall). 6000—9200 ft. Large shrub or tree. May—June.

Chamaebatiaria

C. millefolium (Torr.) Maxim. (FERN-BUSH). Rare in *Artemisia*—Pinyon—Juniper of region. NTS: Local, e. Pahute Mesa. NYE CO.: Crest of Grapevine Mtns (Phinney Cyn), cent. Belted Range (local, Wheelbarrow Peak area). 6600—7800 ft. Shrub. July—Aug.

Coleogyne

C. ramosissima Torr. (BLACKBRUSH). Dominant shrub or, across ecotones, codominant with *Larrea*, *Grayia*, *Lycium andersonii*, or *Artemisia;* often in nearly pure stands on upper bajadas of closed basins, or over basin floors of open basins, at middle elevations and usually associated with calcareous soil materials. NTS: Upper bajadas of Mercury Valley, Rock Valley, Jackass Flats, Frenchman Flat, Yucca Flat, se. Groom Lake, and nearly all the basin floors of Topopah Valley and Mid Valley; local areas in Forty-Mile Cyn. CLARK CO.: Dominant shrub of upper bajadas on S slope of Spring Mtns (n. Amargosa and Pahrump Valley drainages). LINCOLN CO.: With *Grayia* on upper bajadas below W face of cent. Groom Range, where at its known n. limits. 4000—5000 (—6000) ft. Shrub. Apr—May.

Cowania

C. mexicana D. Don var. **stansburiana** (Torr.) Jeps. Widely distributed in the region, usually as small local populations which are often in washes or depressions, most common in areas of limestone mountains or outcrops; *Coleogyne*, *Artemisia*, *Artemisia*—Pinyon—Juniper. NTS: W. Spotted Range (Red Mtn), Buried Hills, Halfpint Range (Banded Mtn and hills to E), CP Hills, Mine Mtn, below Timber Mtn, Eleana Range, s. Belted Range (Rainier Mesa, Oak Spring Butte), Pahute Mesa (where attains tree size) and below N rim. CLARK CO.: Nw. Spring Mtns (cyns of both slopes). NYE CO.: Nw. Yucca Mtn, Bare Mtn, nw. Stonewall Mtn, Cactus Range, s. Kawich Range (to tree size in Cedar Pass area), cent. Belted Range. 4200—7000 ft. Large shrub to small tree. May—June.

Fallugia

F. paradoxa (D. Don) Endl. (APACHE PLUME). Restricted to major washes, in *Coleogyne*, *Artemisia*, *Artemisia*—Pinyon—Juniper, *Atriplex*, usually below limestone mountain ranges or limestone outcrops in volcanic areas. NTS: W. Spotted Range (Red Mtn), Ranger Mtns; local, limestone outcrop of nw. Papoose Range. CLARK CO.: Nw. Spring Mtns (Wheeler Wash); abundant below N-S axis of Spotted Range (e. Frenchman Flat). LINCOLN CO.: Washes below Groom Range. NYE CO.: Nw. Spring Mtns (Crystal Spg cyn, and N slope below Mt. Stirling); local, limestone area of nw. Belted Range. 3700—6400 ft. Shrub. May—June.

Geum

G. macrophyllum Willd. NYE CO.: Common locally in moist to wet soils in *Artemisia*—Pinyon—Juniper, n. Kawich Range (Eden Creek and Longstreet cyns). 7100—7200 ft. Perennial. June.

Holodiscus

H. microphyllus Rydb. var. **microphyllus.** Common in canyons of volcanic mountains and mesas, bases and ledges of cliffs and around boulders, *Artemisia*—Pinyon—Juniper, Fir—Pinyon. NTS: Shoshone Mtn, Timber Mtn, s. Belted Range (Rainier Mesa), Pahute Mesa. CLARK CO.: Nw. Spring Mtns (upper Clark Cyn). LINCOLN CO.: Cent. Groom Range (Bald Mtn). NYE CO.: Nw. Spring Mtns (N slope below Mt. Stirling), Stonewall Mtn, n. Kawich Range (Eden Creek and Longstreet cyns), cent. Belted Range (cyns of both slopes). 5300—9000 ft. Shrub. Late June—Aug.

Ivesia

I. eremica (Cov.) Rydb. [*Potentilla e.* Cov.]. NYE CO.: Collected in 1891 and described by Coville (Coville, F. V., *Contr. U. S. Nat. Herb.*, 4, 1893) from non-flowering material from S of Devils Hole, n. Ash Meadows; an endemic apparently restricted to Ash Meadows, on light-colored clay uplands, where it occurs with other rare and locally endemic species in the spg areas of n. and e. parts, in *Atriplex* and *Distichlis* communities. 2200—2300 ft. Perennial. Sept.—Oct.

I. sabulosa (M. E. Jones) Keck. Uncommon species, but usually abundant where it occurs. NTS: *Artemisia*—Pinyon—Juniper, usually at bases of cliffs, around boulders, and along cyn washes; certain cyns of Silent Canyon and South Silent Canyon drainages of n. Pahute Mesa, and flatrock area near SE rim. NYE CO.: Around rock outcrops, White Blotch Spg W of n. Groom Range; locally common on steep talus slopes and high ridges of n. Kawich Range (Kawich Peak area of upper Eden Creek cyn), in *Artemisia*—*Cercocarpus* and *Artemisia*—Pinyon—Juniper. 6400—9100 ft. Perennial. Late June—Aug.

Peraphyllum

P. ramosissimum Nutt. (SQUAW-APPLE). NTS: Local populations on slopes and along washes in *Coleogyne* or *Artemisia*, area of Tippipah Spg (se. Forty-Mile Cyn), and cyn bottoms of n. Halfpint Range (W side of "The Hump", and Cockeyed Ridge). CLARK CO.: Common species of both slopes of nw. Spring Mtns (incl. Wheeler Wash—Clark Cyn area, and Cold Creek cyn), *Artemisia*—Pinyon—Juniper. NYE CO.: Occasional to common, *Artemisia*—Pinyon—Juniper, nw. Spring Mtns (Gold Spg and elsewhere on N slope below Mt. Stirling). 5000—7000 ft. Shrub. Apr—June.

Petrophytum

P. caespitosum (Nutt.) Rydb. Forming large mats on cliff-faces in limestone mountain ranges; *Atriplex*, *Larrea/Coleogyne*, *Coleogyne*, *Artemisia*—Pinyon—Juniper, Yellow Pine—Pinyon. NTS: W. Spotted Range (Mercury Ridge), Buried Hills (hills near head of Nye Cyn), CP Hills. CLARK CO.: Nw. Spring Mtns (upper Clark Cyn, along Wheeler Wash); N-S axis of Spotted Range. NYE CO.: Nw. Spring Mtns (N slope below Mt. Stirling); local limestone outcrop, nw. Belted Range. 4300—7200 ft. Shrub. Aug—Sept.

Potentilla

P. biennis Greene. Moist soils near springs, base of cliffs or along washes in usually volcanic mountains and mesas; in *Coleogyne*, *Artemisia*, most often *Artemisia*—Pinyon—Juniper. NTS: Near cyn bottom below Tippipah Reservoir (se. Forty-Mile Cyn), Whiterock Spg (nw. Yucca Flat); nw. Pahute Mesa. CLARK CO.: Nw. Spring Mtns (Cold Creek Spg). NYE CO.: Stonewall Mtn Spg; n. Kawich Range (Eden Creek and Longstreet cyns). 5000—8100 ft. Biennial. June—July.

P. gracilis Dougl. ex Hook. ssp. **nuttallii** (Lehm.) Keck. NYE CO.: Locally common, moist soils of spg areas and mountain meadows in *Artemisia*—Pinyon—Juniper, n. Kawich Range (Eden Creek and Longstreet cyns). 6800—8100 ft. Perennial. June—July.

P. propinqua (Rydb.) Rydb. CLARK CO.: Common on lower slopes of upper Clark Cyn, nw. Spring Mtns, in Yellow Pine—Fir. 8500 ft. Perennial. July—Aug.

Prunus

P. andersonii A. Gray (DESERT PEACH). NYE CO.: Local populations around volcanic boulders and cliff bases, or in ravines, in *Artemisia*, *Artemisia*—Pinyon—Juniper; Grapevine Mtns (upper Phinney Cyn) and foothills of s. Monitor Range. 6000—7000 ft. Shrub. May.

P. fasciculata (Torr.) A. Gray (DESERT ALMOND). Widely distributed and common shrub of the region, restricted to washes; *Larrea*, *Atriplex*, *Grayia—Lycium*, *Coleogyne*, *Artemisia tridentata*, uncommon in *Artemisia*—Pinyon—Juniper. NTS: N. Amargosa Valley, Mercury Valley, Rock Valley, Jackass Flats, Topopah Valley, Frenchman Flat, Mid Valley, Yucca Flat, Forty-Mile Cyn, Thirsty Cyn, w. Emigrant Valley. CLARK CO.: Below nw. Spring Mtns (Cold Creek Spg). NYE CO.: Nw. Spring Mtns (common on N slope below Mt. Stirling; Crystal Spg cyn); Crater Flat, Tolicha Peak area; Reveille Range. 3300—6000 ft. Shrub. Apr—May.

P. virginiana L. var. **melanocarpa** (A. Nels.) Sarg. (WESTERN CHOKE CHERRY). NYE CO.: Known in the region only from n. Kawich Range (Eden Creek and Longstreet cyns), in *Artemisia*—Pinyon—Juniper, where it attains tree-size; also in cent. Hot Creek Range to N. 6800—8000 ft. Small tree. June.

Purpusia

P. saxosa Bdg. Locally common, crevices of volcanic cliffs, *Artemisia*—Pinyon—Juniper. NTS: Sw. Shoshone Mtn, and South Silent Canyon drainage of n. Pahute Mesa; all populations with white corollas. NYE CO.: Cyns of both slopes of cent. Belted Range (Johnnies Water cyn, Indian Spg); all populations with bright yellow corollas. 6000—6800 ft. Perennial. June—July.

Purshia

P. glandulosa Curran. Widely distributed and common in washes, most frequent at or near lower limits of *Artemisia*—Pinyon—Juniper; also washes in *Artemisia*, *Coleogyne*, *Grayia—Lycium*, *Atriplex—Ceratoides*, rarely *Larrea.* NTS: Area of Shoshone Mtn, Timber Mtn, Buckboard Mesa, Yucca Mtn (nw. Jackass Flats), Eleana Range, Halfpint Range (Scarp Cyn and sw. Emigrant Valley), foothills of s. Belted Range (n. Yucca Flat). LINCOLN CO.: Foothills of cent. Groom Range. NYE CO.: Nw. Spring Mtns (N slope below Mt. Stirling), n. Yucca Mtn (Beatty Wash, Fluorspar Cyn), Bullfrog Hills, Grapevine Mtns, White Blotch Spg W of n. Groom Range. 3200—6000 ft. Shrub. Late Apr—early June.

P. tridentata (Pursh) DC. (BITTERBRUSH). Occasional to common, *Artemisia*—Pinyon—Juniper. NTS: Eleana Range, s. Belted Range (Rainier Mesa and slopes, Gold Meadows), s. Pahute Mesa and S face. NYE CO.: Stonewall Mtn, n. Kawich Range (Eden Creek and Longstreet cyns). 6000—7600 ft. Shrub, some populations low, spreading, and nearly prostrate. May—June.

Rosa

R. woodsii Lindl. Scattered, usually small populations, in seepage areas or usually in washes below springs, sometimes as dense thickets, *Artemisia*—Pinyon—Juniper, Yellow Pine—Fir. NTS: Nw. Pahute Mesa (headwaters of East Thirsty Cyn). CLARK CO.: Common, nw. Spring Mtns (upper Clark Cyn, Trough Spg, Willow Creek Spg). LINCOLN CO.: Cent. Groom Range (Bald Mtn Spg). NYE CO.: Nw. Spring Mtns (Rock Spg, Gold Spg, and Crystal Spg, all of Mt. Stirling area); grotto at Stonewall Mtn Spg, and near "Ruins"; n. and s. Kawich Range (along streams of Eden Creek and Longstreet cyns, and at Rose Spg); cent. Belted Range (Cliff Spg). 5400—9000 ft. Shrub. June—July.

Rubus

R. procerus P. J. Muell. CLARK CO.: Common around lake margin, Indian Springs townsite. 3100 ft. Introduced shrub. July.

RUBIACEAE. Madder Family

Galium

G. aparine L. Local on moist soils near springs or streams, *Artemisia*—Pinyon—Juniper. NTS: S. Shoshone Mtn (Topopah Spg), s. Belted Range (Oak Spg). NYE CO.: N. Kawich Range (Eden Creek cyn). 5800—7300 ft. Annual, presumably introduced. May—June.

G. bifolium S. Wats. Local in *Artemisia*—Pinyon—Juniper. NTS: Base of boulder-cliffs, S rim of Pahute Mesa; LINCOLN CO.: Cent. Groom Range (under Pinyons, W slope of Bald Mtn). 7300—8000 ft. Annual. June.

G. hilendiae Demp. & Ehrend. ssp. **hilendiae.** NTS: Sw. Shoshone Mtn, *Artemisia*—Pinyon—Juniper. 5700 ft. Perennial, woody at base. June.

ssp. **kingstonense** (Demp.) Demp. & Ehrend. NTS: S. Belted Range (cliff at head of Butte Wash[33] below Oak Spring Butte); *Artemisia*—Pinyon—Juniper. 5600 ft. Perennial. June.

G. hypotrichium A. Gray ssp. **nevadense** Demp. & Ehrend. Usually common on sites of occurrence, commonly on loose talus and nearly always under trees, in *Artemisia*—Pinyon—Juniper, *Artemisia*—*Cercocarpus*, Fir, and Fir—Pinyon. CLARK CO.: Nw. Spring Mtns (upper Clark Cyn). LINCOLN CO.: Cent. Groom Range (Bald Mtn). NYE CO.: Grapevine Mtns (Phinney Cyn), ne. Stonewall Mtn, n. Kawich Range (Eden Creek and Longstreet cyns). 6400—9000 ft. Perennial, woody near base. June—Aug.

G. magnifolium Demp. NTS: Local in limestone crevices in *Coleogyne*, cyns of W slope of Buried Hills. 4200–5000 ft. Perennial, woody near base. May.

G. multiflorum Kell. CLARK CO.: Nw. Spring Mtns (slopes along Wheeler Wash), *Artemisia*–Pinyon–Juniper. 6600 ft. Perennial, woody near base. June.

G. parishii Hilend & Howell. CLARK CO.: Nw. Spring Mtns (common in Yellow Pine–Fir on slopes and floor of upper Clark Cyn, ledges near Trough Spg). NYE CO.: *Artemisia*–Pinyon–Juniper of N slope below Mt. Stirling. 7200–8200 ft. Perennial, strongly woody at base. June–July.

G. proliferum A. Gray CLARK CO.: Locally common at base of cliffs, in *Coleogyne* and *Larrea/Coleogyne*, N-cent. part of N-S axis of Spotted Range. 5000–5400 ft. Annual. Apr–early May.

G. stellatum Kell. Locally frequent in limestone mountain ranges, esp. in cliff crevices and on rock ledges, *Atriplex*, *Larrea–Atriplex*, *Larrea–Ambrosia*. NTS: Specter Range, w. Spotted Range (Red Mtn). NYE CO.: Nw. Spring Mtns (Johnnie Mine area), Bare Mtn (both slopes). 3200–4200 ft. Perennial, strongly woody toward base. Apr–May.

Kelloggia

K. galioides Torr. CLARK CO.: Locally common on moist soils near spg, upper Clark Cyn of nw. Spring Mtns, in Yellow Pine–Fir. 9000 ft. Perennial. June–July.

RUTACEAE. Rue Family

Thamnosma

T. montana Torr. & Frém. (TURPENTINE-BROOM). Washes or other disturbed soils of bajadas of s. part of region, in *Larrea*, *Grayia–Lycium*, *Coleogyne*, *Coleogyne/Artemisia*, *Atriplex*. NTS: Mercury Valley, Rock Valley, Jackass Flats, Frenchman Flat, Mid Valley; esp. common along Cane Spg wash of w. Frenchman Flat and E slope of Jackass Flats. NYE CO.: Common, washes and slopes of N face of nw. Spring Mtns, below Mt. Stirling. 4000–6000 ft. Shrub. Mar–Apr, occasional any month Jan–Aug.

SALICACEAE. Willow Family

Populus

P. fremontii S. Wats. var. **fremontii** (COTTONWOOD). Native tree associated with spring areas or other areas of moist soils, but usually planted (and escaping) around old homesteads and townsites. CLARK CO.: Indian Springs townsite; Willow Spg at base of N slope of nw. Spring Mtns, in *Artemisia*—Pinyon—Juniper. NYE CO.: N. Ash Meadows, with Mesquite; Crystal Spg (S slope of Spring Mtns); Antelope Spg (s. Cactus Range); along stream below Warm Spgs (s. Hot Creek Range); commonly planted in Pahrump Valley (for shade around habitations and windbreak along fencerows) and Beatty townsite. 2200—6200 ft. Large tree (to 30+ m tall and 1 m diam.). Mar—Apr.

P. tremuloides Michx. (QUAKING ASPEN). NYE CO.: Cyn bottom in *Artemisia*—Pinyon—Juniper, cent. Hot Creek Range; to be expected in n. Kawich Range. 8700 ft. Small tree. June.

P. trichocarpa Torr. & Gray. var. **trichocarpa** (BLACK COTTONWOOD). NYE CO.: Locally common with *Salix*, in *Artemisia*—Pinyon—Juniper, in cyns with perennial streams in n. Kawich Range (Eden Creek and Longstreet cyns). 6600—7000 ft. Medium-size to tall tree, to 30+ m high. Apr.

Salix

S. exigua Nutt. ssp. **exigua** var. **stenophylla** (Rydb.) Schneid. (NARROW-LEAF WILLOW). Widely distributed in moist to wet soils, usually common locally, with Cottonwood, Ash—Screwbean, and in *Atriplex*, *Coleogyne*, *Artemisia*, and *Artemisia*—Pinyon—Juniper areas. NTS: Seepage sites at Whiterock Spg, Oak Spg, n. Pahute Mesa, and water impoundments of n. Yucca Flat. NYE CO.: Ash Meadows; nw. Spring Mtns (Rock Spg, Crystal Spg cyn), e. Bullfrog Hills (Indian Spgs), Stonewall Mtn (grotto at Spg), e. Goldfield Hills (Wildhorse Spg), n. Kawich Range (Eden Creek and Longstreet cyns). 2300—8000 ft. Shrub. Apr—May.

S. gooddingii Ball var. **gooddingii** (BLACK WILLOW). The tree-size native Willow of usually lower elevations of region, near springs or streams. NTS: Several large trees and numerous small ones at Cane Spg (w. Frenchman Flat), in *Atriplex canescens;* small trees at water impoundment, cent. Jackass Flats, in *Larrea*. CLARK CO.: Two large trees below Cold Creek Spg (base of nw. Spring Mtns), in *Artemisia*—Pinyon—Juniper, and trees of various size at Indian Springs townsite. NYE CO.: N. Oasis Valley, several large trees in *Larrea*—*Sarcobatus*, thickets elsewhere in *Juncus* swamps. 3700—6200 ft. Large tree. Apr—early May.

S. lasiolepis Benth. var. **lasiolepis** (ARROYO WILLOW). CLARK CO.: Extensive thickets at Willow Spg, in *Artemisia*—Pinyon—Juniper, and around pond at Cold Creek Field Sta., in *Artemisia*, nw.

Spring Mtns. NYE CO.: Thickets above Crystal Spg, in Pinyon—Juniper, nw. Spring Mtns. 5400—6100 ft. Shrub. Apr—May.

S. rigida Muhl. var. **watsonii** (Bebb) Cronq. NYE CO.: The Willow (with *S. exigua*) in *Artemisia*—Pinyon—Juniper of cyns of n. and s. Kawich Range (abundant in Eden Creek and Longstreet cyns, and thickets at Rose Spg of Cedar Pass area). 6400—8000 ft. Apr—May.

SANTALACEAE. Sandalwood Family

Comandra

C. umbellata (L.) Nutt. ssp. **californica** (Eastw. ex Rydb.) Piehl. Frequent in *Artemisia*—Pinyon—Juniper of both slopes of nw. Spring Mtns (Wheeler Cyn, below Trough Spg, and Cold Creek area). 6000—8000 ft. Perennial. May—June.

SAURURACEAE. Lizard-Tail Family

Anemopsis

A. californica Hook. Locally common on seasonally or year-round wet soils, esp. near springs and in *Juncus* or *Distichlis* meadows. CLARK CO.: Near lake, Indian Springs townsite. NYE CO.: Many local areas in Ash Meadows, Oasis Valley, and near spgs in Bullfrog Hills. 2200—4000 ft. Perennial. May—Oct.

SAXIFRAGACEAE. Saxifrage Family

Fendlerella

F. utahensis (S. Wats.) Heller. Locally common on canyon slopes in limestone mountain ranges, in *Atriplex*, *Coleogyne*, or *Artemisia*—Pinyon—Juniper. NTS: E. Specter Range, w. Spotted Range, Buried Hills (esp. hills near head of Nye Cyn). CLARK CO.: Nw. Spring Mtns (ledges of cliff, lower Clark Cyn); N-S axis of Spotted Range. 3900—6500 ft. Shrub. July.

Heuchera

H. rubescens Torr. var. **pachypoda** (Greene) Rosend., Butt. & Lak. Locally common, bases of cliffs, around rock ledges, and steep talus slopes, *Artemisia*—Pinyon—Juniper, *Artemisia*—*Cercocarpus*, Yellow Pine—Fir. CLARK CO.: Nw. Spring Mtns (upper Clark Cyn). NYE CO.: On cliffs and in waterfall at grotto of Stonewall Mtn Spg;

n. Kawich Range (upper Eden Creek and Longstreet cyns); cent. Belted Range (below Wheelbarrow Peak). 5600—9000 ft. Perennial. June—July.

Jamesia

J. americana Torr. & Gray var. **californica** (Small) Jeps. CLARK CO.: Frequent along cliffs in upper Clark Cyn, nw. Spring Mtns, in Yellow Pine—Pinyon and Yellow Pine—Fir. 7000—9200 ft. Shrub. July.

Lithophragma

L. tenellum Nutt. in Torr. & Gray. NTS: Rare, but locally common species of volcanic flatrock areas in *Artemisia*—Pinyon—Juniper; S rim of Pahute Mesa, and Rainier Mesa (s. Belted Range). 7200—7500 ft. Perennial. May—June.

Philadelphus

P. microphyllus A. Gray ssp. **stramineus** (Rydb.) C. L. Hitchc. (MOCK ORANGE). CLARK CO.: More or less common on slopes of upper Clark Cyn, nw. Spring Mtns, in Yellow Pine—Pinyon and Yellow Pine—Fir. 7200—8000 ft. Shrub. July.

Ribes

R. aureum Pursh var. **aureum** (GOLDEN CURRANT). Common in canyon bottoms with perennial streams, usually in Willow thickets in *Artemisia*—Pinyon—Juniper. CLARK CO.: Nw. Spring Mtns (Cold Creek—Willow Creek area). NYE CO.: N. Kawich Range (Eden Creek and Longstreet cyns). 6100—8000 ft. Shrub. Apr—June.

R. cereum Dougl. var. **cereum** (SQUAW CURRANT). Local around boulders and bases of cliffs in canyons of mountain ranges, in *Artemisia*—Pinyon—Juniper, Yellow Pine—Pinyon, Yellow Pine—Fir, Fir—Pinyon, or Fir. NTS: N. (with *R. velutinum*) and s. Pahute Mesa. CLARK CO.: Nw. Spring Mtns (common on both slopes, incl. upper Clark Cyn, Wheeler Pass, and Cold Creek areas). LINCOLN CO.: Cent. Groom Range (common, with *R. velutinum*, on Bald Mtn). NYE CO.: White Blotch Spg W of n. Groom Range. 6000—9200 ft. Shrub. May—July.

R. niveum Lindl. NYE CO.: Local, *Artemisia*—Pinyon—Juniper, n. Kawich Range (upper Eden Creek cyn). 7100—7200 ft. Shrub. May—June.

R. velutinum Green var. **velutinum** (GOOSEBERRY). The widely distributed and common *Ribes* of volcanic mountain ranges, often abundant on canyon slopes and bottoms, along cliffs and around boulders; *Artemisia*—Pinyon—Juniper, Fir—Pinyon, uncommonly *Artemisia* or *Coleogyne/Artemisia.* NTS: Shoshone Mtn, Timber Mtn, Eleana Range, s. Belted Range (Rainier Mesa and slopes, Gold Meadows, Oak Spring Butte), Pahute Mesa, Halfpint Range (Cockeyed Ridge). LINCOLN CO.: Groom Range (Bald Mtn) and e. foothills. NYE CO.: Nw. Spring Mtns (N face below Mt. Stirling), Grapevine Mtns, Stonewall Mtn, cent. Belted Range (cyns of both slopes). 5300—8000 ft. Shrub. May—June.

SCROPHULARIACEAE. Figwort Family

Antirrhinum

A. filipes A. Gray. Occasional, twining within shrubs and inconspicuous; *Larrea—Ambrosia*, *Larrea—Atriplex*, *Atriplex*, bajadas below limestone mountain ranges or calcareous volcanic hills. NTS: Below Specter Range, w. Spotted Range (incl. Red Mtn), E end of Skull Mtn, Ranger Mtns. NYE CO.: Bare Mtn. 2900—4000 ft. Winter annual. Apr—early May.

A. kingii S. Wats. Occasional, esp. in washes and on disturbed soils, *Coleogyne*, *Artemisia*, *Artemisia*—Pinyon—Juniper, infrequently *Larrea*, *Atriplex—Kochia.* NTS: N. Jackass Flats, sw. Frenchman Flat, e. Mid Valley, sw. Yucca Flat (area of Mine Mtn and CP Hills) and nw. Yucca Flat (Eleana Range), n. Forty-Mile Cyn; n. Pahute Mesa. CLARK CO.: N-S axis of Spotted Range. NYE CO.: N. Crater Flat. 4100—6500 ft. Annual. Apr—Sept.

Castilleja

C. chromosa A. Nels. The widely distributed Paint-Brush of the middle and lower elevations, often growing under shrubs, in *Larrea*, *Coleogyne*, *Atriplex*, *Artemisia*, sometimes *Artemisia*—Pinyon—Juniper or Yellow Pine—Pinyon; around rock ledges and along washes or lower canyon walls, esp. common and characteristic of areas of limestone mountains or foothills. NTS: In or below Specter Range, w. Spotted Range (Red Mtn, Mercury Ridge), Ranger Mtns, Buried Hills, Halfpint Range (incl. French Peak mtn, Banded Mtn), nw. Papoose Range, CP Hills, Eleana Range, s. Belted Range (Rainier Mesa and uplands to E), Shoshone Mtn (Topopah Spg, where there are pubescence characters from *C. martinii* and conspicuous color phases); no colls. from Pahute Mesa. CLARK CO.: N-S axis of Spotted Range; nw. Spring Mtns (Clark Cyn area, Trough Spg). NYE CO.: Nw. Spring Mtns (N slope below Mt. Stirling), nw. Yucca Mtn (Fluorspar Cyn, Tates Wash, Beatty Wash), Bullfrog Hills, Grapevine

Mtns, cent. Reveille Range, n. Belted Range. 3600—6000 (—8200) ft. Perennial. Apr.—June.

C. **linariaefolia** Benth. var. **linariaefolia.** The linear-leaved Paint-Brush of the higher elevations of the region; *Artemisia*—Pinyon—Juniper, *Artemisia—Cercocarpus*, Yellow Pine—Pinyon. NTS: S. Belted Range (Rainier Mesa and slopes, Gold Meadows, and s. Kawich Valley), Pahute Mesa. CLARK CO.: Nw. Spring Mtns (upper Clark Cyn, Trough Spg). LINCOLN CO.: Cent. Groom Range (Bald Mtn). NYE CO.: Cent. Belted Range (cyns of both slopes), n. and s. Kawich Range (Eden Creek cyn and Cedar Pass area). The local populations common in moist soils in Ash—Screwbean and *Atriplex—Haplopappus* of the spg areas in n. and e. Ash Meadows at 2200—2300 ft, and with *Baccharis* and other Ash Meadows species locally at 5000 ft in Crystal Spg cyn of S slope of nw. Spring Mtns, appear to be ecologically distinct and also worthy of taxonomic distinction. (2200—) 6200—9100 ft. Perennial. June—Aug (Ash Meadows populations into October).

C. **martinii** Abrams var. **clokeyi** (Pennell) N. Holmgren. The common Paint-Brush of higher elevations, in *Artemisia*—Pinyon—Juniper, Yellow Pine—Pinyon, Yellow Pine—Fir. NTS: Shoshone Mtn, s. Belted Range (Rainier Mesa, Gold Meadows), Pahute Mesa. CLARK CO.: Nw. Spring Mtns (upper Clark Cyn, Trough Spg). NYE CO.: Cent. Grapevine Mtns, s. Kawich Range (E of Cedar Pass), cent. Belted Range (E slope). 6200—9000 ft. Perennial. May—July.

C. **viscidula** A. Gray. NYE CO.: Common on steep treeless slopes of Kawich Peak area, n. Kawich Range (upper Eden Creek and Longstreet cyns), in *Artemisia tridentata* and *Artemisia—Cercocarpus*. 7600—9100 ft. Perennial. June—July.

Collinsia

C. **callosa** Parish. NYE CO.: In pine litter on loose talus, *Artemisia*—Pinyon, cent. Grapevine Mtns (head of Phinney Cyn). 7000 ft. Annual. Late Apr—May.

C. **parviflora** Dougl. ex Lindl. Common locally in volcanic flatrock areas, washes, or base of boulder-cliffs, *Artemisia*—Pinyon—Juniper. NTS: Eleana Range, s. Belted Range (Rainier Mesa and slopes), S rim and cyns of n. Pahute Mesa. NYE CO.: Cent. Belted Range (E slope), n. and s. Kawich Range (Eden Creek cyn and Cedar Pass area). 5800—8000 ft. Annual. May—June.

Cordylanthus

C. **tecopensis** Munz & Roos. NYE CO.: Common locally, restricted to more or less salt-encrusted clay soils of certain spg areas of n. and e. Ash Meadows, with *Distichlis* and *Atriplex*. 2200—2300 ft. Annual. July—Oct.

Keckiella

K. rothrockii (A. Gray) Straw var. **rothrockii** [*Penstemon r.* A. Gray]. Common locally, *Artemisia*—Pinyon—Juniper, volcanic mountains and mesas. NTS: S rim of Pahute Mesa. NYE CO.: Grapevine Mtns (upper Phinney Cyn), Stonewall Mtn (common, esp. at 6300—6500 ft in cyn with "Ruins"), cent. Belted Range (W slope). 6300—7300 ft. Shrub. July—Aug.

Mimulus

M. bigelovii (A. Gray) A. Gray var. **bigelovii.** Common locally in *Larrea—Ambrosia*, *Larrea—Atriplex*, *Coleogyne*. NTS: Loose talus and gravel terraces, e. Specter Range (Amargosa Valley). NYE CO.: Banks of major wash, N slope of nw. Spring Mtns below Mt. Stirling. 3300—5000 ft. Probably winter annual. Apr—May.

M. densus A. L. Grant. Locally common in *Artemisia*—Pinyon—Juniper, less common in *Artemisia*, esp. volcanic flatrock areas and shallow soils. Within a population, plants with red-purple flowers usually greatly outnumber those with yellow flowers. NTS: S. Belted Range (Rainier Mesa and slopes, Gold Meadows), Pahute Mesa and below NE face in s. Kawich Valley. NYE CO.: Cent. Belted Range (Johnnies Water), s. Kawich Range (Cedar Pass area). 6100—7500 ft. Annual. Late May—July.

M. guttatus Fisch. ex DC. var. **guttatus.** In water and on wet soils, seepage areas near most major springs of region, in *Atriplex*, *Coleogyne*, *Artemisia*, *Artemisia*—Pinyon—Juniper, Yellow Pine—Pinyon. NTS: N slope of Skull Mtn (Cane Spg), s. Shoshone Mtn (Topopah Spg), Eleana Range (Tippipah Spg), s. Belted Range (Whiterock Spg and Oak Spg). CLARK CO.: Nw. Spring Mtns (Trough Spg, Cold Creek Spg, Willow Spg). LINCOLN CO.: Cent. Groom Range (Bald Mtn Spg). NYE CO.: Ash Meadows, Oasis Valley; nw. Spring Mtns (Crystal Spg), Bullfrog Hills (Indian Spgs), n. Kawich Range (Eden Creek and Longstreet cyns), cent. Belted Range (Johnnies Water), hill W of n. Groom Range (White Blotch Spg). 2300—8200 ft. Perennial. Apr—July.

M. montioides A. Gray. NTS: Locally abundant, *Larrea—Ambrosia—Atriplex*, foothills N of w. Spotted Range (e. Mercury Valley). 4000 ft. Probably winter annual. Apr.

M. parryi A. Gray. CLARK CO.: Common on bajadas, and on loose talus below cliffs, in *Coleogyne*, cent. part of N-S axis of

Spotted Range (e. Frenchman Flat). 3800—5200 ft. Probably winter annual. Apr—May.

M. pilosus (Benth.) S. Wats. [*Mimetanthe p.* Greene]. Locally common, usually along washes or other disturbed sites, *Artemisia*—Pinyon—Juniper. NTS: S. Shoshone Mtn (Topopah Spg), n. and e. Pahute Mesa. NYE CO.: Cent. Belted Range (Johnnies Water), n. and s. Kawich Range (Eden Creek cyn and Cedar Pass area). 5800—7200 ft. Annual. Late May—July.

M. rubellus A. Gray in Torr. NYE CO.: Local, moist soils in *Artemisia*—Pinyon—Juniper; cent. Grapevine Mtns (upper Phinney Cyn), cent. Belted Range (near Johnnies Water). 6500—7000 ft. Annual. Apr—July.

M. spissus A. L. Grant. The most widely distributed and common *Mimulus* of the region, in sandy washes or on steep slopes with loose talus, in *Atriplex*, *A. hymenelytra*, *A. canescens*, *Larrea*, *Lycium pallidum*—*Grayia*, *Coleogyne*, *Artemisia*, *Artemisia*—Pinyon—Juniper. NTS: In or below Specter Range, Skull Mtn, w. Spotted Range (incl. Mercury Ridge), Ranger Mtns, Halfpint Range (French Peak mtn, Banded Mtn), Shoshone Mtn, Timber Mtn, Buckboard Mesa, Mine Mtn, Eleana Range, Pahute Mesa; Thirsty Cyn, Emigrant Valley. NYE CO.: Nw. Spring Mtns (Crystal Spg cyn), E-W axis of Spotted Range, cent. Bare Mtn, Beatty Mtn, Obsidian Butte area. 3100—6600 (most frequent at 4000—5000) ft. Annual. May—Sept.

M. suksdorfii A. Gray. Common locally, esp. around volcanic flatrock areas, *Artemisia* and *Artemisia*—Pinyon—Juniper. NTS: Below se. Skull Mtn (Mara Wash), s. Shoshone Mtn (Topopah Spg), s. Belted Range (Rainier Mesa and slopes), Pahute Mesa. NYE CO.: Both slopes of cent. Belted Range (Johnnies Water, Cliff Spg), W of n. Groom Range (White Blotch Spg), s. Kawich Range (Cedar Pass area). (3700—) 5700—7500 ft. Annual. May—June.

Mohavea

M. breviflora Cov. Local in washes or on loose talus of limestone mountain ranges, in *Larrea*—*Ambrosia*, *Larrea*—*Atriplex*. NTS: Cent. and e. Specter Range (n. Amargosa Valley), w. Spotted Range (e. Mercury Valley), sw. Ranger Mtns (se. Frenchman Flat). NYE CO.: Bare Mtn (n. Amargosa Valley). 3000—4200 ft. Probably winter annual.

Pedicularis

P. semibarbata A. Gray. CLARK CO.: Nw. Spring Mtns (locally abundant, upper Clark Cyn in Yellow Pine—Fir). LINCOLN CO.:

Cent. Groom Range (common under Pinyons, in *Artemisia*—Pinyon—Juniper, Bald Mtn). 7000—8000 ft. Perennial. May—June.

Penstemon

P. albomarginatus M. E. Jones. NTS: Abundant in sandy soils below w. Specter Range (n. Amargosa Valley) in *Larrea—Ambrosia*. 2800 ft. Perennial. Late Apr—early May.

P. arenarius Greene. NYE CO.: Locally common, in *Atriplex—Tetradymia glabrata*, sands below Tolicha Peak (s. Sarcobatus Flat). 4000 ft. Perennial. May.

P. bridgesii A. Gray. Widely distributed and common in volcanic mountains, in *Artemisia*, *Artemisia*—Pinyon—Juniper, Fir—Pinyon. NTS: Common in cyns of Shoshone Mtn, Timber Mtn, s. Belted Range (Rainier Mesa and slopes, Gold Meadows), Pahute Mesa. LINCOLN CO.: Cent. Groom Range (common on Bald Mtn). NYE CO.: Common, ne. Stonewall Mtn; s. Kawich Range (uncommon, Rose Spg cyn); cent. Belted Range (uncommon in cyns of both slopes). 5600—8500 ft. Perennial. June—Aug.

P. confusus M. E. Jones var. **confusus**. Uncommon as local populations, *Artemisia nova*—Pinyon—Juniper. LINCOLN CO.: Below ne. Groom Range (nw. Desert Valley). NYE CO.: Local limestone outcrop W of Belted Peak of n. Belted Range. 6200 ft. Perennial. Late May—early June.

P. eatonii A. Gray var. **eatonii**. *Artemisia*—Pinyon—Juniper of certain mountain canyons. CLARK CO.: Nw. Spring Mtns (occasional in cyns of both slopes). LINCOLN CO.: Cent. Groom Range (common in area of Bald Mtn). NYE CO.: Nw. Spring Mtns (local populations on N face below Mt. Stirling), cent. Belted Range (common in cyns of both slopes, but esp. in Johnnies Water cyn), occasional in n. and s. Kawich Range (Eden Creek and Longstreet cyns, and Rose Spg cyn). 5800—7500 ft. Perennial. June—July.

P. floridus Bdg. ssp. **austinii** (Eastw.) Keck. Widely distributed as small local populations in *Artemisia*—Pinyon—Juniper of canyons in volcanic mountain ranges, and esp. in washes in *Artemisia tridentata* at lower elevations. NTS: Cyns of Shoshone Mtn, Timber Mtn, Eleana Range, s. Belted Range (Rainier Mesa and slopes), Pahute Mesa. NYE CO.: Washes of Tolicha Peak—Obsidian Butte area, and s. San Antonio Mtns. (3600—) 5200—7200 ft. Perennial. June—July.

P. fruticiformis Cov. ssp. **amargosae** Keck. NTS: Local in washes in *Larrea—Ambrosia—Atriplex*, lower cyns of S face of Specter Range (Amargosa Valley). NYE CO.: Occasional along washes in *Coleogyne*, nw. Spring Mtns (Crystal Spg cyn). 3300—5200 ft. Perennial. Late Apr—June.

P. humilis Nutt. ex Gray var. **humilis**. Rather widely distributed, occasional to common locally, *Artemisia*—Pinyon—Juniper. NTS: Timber Mtn, s. Belted Range (Rainier Mesa and slopes, Gold Meadows), Pahute Mesa. NYE CO.: N. Belted Range, n. Reveille Range, n. and esp. s. Kawich Range (upper Eden Creek cyn; Cedar Pass area, where grows with *P. kingii* which it resembles). 6000—8000 ft. Perennial. May—June.

P. kingii S. Wats. NYE CO.: *Artemisia*—Pinyon—Juniper, locally common on local limestone area, s. Kawich Range (Cedar Pass area), where occurs with *P. humilis* which it resembles. 7000 ft. Perennial. June.

P. pahutensis N. Holmgr. (Holmgren, N. H., *Aliso*, 7: 351. 1971). Common in areas of occurrence, *Artemisia*—Pinyon—Juniper. NTS: Cyns of s. Shoshone Mtn, Eleana Range, s. Belted Range (Rainier Mesa and slopes (type locality below NW face), Gold Meadows), s. Pahute Mesa. NYE CO.: Stonewall Mtn (cyn with "Ruins"). 5800—7500 ft. Perennial. June.

P. palmeri A. Gray. *Artemisia*—Pinyon—Juniper, Yellow Pine—Fir, where large populations may occur locally, esp. on disturbed sites and canyon bottoms; also scattered along washes in canyons of limestone mountain ranges at lower elevations. NTS: S. Belted Range (Rainier Mesa), s. Pahute Mesa; w. Spotted Range (Red Mtn, Mercury Ridge), Ranger Mtns. CLARK CO.: Both slopes of nw. Spring Mtns (Willow Creek area, and the common *Penstemon* of 5000—8000 ft in Wheeler Wash and Clark Cyn); N-S axis of Spotted Range. LINCOLN CO.: Occasional in washes below n. and cent. Groom Range. NYE CO.: Both slopes of cent. Belted Range (common in Johnnies Water cyn, and cyn leading to Indian Spg), n. Kawich Range (Eden Creek cyn, where hybridizes with *P. eatonii*). 3300—8000 ft. Perennial. May—July.

P. petiolatus Bdg. The common *Penstemon* of certain limestone mountain ranges; *Larrea*—*Atriplex*, *Coleogyne*—*Atriplex*, *Atriplex*. NTS: W. Spotted Range (incl. Red Mtn, Mercury Ridge), Ranger Mtns, Buried Hills. CLARK CO.: N-S axis of Spotted Range. 3800—5200 ft. Shrub. Late Apr—early June.

P. pudicus Reveal & Beatley (Reveal, J. L., and J. C. Beatley, *Bull. Torr. Bot. Club*, 98: 332. 1971). NYE CO.: Common, *Artemisia*—*Cercocarpus* and adjacent *Artemisia*—Pinyon—Juniper, steep mountain slopes and ridges of Kawich Peak area, n. Kawich Range (upper Eden Creek cyn); type locality. 7600—9200 ft. Perennial. Late June—early July.

P. thompsoniae (A. Gray) Rydb. CLARK CO.: Local, *Artemisia*—Pinyon—Juniper, Clark Cyn area of nw. Spring Mtns. 6000 ft. Perennial. Late May—June.

P. thurberi Torr. var. **anestius** Reveal & Beatley (Reveal, J. L., and J. C. Beatley, *Great Basin Natur.*, 34: 230. 1974). NTS: Common over upper bajada below sw. Buried Hills, on deep loose sands in *Larrea—Ambrosia*; type locality. 3800—4100 ft. Shrub. June.

P. utahensis Eastw. CLARK and NYE COS.: Occasional to locally common on slopes and in washes in *Coleogyne*, *Coleogyne/Artemisia*, and lower *Artemisia*—Pinyon—Juniper, both slopes of nw. Spring Mtns (areas of Wallace—Clark Cyns, and N face below Mt. Stirling). 5200—6000 ft. Perennial. Late Apr—early June.

P. venosus (Keck ex Kearn. & Peebl.) Reveal [*P. angustifolius* Nutt. ssp. *v.* Keck]. NTS: Common in *Atriplex* on sands of s. Groom Lake, below se. Belted Range. NYE CO.: Common in *Atriplex canescens*, loose sands E of upper playa of Kawich Valley, below W slope of N-cent. Belted Range. 4500—5400 ft. Perennial. Late May—early June.

Scrophularia

S. desertorum (Munz) Shaw. Occasional, *Artemisia*—Pinyon—Juniper, base of cliffs, cliff crevices, around boulders on talus slopes and in washes. NTS: Shoshone Mtn, s. Belted Range (E rim of Rainier Mesa), Pahute Mesa. NYE CO.: Grapevine Mtns, Stonewall Mtn, n. Kawich Range (Eden Creek cyn), cent. Belted Range (cyns of both slopes). 5400—7800 ft. Perennial. May—July.

Verbascum

V. thapsus L. (COMMON MULLEIN). CLARK CO.: Local on moist disturbed site in *Artemisia*, below N face of nw. Spring Mtns. 6000 ft. Introduced biennial. June—July.

Veronica

V. americana (Raf.) Schw. Uncommon, wet soils of spring areas or around water impoundments, in *Coleogyne*, *Artemisia*—Pinyon—Juniper, Yellow Pine—Fir, and Ash—*Baccharis*. NTS: W. Frenchman Flat (Cane Spg), water impoundment of nw. Yucca Flat. CLARK CO.: Common ground cover along stream near spg of upper Clark Cyn, nw. Spring Mtns. NYE CO.: Ash Meadows; nw. Kawich Range (Eden Creek cyn). 2300—9000 ft. Perennial. July—Oct.

V. anagallis-aquatica L. Widely distributed and predictable near springs throughout region, in *Atriplex*, *Coleogyne*, *Artemisia*, *Artemisia*—Pinyon—Juniper. NTS: Base of Skull Mtn (Cane Spg); Tippipah Reservoir. CLARK CO.: Nw. Spring Mtns (Willow Spg).

LINCOLN CO.: Cent. Groom Range (Bald Mtn Spg, Cattle Spg). NYE CO.: Ash Meadows; nw. Spring Mtns (Crystal Spg), e. Goldfield Hills (Willow Spg), Cactus Range (Cactus Spg), spgs of n. and cent. Kawich Range. 2200—6700 ft. Introduced perennial. May—Sept.

V. peregrina L. ssp. **xalapensis** (HBK.) Penn. NTS: Common locally, seepage area in *Coleogyne* below Whiterock Spg (nw. Yucca Flat). 5000 ft. Annual. May—June.

SIMAROUBACEAE. Quassia Family

Ailanthus

A. altissima (Mill.) Swingle (TREE OF HEAVEN). NYE CO.: Common planted and escaped small tree, Beatty townsite. 3300 ft. Introduced tree. June.

SOLANACEAE. Nightshade Family

Datura

D. meteloides A. DC. Uncommon in region except along certain major washes in *Larrea*, *Atriplex*, *A. canescens.* NTS: Scattered plants along most of length of Forty-Mile Cyn wash, through Shoshone Mtn usually at bases of cliffs. NYE CO.: Local, Ash Meadows, Oasis Valley and washes of Amargosa Valley drainage to S and E; weed of cotton fields in Pahrump Valley. 2200—4200 ft. Perennial. May—Oct.

Lycium

L. andersonii A. Gray (DESERT-THORN). Common and widely distributed shrub, dominant or associated species in many kinds of plant communities of the bajadas of the region, esp. *Larrea* and *Coleogyne*; replaced by *L. shockleyi* in some areas, and largely replaced by *L. pallidum* in others; with *Grayia* dominant over large areas transitional between the Mojave and Great Basin deserts. NTS: Ne. Amargosa Valley, Mercury Valley, Rock Valley, Jackass Flats, Frenchman Flat (replaced in se. part by *L. shockleyi* below Ranger Mtns, and by *L. pallidum* in sw. part below e. Skull Mtn), Mid Valley, Yucca Flat (incl. Plutonium Valley), Forty-Mile Cyn, Thirsty Cyn, s. Groom Lake. NYE CO.: Local in n. Ash Meadows; ne. Amargosa Valley, sw. Penoyer Valley. 2200—5000 (—6000) ft. Shrub. Late Mar—Apr; commonly leafing-out following rains in late summer or early autumn, rarely with occasional flowers.

L. pallidum Miers var. **oligospermum** C. L. Hitchc. Dominant or associated shrub, restricted to certain areas of either limestone or volcanic mountain ranges or hills, usually associated with *Larrea* or *Atriplex*. NTS: S slope of Skull Mtn and below N slope of w. Spotted Range (Red Mtn and Mercury Ridge), SW across Rock Valley and Specter Range to Amargosa Valley; below N slope of Little Skull Mtn NW to conspicuous hill outlier in w. Jackass Flats, and below S end of Yucca Mtn in extreme w. Jackass Flats [absent below s. Shoshone Mtn (Calico Hills)]; with *Larrea* below SE end of Skull Mtn (sw. Frenchman Flat), and the dominant species (without *Larrea* and other shrub species of the nearby *Larrea* communities, except *Grayia*) E across the lowlands of s. Frenchman Flat to playa. NYE CO.: Local at edge of Mesquite thicket in s. Stewart Valley, local in n. Ash Meadows, lower slope of mtn N end of Pahrump Valley, ne. Amargosa Valley (near playa N of Ash Meadows), below s. Bare Mtn (Steves Pass), s. Beatty Mtn, se. Bullfrog Hills. 2300—4200 ft. Shrub. Late Mar—Apr; some yrs leafing-out following heavy summer or early autumn rains.

L. shockleyi A. Gray (*L. rickardii* C. H. Mull.). Dominant species, with *Atriplex* and with or without *Larrea*, restricted to certain bajadas below either limestone or volcanic mountain ranges or hills, on soils of widely differing physical characteristics; a species primarily of the transition between the Mojave and Great Basin deserts, known from Clark, Nye, Esmeralda, and Mineral (type locality near Candelaria) Cos.; reported also from Churchill Co. (pers. comm., C. L. Hitchcock, 1973). NTS: Bajada below Ranger Mtns (se. Frenchman Flat) and foothills and bajada below W face of w. Spotted Range (e. Mercury Valley). CLARK CO.: Dominant on bajada below s. Spotted Range (w. Indian Springs Valley). NYE CO.: With *Atriplex*, S of Skeleton Hills, extreme nw. Ash Meadows (n. Amargosa Valley); in *Sarcobatus—Atriplex* over large area of N slope of Stonewall Flat below Goldfield Hills and Cactus Range; common, esp. on low sand ridges N of Mud Lake (Ralston Valley). 2400—5200 ft. Shrub. Apr, infrequently leafing-out in summer or autumn.

Lycopersicon

L. esculentum Mill. (TOMATO). NTS: Small roadside population on moist site, s. Frenchman Flat, and an irrigated site of s. Jackass Flats. 3200—3600 ft. Introduced annual. Aug—Sept, frs. maturing Sept—Nov.

Nicotiana

N. attenuata Torr. Common on disturbed sites, but belonging to canyons, usually on ledges or at bases of cliffs of usually volcanic mountains and mesas, in *Artemisia* and *Artemisia*—Pinyon—Juniper, sometimes *Atriplex—Ceratoides* and *A. canescens*. NTS: Shoshone Mtn, Timber Mtn, s. Belted Range (Rainier Mesa), Pahute Mesa; Thirsty Cyn; local population on a temporarily moist site in *Larrea*, n. Frenchman Flat. CLARK CO.: Nw. Spring Mtns (disturbed sites, lower Clark Cyn area). NYE CO.: In or below Grapevine Mtns, Cactus Range, n. and s. Kawich Range, cent. Reveille Range, cent. Belted Range, White Blotch Spg W of n. Groom Range. (3500—) 4900—7500 ft. Annual. (May—) July—Sept.

N. trigonophylla Dunal in A. DC. var. **trigonophylla**. The tobacco of limestone mountains, where occasional to common on ledges, at bases of cliffs, or in washes, in *Larrea—Ambrosia, Atriplex*. NTS: Specter Range, w. Spotted Range (incl. Red Mtn, Mercury Ridge), Ranger Mtns, Buried Hills and limestone butte to W, Halfpint Range (below French Peak mtn), Yucca Mtn. CLARK CO.: S. Spotted Range. NYE CO.: Mtn E side of Stewart Valley, washes below W end of Spring Mtns, below n. Yucca Mtn and Bare Mtn. 2600—4800 ft. Perennial. Apr—May, some yrs through summer and autumn (Nov).

Physalis

P. crassifolia Benth. Locally common in washes, or cliffs, crevices, and ledges of limestone mountain ranges, in *Atriplex*, *A. hymenelytra*, *Larrea—Atriplex*, *Larrea—Ambrosia*. NTS: Specter Range, Ranger Mtns. CLARK CO.: limestone hill below N face of nw. Spring Mtns, and s. Spotted Range (w. Indian Springs Valley). NYE CO.: Mtn E side of Stewart Valley, mtn N end of Pahrump Valley, below W end of Spring Mtns, and Bare Mtn. 2600—4300 ft. Perennial. Apr—Aug.

P. hederaefolia A. Gray var. **cordifolia** (A. Gray) Waterfall [*P. fendleri* A. Gray]. Local in *Artemisia*—Pinyon—Juniper. CLARK CO.: Slopes near Cold Creek and Willow Creek, nw. Spring Mtns. NYE CO.: Gold Spg of N slope, nw. Spring Mtns. 5900—6200 ft. Perennial. June—July.

Solanum

S. elaeagnifolium Cav. NYE CO.: Local large population on disturbed site in *Atriplex*, *Distichlis*, and Ash thicket, s. Ash Meadows. 2200 ft. Introduced perennial. May—July.

S. sarrachoides Sendt. ex Mart. NYE CO.: Edge of grazed meadow in *Artemisia*—Pinyon—Juniper, n. Kawich Range (Longstreet cyn). 7200 ft. Introduced annual. Sept.

S. triflorum Nutt. Seepage areas or other moist, usually disturbed soils in *Artemisia tridentata* or *Artemisia*—Pinyon—Juniper.

LINCOLN CO.: Groom Range (Bald Mtn Spg and Cattle Spg). NYE CO.: N. Kawich Range (Eden Creek and Longstreet cyns). 5900—7100 ft. Annual. June—Sept.

TAMARICACEAE. Tamarisk Family

Tamarix

T. africana Poir. CLARK CO.: Indian Springs townsite. 3100 ft. Introduced shrub. Apr—May.

T. aphylla (L.) Karst. (ATHEL). NYE CO.: Occasional escape, often planted around homesteads where it reaches 15 m height; Ash Meadows. 2300 ft. Introduced large tree. Fl. time not known.

T. parviflora DC. Occasional escape, usually planted. CLARK CO.: Indian Springs townsite. NYE CO.: Ash Meadows; Beatty townsite. 2200—3300 ft. Introduced small tree. Apr—May.

T. ramosissima Ledeb. The common Tamarisk of the region, on wet or moist disturbed soils, usually as occasional plants or small populations. NTS: W. Jackass Flats (near reservoir), Frenchman Flat (Cane Spg and on playa), Yucca Flat (near playa), ne. Forty-Mile Cyn; Pahute Mesa. CLARK CO.: Indian Springs townsite. NYE CO.: Uncommon in Ash Meadows, but common in Pahrump Valley where it sometimes forms extensive thickets as an escape. 2200—6900 ft. Introduced shrub or small tree. May—Oct.

ULMACEAE. Elm Family

Ulmus

U. minor Mill. [*U. carpinifolia* Gleditsch] (SMOOTH-LEAVED ELM). Planted near homesteads, persisting near springs, and occasional escapes. NTS: Mercury townsite; cent. Jackass Flats. NYE CO.: Cent. Penoyer Valley. 3700—6300 ft. Introduced small tree. Mar.

U. parvifolia Jacq. (CHINESE ELM). NTS: Occasional along disturbed roadsides of Pahute Mesa. 6500—7200 ft. Introduced small tree. Fl. time not known.

URTICACEAE. Nettle Family

Urtica

U. holosericea Nutt. Abundant in Willow thickets along mountain streams, or near springs, in *Artemisia nova* or usually

Artemisia—Pinyon—Juniper. CLARK CO.: Nw. Spring Mtns (Willow Creek). NYE CO.: Stonewall Mtn Spg, n. Kawich Range (Eden Creek and Longstreet cyns). 5600—7100 ft. Perennial. June—July.

VALERIANACEAE. Valerian Family

Valeriana

V. puberulenta Rydb. CLARK CO.: Locally common in Yellow Pine—Fir, on steep slope at head of Clark Cyn, nw. Spring Mtns. 9200 ft. Perennial. July.

VERBENACEAE. Vervain Family

Lippia

L. incisa (Small) Tidestr. CLARK CO.: Local on moist bank of lake, Indian Springs townsite. 3100 ft. Perennial. June—Aug.

Verbena

V. bracteata Lag. & Rodr. Local on moist, usually disturbed sandy sites, in *Larrea—Ambrosia*, *A. canescens*, *Coleogyne*, *Artemisia*, *Artemisia*—Pinyon—Juniper. NTS: Cane Spg (w. Frenchman Flat). CLARK CO.: Common along Wheeler Wash of nw. Spring Mtns. LINCOLN CO.: Foothills of Groom Range. NYE CO.: With Mesquite, n. Pahrump Valley; s. foothills of Monitor Range, s. Kawich Range (Cedar Pass), spg in n. Belted Range. 2600—7000 ft. Perennial. May—Sept.

V. gooddingii Briq. Washes or other disturbed sites in *Larrea—Ambrosia*, *Coleogyne*, *Artemisia*. CLARK CO.: Occasional along Wheeler Wash, nw. Spring Mtns. NYE CO.: Scattered populations or locally common along washes of foothills and upper bajadas below W end of Spring Mtns. 2500—5200 ft. Perennial. Apr—Sept.

VIOLACEAE. Violet Family

Viola

V. nephrophylla Greene. Wet meadows or along streams in *Artemisia*—Pinyon—Juniper. CLARK CO.: Nw. Spring Mtns (Willow Creek area). NYE CO.: N. Kawich Range (Eden Creek and Longstreet cyns). 6000—7200 ft. Perennial. June.

VISCACEAE. Mistletoe Family

Arceuthobium

A. abietinum (Engelm.) Hawksworth & Wiens. LINCOLN CO.: Occasional on *Abies concolor*, cent. Groom Range (Bald Mtn). 8500 ft. Perennial. Summer.

A. divaricatum Engelm. Occasional, on *Pinus monophylla*, more common some years than others, esp. on damaged trees; throughout the range of the host species. NTS: Shoshone Mtn, s. Belted Range (Rainier Mesa), Pahute Mesa. CLARK CO.: Nw. Spring Mtns (Clark Cyn). LINCOLN CO.: Cent. Groom Range (W slope of Bald Mtn). NYE CO.: Cent. Belted Range. 6300—7500 ft. Perennial. Summer.

Phoradendron

P. californicum Nutt. Uncommon, on *Prosopis glandulosa*. CLARK CO.: Cactus Spgs and Indian Spgs. NYE CO.: Ash Meadows. 2200—3200 ft. Perennial. July.

P. juniperinum Engelm. Occasional, on *Juniperus osteosperma*; abundant some years, when present on most trees of some areas; range of the host species. NTS: Shoshone Mtn, s. Belted Range (Rainier Mesa), Pahute Mesa. CLARK CO.: Nw. Spring Mtns (lower Clark Cyn). NYE CO.: Cent. Belted Range. 6200—6700 ft. Perennial. Summer.

VITACEAE. Grape Family

Parthenocissus

P. inserta (Kerner) Fritsch. Planted and escaped. CLARK CO.: Indian Springs townsite. NYE CO.: Beatty townsite. 3100—3300 ft. Introduced perennial. May.

Vitis

V. arizonica Engelm. Wet to moist soils near springs. CLARK CO.: Indian Springs townsite. NYE CO.: Spg areas of Ash Meadows, in Ash—Screwbean or *Atriplex*; Crystal Spg cyn of nw. Spring Mtns. 2200—5000 ft. Woody vine. Apr—May.

ZYGOPHYLLACEAE. Caltrop Family

Kallstroemia

K. californica (S. Wats.) Vail. NTS: Disturbed site in *Larrea—Ambrosia*, Mercury Valley. 3600 ft. Annual. Aug—Oct.

Larrea

L. tridentata (Sessé & Moç. ex DC.) Cov. [*L. divaricata* Cav. of N. Amer. auths.] (CREOSOTE-BUSH). Dominant shrub of most Mojave Desert communities, reaching its n. limits within and across the region from E to W: Locally in w. Emigrant Valley at 5100 ft; ne. Yucca Flat, at 4500 ft; several widely scattered large plants in s. Mid Valley; N slope of Jackass Flats, s. Topopah Valley, and washes below Yucca Mtn, all at *ca* 4500 ft; uplands of Thirsty Cyn at 5200 ft; n. Crater Flat at 4200 ft; along Beatty Wash, Thirsty Cyn drainage and other drainages S or W into Oasis Valley, at 4200—5200 ft; Sarcobatus Flat below Tolicha Peak at 4600 ft, NW nearly to saddle into Stonewall Flat; mostly replaced by *Atriplex* communities in Ash Meadows, but on uplands in n. part and in other areas locally. NTS: N. Amargosa Valley, Mercury Valley, Rock Valley, Jackass Flats, Topopah Valley, Frenchman Flat, Yucca Flat, Thirsty Cyn, w. Emigrant Valley. CLARK CO.: E-W axis of Spotted Range. NYE CO.: Ash Meadows, Crater Flat, Oasis Valley, n. Amargosa Valley (Fluorspar Cyn), Sarcobatus Flat. 2200—4500 (—5200) ft. Shrub. May—June, some yrs scattered fls. Sept—Oct.

Tribulus

T. terrestris L. (PUNCTURE VINE). Uncommon in the region, always on disturbed moist soils; *Larrea—Ambrosia*, *Atriplex*, *Artemisia*. NTS: Roadside, Mercury Valley. CLARK CO.: Indian Springs townsite. LINCOLN CO.: Near spg, foothills of Groom Range. NYE CO.: In or near Lathrop Wells, Beatty, Tonopah, and Warm Springs townsites; Stonewall Mtn Spg; Pahrump Valley. 2600—6000 ft. Annual. May—Oct.

Class **LILIATAE.** Monocotyledons

AGAVACEAE. Agave Family

Agave

A. utahensis Engelm. var. **eborispina** (Hester) Breit. Local populations in certain canyons of limestone mountain ranges, usually in *Atriplex*. NTS: W. Spotted Range (incl. Mercury Ridge), Ranger Mtns, and Buried Hills. CLARK CO.: N-S axis of Spotted Range. 3900—4800 ft. Perennial. May—June.

Yucca

Y. baccata Torr. var. **vespertina** McKelvey. Occasional to common locally, identified with upper *Coleogyne*, *Artemisia*, and lower

Artemisia—Pinyon—Juniper. NTS: In and below Eleana Range (nw. Yucca Flat), s. Belted Range (Rainier Mesa), cyns of Halfpint Range (Banded Mtn). CLARK CO.: Abundant on upper bajada, with *Y. brevifolia* and in some places *Y. schidigera*, below S face of Spring Mtns. NYE CO.: Cent. Belted Range (Johnnies Water cyn). 4500—6500 (—7500) ft. Shrub. May—June, to some extent most yrs.

Y. brevifolia Engelm. in Wats. (JOSHUA TREE). Occasional plants to locally extensive populations, upper *Larrea*, *Atriplex*, *Grayia—Lycium*, lower *Coleogyne* and less commonly *Artemisia*. NTS: N. and e. Jackass Flats, Mid Valley (esp. s. half), n. and e. Frenchman Flat, most parts of Yucca Flat (esp. s. one-third), uplands of Thirsty Cyn, s. Groom Lake (where reaches its n. limits at *ca* 6000 ft in n. Papoose Range). CLARK CO.: Below both faces of nw. Spring Mtns, often with *Y. schidigera* and/or *Y. baccata*, in forest-like stands of large trees on S-facing bajada (Pahrump Valley drainage); extensive and dense stand below N-S axis of Spotted Range (middle and upper bajadas of e. Frenchman Flat). ESMERALDA CO.: Cent. Goldfield Hills. NYE CO.: Many trees on E slope of cent. and s. Cactus Range, where ascends to ridges at *ca* 7000 ft (sight record); Crater Flat (below Bare Mtn); Stonewall Flat (common in Goldfield Hills). 3500—7000 ft. Small tree. Apr—May, to some extent every yr, but rarely the majority of plants the same yr.

Y. schidigera Roezl ex Ortgies (MOJAVE YUCCA). The common *Yucca* of lower elevations, usually in *Larrea—Atriplex* or *Larrea—Ambrosia*, less frequent in *Coleogyne*. NTS: S. Frenchman Flat and Mercury Valley below w. Spotted Range (Red Mtn, Mercury Ridge), and w. Frenchman Flat (below N slope of e. half of Skull Mtn). CLARK CO.: Below both faces of nw. Spring Mtns, where often occurs with *Y. brevifolia* and/or *Y. baccata*. 3100—5000 ft. Shrub or small tree. Apr—May, to some extent nearly every yr, but rarely the majority of plants the same yr.

AMARYLLIDACEAE. Amaryllis Family

Allium

A. atrorubens S. Wats. LINCOLN CO.: Common in *Artemisia* and *Artemisia*—Pinyon—Juniper, W slope of cent. Groom Range (below Bald Mtn). 6000—7000 ft. Perennial. June.

A. bisceptrum S. Wats. NYE CO.: Local populations, usually under Willow trees (in *Artemisia*—Pinyon—Juniper), cyn bottoms of

n. Kawich Range (Eden Creek and Longstreet cyns). 6700—7200 ft. Perennial. June.

A. nevadense S. Wats. Widely distributed and locally often common, esp. along washes or around flatrock areas, volcanic and limestone mountain ranges; usually *Coleogyne*, *Artemisia nova*, *Artemisia*—Pinyon—Juniper, occasionally *Larrea* or *Atriplex*. NTS: E. Specter Range and below Skull Mtn (Rock Valley), w. Spotted Range (Red Mtn, Mercury Ridge), Mine Mtn, s. Shoshone Mtn (Topopah Spg), Eleana Range (Capt. Jack Spg area), s. Belted Range (Rainier Mesa, Gold Meadows), most areas of Pahute Mesa. CLARK CO.: N-S axis of Spotted Range. NYE CO.: Nw. Spring Mtns (N slope below Mt. Stirling), Bullfrog Hills (summit of Sawtooth Mtn), Stonewall Mtn (near Spg), s. Kawich Range (Cedar Pass), n. Belted Range. 4000—7400 ft. Perennial. Apr—June.

A. scorodoprasum L. NTS: S. Yucca Flat, banks of reservoir. 3900 ft. Introduced perennial. June.

Androstephium

A. breviflorum S. Wats. Usually in *Atriplex*, with various other codominants (incl. *Larrea*, *Ambrosia*, *Coleogyne*, *Lycium pallidum*, *L. shockleyi*) and on bajadas below limestone hills or mountain ranges, sometimes in lowlands near playas. NTS: Amargosa Valley and Rock Valley (cent. Specter Range); Mercury Valley (below w. Spotted Range and e. Specter Range); s. and se. Frenchman Flat (below Spotted Range—Ranger Mtns complex), in *Atriplex canescens* N of playa (French Peak area) and *Larrea—Ambrosia* in sands W of Buried Hills; sw. Yucca Flat (CP Hills). CLARK CO.: N-S axis of Spotted Range. 2900—4400 ft. Perennial. Apr—early May.

Dichelostemma

D. pulchellum (Salisb.) Heller [*Brodiaea p.* Greene; *D. capitata* A. Wood]. Plants rarely occurring singly, in populations sometimes covering extensive areas, in *Larrea—Grayia—Lycium*, *Grayia—Lycium*, *Coleogyne*, *Artemisia* (usually *A. nova*) and *Artemisia*—Pinyon—Juniper. NTS: Mercury Valley, Rock Valley, n. Jackass Flats, w. Frenchman Flat, nw. Yucca Flat, Forty-Mile Cyn; s. Belted Range (below W face of Rainier Mesa). CLARK CO.: Nw. Spring Mtns (Willow Spg area). NYE CO.: Nw. Spring Mtns (N slope below Mt. Stirling), below nw. Yucca Mtn (common length of Tates Wash), ne. Bullfrog Hills. 3500—6800 ft. Perennial. Apr—June, fl. abundantly some yrs; in other yrs absent, or with leaves only.

CYPERACEAE. Sedge Family

Carex

C. **alma** Bailey. NTS: Near Oak Spg of Oak Spring Butte (s. Belted Range), *Artemisia*—Pinyon—Juniper. 5900 ft. Perennial. June—July.

C. **aurea** Nutt. NYE CO.: Locally common, wet soils of meadow in *Artemisia*—Pinyon—Juniper, n. Kawich Range (Eden Creek cyn). 7200 ft. Perennial. June.

C. **douglasii** Boott. Local populations, talus slopes, base of cliffs, flatrock areas, mountain meadows, or spring areas, *Artemisia* or *Artemisia*—Pinyon—Juniper of volcanic mountains and mesas. NTS: Flatrock area of sw. Rainier Mesa (s. Belted Range), s. and e. Pahute Mesa. NYE CO.: Near spgs or in moist meadows, W slope of cent. and n. Kawich Range (Stinking Spg, Longstreet cyn). 6600—7400 ft. Perennial. May—June.

C. **lanuginosa** Michx. NYE CO.: Common, banks of stream in mountain meadow in *Artemisia*—Pinyon—Juniper, n. Kawich Range (Longstreet cyn). 6800 ft. Perennial. June.

C. **microptera** Mkze. Moist or wet soils, *Artemisia*—Pinyon—Juniper. CLARK CO.: Nw. Spring Mtns (common, spg near head of Clark Cyn and Cold Creek Spg). NYE CO.: Locally common near stream, n. Kawich Range (Eden Creek cyn). 6200—9000 ft. Perennial. June—July.

C. **nebrascensis** Dewey. NYE CO.: Locally abundant in wet soils, *Artemisia*—Pinyon—Juniper, n. Kawich Range (Eden Creek and Longstreet cyns). 6800—7100 ft. Perennial. June.

C. **occidentalis** Bailey. Local populations, volcanic boulders in washes or around boulder-cliffs, cliff crevices, or steep talus slopes; *Artemisia*—Pinyon—Juniper. NTS: S rim of Pahute Mesa. NYE CO.: Wet cliff face at Stonewall Mtn Spg, n. and s. Kawich Range (Eden Creek cyn, Cedar Pass), cent. Belted Range (Cliff Spg). 5600—7600 ft. Perennial. Late May—June.

C. **praegracilis** W. Boott. The most common and widely distributed *Carex* of the region, on wet soils in *Atriplex*, *A. canescens*, *Distichlis* or *Juncus* meadows, and Ash—Screwbean; in *Artemisia*—Pinyon—Juniper or Yellow Pine—Fir at higher elevations. NTS: Cane Spg (w. Frenchman Flat). CLARK CO.: Irrigation channels, Indian Springs townsite; nw. Spring Mtns (seepage area below N slope, and upper Clark Cyn near stream). NYE CO.: Common in spg areas of Ash Meadows and Oasis Valley; spgs of Bullfrog Hills and both slopes of Kawich Range (Stinking Spg, Eden Creek and Longstreet cyns), base of s. Hot Creek Range (Warm Spgs). 2200—9000 ft. Perennial. Apr—June.

C. scopulorum Holm. NYE CO.: Locally common, wet soils along Eden Creek in *Artemisia*—Pinyon—Juniper, n. Kawich Range. 7100 ft. Perennial. June.

Cladium

C. californicum (S. Wats.) O'Neill in Tidestr. & Kittell. NYE CO.: Local, usually small populations, with *Carex*, *Juncus*, or *Distichlis*, known from 3 spring areas of n. and e. Ash Meadows. 2200—2300 ft. Perennial. June—July.

Eleocharis

E. macrostachya Britt. in Small. Frequent in region around springs, water impoundments, or seepage sites, in *Atriplex*, *Artemisia*, *Artemisia*—Pinyon—Juniper. NTS: Margin of ponds in s. Frenchman Flat (Well 5-B), Forty-Mile Cyn (Well 8), and nw. Yucca Flat below Rainier Mesa. LINCOLN CO.: Bald Mtn Spg (cent. Groom Range). NYE CO.: Willow Spg (e. Goldfield Hills), and pond margin at Reveille Mill (Reveille Valley). 3100—7000 ft. Perennial. June—Aug.

E. parishii Britt. [*E. montevidensis* Kunth var. *p.* (Britt.) V. Grant]. Common, moist soils of region, usually in large or local areas of *Juncus* meadows (or Ash—Screwbean), *Atriplex*, or *Artemisia*—Pinyon—Juniper. NTS: Cane Spg (w. Frenchman Flat). CLARK CO.: Along stream, Indian Springs townsite; Willow Creek (nw. Spring Mtns). NYE CO.: Spg areas of Ash Meadows, Oasis Valley; e. Bullfrog Hills, n. Kawich Range (Eden Creek cyn), s. Hot Creek Range (Warm Spgs). 2200—8100 ft. Perennial. Apr—Sept.

E. rostellata (Torr.) Torr. NYE CO.: Common in wet soils of spg areas of Ash Meadows, often with *E. parishii;* Ash—Screwbean, *Juncus* and *Distichlis* meadows. 2200—2300 ft. Perennial. Apr—July.

Fimbristylis

F. thermalis S. Wats. NYE CO.: Known from 3 spg areas of ne. Ash Meadows, in wet soils with *Juncus* or Ash—Screwbean; also from spgs NE of Beatty (*Kral 21833*, RSA). 2200—3300 ft. Perennial. June—July.

Schoenus

S. nigricans L. NYE CO.: Known from 3 small populations in n. Ash Meadows, in *Atriplex* or Ash—Screwbean. 2200—2300 ft. Perennial. Apr—June.

Scirpus

S. acutus Muhl. Usually small populations, near water impoundments or along stream. NTS: Reservoirs in s. Yucca Flat (Well C), s. Frenchman Flat (Well 5-B), ponds of W-cent. Jackass Flats. CLARK CO.: Indian Springs townsite. 3100—3900 ft. Perennial. May—June.

S. americanus Pers. NYE CO.: Margin of pond at Reveille Mill (Reveille Valley) and banks of drainage channel at Warm Spgs (s. Hot Creek Valley), disturbed *Atriplex*. 5400—6000 ft. Perennial. May—July.

S. microcarpus Presl. NYE CO.: Local in grotto at Stonewall Mtn Spg; *Artemisia*—Pinyon—Juniper. 5600 ft. Perennial. July.

S. olneyi A. Gray. The most common Bulrush of the region. CLARK CO.: Lake margin, Indian Springs townsite. NYE CO.: Dense stands in swamp areas of Ash Meadows and Oasis Valley. 2200—3600 ft. Perennial. Apr—June.

S. robustus Pursh. Uncommon, small populations in perennially wet soils or standing water. NTS: Edge of Groom Lake playa. NYE CO.: Local populations in former Carson Slough area and e. Ash Meadows; Oasis Valley. 2200—5000 ft. Perennial. July—Oct.

IRIDACEAE. Iris Family

Iris

I. missouriensis Nutt. NYE CO.: Locally abundant, *Carex* wet meadows in *Artemisia*—Pinyon—Juniper, Longstreet cyn of nw. Kawich Range. 6600—7200 ft. Perennial. June.

Sisyrinchium

S. demissum Greene [*S. funereum* Bickn.]. NYE CO.: Locally often common in *Juncus*—*Distichlis* wet meadows of spg areas, n. and e. Ash Meadows; local in Crystal Spg cyn in Pinyon—Juniper at 4800 ft, S slope of nw. Spring Mtns. 2200—4800 ft. Perennial. Apr—June.

S. halophilum Greene. NYE CO.: Local in *Carex* wet meadows in *Artemisia*—Pinyon—Juniper, n. and cent. Kawich Range (Eden Creek and Longstreet cyns, Stinking Spg). 6600—6800 ft. Perennial. June.

JUNCACEAE. Rush Family

Juncus

J. balticus Willd. The most predictable species of more or less permanently wet alkaline soils of the region, usually locally the dominant species and occurring in dense stands in areas otherwise *Atriplex*, *A. canescens*, *Coleogyne*, *Artemisia*, or *Artemisia*—Pinyon—Juniper; populations extremely variable, and here included are plants variously with characters of *J. mexicanus* Willd., whose distribution in this region is not distinguishable from that of *J. balticus*. NTS: Cane Spg (w. Frenchman Flat), Topopah Spg (n. Topopah Valley), Whiterock Spg (nw. Yucca Flat). CLARK CO.: Local on sand dune, Cactus Spgs. LINCOLN CO.: Bald Mtn Spg (base of cent. Groom Range). NYE CO.: Locally common in many wet areas of Ash Meadows, Oasis Valley; Crystal Spg cyn (nw. Spring Mtns); wet areas of Bullfrog Hills; wet meadows of Eden Creek cyn, Stinking Spg, and Cedar Spg (n., cent., and s. Kawich Range). 2200—6800 ft. Perennial. Late Apr—July.

J. bufonius L. NYE CO.: Locally common on wet soils near streams in *Artemisia*—Pinyon—Juniper, n. Kawich Range (Eden Creek cyn). 7200—8100 ft. Annual. June—Sept.

J. cooperi Engelm. NYE CO.: Known in region from one large plant near pond, w. Ash Meadows, with *Distichlis* and *Baccharis*. 2300 ft. Perennial. May.

J. ensifolius Wikst. var. **montanus** (Engelm.) C. L. Hitchc. [incl. *J. saximontanus* A. Nels. and f. *brunnescens* (Rydb.) F. J. Herm., and *J. xiphioides* E. Mey.]. Locally common in spg areas and along streams, *Artemisia*—Pinyon—Juniper. CLARK CO.: Nw. Spring Mtns (Cold Creek Spg, seepage site at Field Sta.). NYE CO.: Nw. Spring Mtns (Crystal Spg cyn), n. Kawich Range (upper Eden Creek cyn). 5000—7200 ft. May—July.

J. longistylis Torr. Occasional, moist soils in *Artemisia*—Pinyon—Juniper, Yellow Pine—Fir. CLARK CO.: Nw. Spring Mtns (Cold Creek, upper Clark Cyn). NYE CO.: N. Kawich Range (upper Eden Creek cyn). 6200—9000 ft. Perennial. July.

J. nodosus L. Local populations near water, in springs, streams, or seepage areas, with *Carex*, *Scirpus*, *Distichlis*, Ash—Screwbean, or in *Artemisia*—Pinyon—Juniper. LINCOLN CO.: Bald Mtn Spg and Cattle Spg (base of Groom Range). NYE CO.: Usually represented to some extent at most spg areas of Ash Meadows; nw. Spring Mtns (near pond at Cold Creek Field Sta.); Oasis Valley. 2200—6500 ft. Perennial. June—Aug.

JUNCAGINACEAE. Arrow-Grass Family

Triglochin

T. concinnum Davy var. **debile** (Jones) J. T. Howell. NYE CO.: Occasional local populations, usually with *Distichlis*, Ash Meadows and Oasis Valley. 2200—3600 ft. Perennial. May—June.

LEMNACEAE. Duckweed Family

Lemna

L. minuta HBK. [*L. minima* Phil.]. NYE CO.: Common in stream in moving water, sedge meadow in *Artemisia*—Pinyon—Juniper, n. Kawich Range (Longstreet cyn). 7200 ft. Flowering time not known.

LILIACEAE. Lily Family

Asparagus

A. officinalis L. Occasional escape from cultivation, as one or a few plants, usually on disturbed moist sites. CLARK CO.: Indian Springs townsite. NYE CO.: N. Ash Meadows, n. Pahrump Valley, Oasis Valley. 2300—3400 ft. Introduced perennial. Late Apr—June.

Calochortus

C. bruneaunis Nels. & Macbr. [*C. nuttallii* Torr. var. *bruneaunis* (Nels. & Macbr.) Ownbey]. NTS: Occasional, *Artemisia*—Pinyon—Juniper, W slope of Rainier Mesa (s. Belted Range) NW to the S rim of Pahute Mesa. 6900—7000 ft. Perennial. June—July.

C. flexuosus S. Wats. Common in *Coleogyne*, *Artemisia*, and lower *Artemisia*—Pinyon—Juniper, occasionally in *Larrea*—*Grayia*—*Lycium* and *Grayia*—*Lycium*. NTS: Mercury Valley, Rock Valley, Jackass Flats, Topopah Valley, Frenchman Flat, Mid Valley, Yucca Flat, n. Forty-Mile Cyn; nw. and s. Pahute Mesa. NYE CO.: Nw. Spring Mtns (N slope below Mt. Stirling), n. Yucca Mtn (w. Crater Flat), Bullfrog Hills, White Blotch Spg W of n. Groom Range. 3300—6500 ft. Perennial. Apr—early June.

Fritillaria

F. atropurpurea Nutt. NTS: Local populations in *Artemisia*—Pinyon—Juniper, s. Belted Range (cent. and n. Rainier Mesa and W slope) NW to flatrock area on S rim of Pahute Mesa. 6000—7400 ft. Perennial. Late May—early June.

F. pinetorum A. Davids. NYE CO.: Local populations in *Artemisia*—Pinyon—Juniper, n. and s. Kawich Range (Eden Creek cyn and Cedar Pass). 7000—7600 ft. Perennial. Late May—early June.

Smilacina

S. stellata (L.) Desf. Locally common on moist sites in *Artemisia*—Pinyon—Juniper and Yellow Pine—Fir. CLARK CO.: Nw.

Spring Mtns (common in upper Clark Cyn, and along Willow Creek). NYE CO.: N. Kawich Range (common locally along stream bank of Longstreet cyn, frequent in Eden Creek cyn). 6000—9000 ft. Perennial. May—June.

Zigadenus

Z. paniculatus (Nutt.) S. Wats. Locally common, *Artemisia nova* and *Artemisia*—Pinyon—Juniper (including flatrock areas), rarely *Coleogyne*. NTS: Timber Mtn (upper Cat Cyn), s. Belted Range (W slope of Rainier Mesa, nw. Gold Meadows), nw. Pahute Mesa. CLARK CO.: Nw. Spring Mtns (near Wheeler Well, Trough Spg area), below N-S axis of Spotted Range (e. Frenchman Flat, Indian Springs Valley). NYE CO.: Nw. Spring Mtns (N slope below Mt. Stirling), s. Kawich Range (Cedar Pass), ne. Groom Range. 4000—7800 ft. Perennial. Late Apr—early June.

NAJADACEAE. Water-Nymph Family

Najas

N. marina L. NYE CO.: Collected by Coville and Funston on the Death Valley Expedition (Coville, F. V., *Contr. U. S. Nat. Herb.*, 4: 1893), in a spring of n. Ash Meadows. 2200 ft. Annual. Summer.

ORCHIDACEAE. Orchid Family

Habenaria

H. sparsiflora S. Wats. CLARK CO.: Locally common near spg, in Yellow Pine—Fir, upper Clark Cyn, nw. Spring Mtns. 9000 ft. Perennial. July.

Spiranthes

S. romanzoffiana Cham. & Schlecht. NYE CO.: Known from two sites in Ash Meadows: Stream bank in Ash thicket, and *Juncus* meadow. 2200 ft. Perennial. July.

POACEAE (GRAMINEAE). Grass Family

Agropyron

A. cristatum (L.) Gaertn. [incl. *A. desertorum* (Fisch.) Schult.]. NTS: Local on burned sites in *Coleogyne*, n. Topopah Valley. CLARK CO.: Common on disturbed sites, esp. burned *Artemisia*—

Pinyon—Juniper, nw. Spring Mtns (Wheeler Pass area). NYE CO.: Disturbed *Artemisia*, Ralston Valley. 5500—7700 ft. Introduced perennial. May—June.

A. elongatum (Host) Beauv. NYE CO.: Local on disturbed site, n. Ash Meadows. 2200 ft. Introduced perennial. July.

A. smithii Rydb. NTS: Local on a mud flat in *Artemisia tridentata*, n. Pahute Mesa. 6400 ft. Perennial. June—July.

A. spicatum (Pursh) Scribn. & J. G. Sm. [incl. var. *inerme* Heller]. Local populations in *Artemisia*—Pinyon—Juniper, around cliffs, on rock outcrops, or talus. CLARK CO.: Nw. Spring Mtns (upper Clark Cyn). NYE CO.: Stonewall Mtn (cyn with "Ruins"), n. Kawich Range (upper Eden Creek cyn). 5600—7700 ft. Perennial. June—July.

A. trachycaulum (Link) Malte. CLARK CO.: Common along stream banks in Yellow Pine—Fir, near spg of upper Clark Cyn, nw. Spring Mtns. NYE CO.: Occasional, *Artemisia*—Pinyon—Juniper, n. Kawich Range (upper Eden Creek cyn). 7800—9000 ft. Perennial. July.

A. vaseyi Scribn. & J. G. Sm. NYE CO.: Common locally along upland mountain washes in *Artemisia*—Pinyon—Juniper, n. Kawich Range (upper Eden Creek cyn). 7800 ft. Perennial. July.

Agrostis

A. exarata Trin. var. **monolepis** (Torr.) A. S. Hitchc. NYE CO.: Common on moist cliff ledges, grotto at Stonewall Mtn Spg; *Artemisia*—Pinyon—Juniper. 5600 ft. Perennial. July.

A. scabra Willd. NYE CO.: Wet soils near spg in *Artemisia*—Pinyon—Juniper, n. Kawich Range (upper Eden Creek cyn). 8100 ft. Perennial. July—Aug.

A. semiverticillata (Forsk.) C. Christ. Occasional to common on disturbed sites near springs throughout region, in Ash—Screwbean, *Distichlis*, *Coleogyne*, *Artemisia*, *Artemisia*—Pinyon—Juniper. NTS: Common, Cane Spg (w. Frenchman Flat), Oak Spg (Oak Spring Butte of s. Belted Range). LINCOLN CO.: Spgs of Groom Range (incl. Cattle Spg, Bald Mtn Spg). NYE CO.: Common in spg or irrigation areas of Ash Meadows, Pahrump Valley, Oasis Valley; Cactus Spg (n. Cactus Range), Eden Creek (n. Kawich Range). 2200—7200 ft. Introduced perennial. May—Sept.

A. stolonifera L. (*A. alba* L.) (REDTOP). Weed of disturbed sites, spring and stream areas, most frequent in *Artemisia*—Pinyon—Juniper. CLARK CO.: Indian Springs townsite; nw. Spring Mtns (abundant near Cold Creek Spg). LINCOLN CO.: Cent. Groom Range (*Juncus* swamp at Bald Mtn Spg). NYE CO.: Cent. and n.

Kawich Range (Stinking Spg and Eden Creek cyn). 3100—7200 ft. Introduced perennial. (May—) June—Aug.

Alopecurus

A. aequalis Sobol. NYE CO.: Moist site in *Artemisia*—Pinyon—Juniper, n. Kawich Range (Eden Creek cyn). 7200 ft. Perennial. July.

Aristida

A. adscensionis L. Local populations on ledges and in rock crevices of lower canyons of limestone mountain ranges, in *Larrea—Ambrosia, Larrea—Atriplex.* NTS: Specter Range, Ranger Mtns. NYE CO.: Foothills below W end of Spring Mtns, Bare Mtn. 2800—4000 ft. Annual. Apr—Oct.

A. fendleriana Steud. Occasional to common locally, *Artemisia, Artemisia*—Pinyon—Juniper, rarely *Grayia—Lycium*. May not be distinguishable from *A. longiseta* in this region. NTS: Forty-Mile Cyn wash (in sw. Forty-Mile Cyn drainage); nw. Pahute Mesa. LINCOLN CO.: Washes and terraces below N end of Groom Range. NYE CO.: Below cent. and n. Belted Range, n. and s. Kawich Range (below Longstreet cyn, Rose Spg cyn of Cedar Pass area), foothills of s. Quinn Canyon Range. 4900—7100 ft. Perennial. June—July.

A. glauca (Nees) Walp. The common *Aristida* of limestone or calcareous volcanic hills, mountains, or outcrops; ledges, crevices, and washes in *Larrea—Ambrosia*, *Larrea—Atriplex*, *Coleogyne*, *Atriplex*. NTS: Cyns of Specter Range, w. Spotted Range (incl. Red Mtn), e. Skull Mtn, Ranger Mtns, Buried Hills and limestone butte to W, Halfpint Range (incl. French Peak mtn, Banded Mtn), CP Hills, Eleana Range (dolomite hills), w. and below s. Shoshone Mtn, Yucca Mtn (w. Jackass Flats). CLARK CO.: Limestone hill below N face of nw. Spring Mtns. NYE CO.: Foothills below W end of Spring Mtns, Bare Mtn, s. Hot Creek Range (slope above Warm Spgs); nw. Reveille Valley. 3000—4500 (—6200) ft. Apr—June, some yrs Sept—Oct.

A. longiseta Steud. var. **robusta** Merr. Common locally, *Coleogyne*, *Artemisia nova*, and esp. *Artemisia*—Pinyon—Juniper; bases of cliffs, washes, or upland volcanic rock outcrops. May not be distinguishable from *A. fendleriana* in this region. NTS: In or below Yucca Mtn (w. Jackass Flats), Shoshone Mtn, Timber Mtn, Eleana Range, se. Belted Range (ne. Yucca Flat), S rim of Pahute Mesa. NYE CO.: Cent. Belted Range (lower Johnnies Water cyn). 4200—7000 ft. Perennial. June—Sept.

A. purpurea Nutt. Washes and rock ledges of limestone mountain ranges, in *Larrea—Ambrosia*, *Larrea—Atriplex*, *Coleogyne*. NTS:

Specter Range, Ranger Mtns, Buried Hills, Halfpint Range (Scarp Cyn). CLARK CO.: N-S axis of Spotted Range, limestone hill below N face of nw. Spring Mtns. 3200—5000 ft. Perennial. Apr—June.

A. wrightii Nash. NTS: Disturbed site in *Larrea/Coleogyne*, w. Spotted Range (Red Mtn). 3800 ft. Perennial. May—June.

Arundo

A. donax L. CLARK CO.: Large reproducing population near stream, Indian Springs townsite. 3100 ft. Introduced perennial. Aug—Sept.

Avena

A. sativa L. (OATS). Occasional plants on disturbed sites. NTS: S. Frenchman Flat. CLARK CO.: Indian Springs townsite. NYE CO.: Ne. Ash Meadows. 2200—3500 ft. Introduced annual. Apr—June.

Blepharidachne

B. kingii (S. Wats.) Hack. Common in *Atriplex*, *A. canescens*, *Atriplex—Ceratoides* of basin floors, and local in *Larrea—Ambrosia*, *Larrea—Atriplex*, *Atriplex*, or *Grayia—Lycium* of slopes and washes in or below limestone mountain ranges and calcareous outcrops elsewhere. NTS: In or below w. Spotted Range (incl. Mercury Ridge), Ranger Mtns, CP Hills; n. Papoose Range, and common in many areas of s. Gold Flat below N face of Pahute Mesa. LINCOLN CO.: Below N end of Groom Range (in *Artemisia tridentata*). NYE CO.: Common in n. Stonewall Flat (abundant, with *Sarcobatus*, in e. Goldfield Hills), w. and s. Cactus Flat, cent. and ne. Kawich Valley, sw. Penoyer Valley. 3100—6200 ft. Perennial. May—early June, some yrs Sept.

Bouteloua

B. barbata Lag. Common locally on rock ledges of lower canyons of usually limestone mountain ranges or hills, and bajadas below, in *Larrea—Ambrosia*, *Larrea—Atriplex*, *Coleogyne*, *Grayia—Lycium*, *Atriplex*, *Atriplex—Ceratoides*. NTS: In or below Specter Range, w. Spotted Range (incl. Mercury Ridge), Mine Mtn, Eleana Range, hills N of Cane Spg; Thirsty Cyn. NYE CO.: Cent. Gold Flat, n. Kawich Valley, and in *Atriplex parryi—Sarcobatus* in s. Hot Creek Valley (Twin Spgs below N end of Reveille Range). 3100—5600 ft. Summer annual. Aug—Sept.

B. gracilis (HBK.) Lag. ex Steudel. Locally abundant in *Artemisia*—Pinyon—Juniper, less frequent in *Artemisia* and

Atriplex—Ceratoides, often sod-forming. NTS: S. Belted Range (Rainier Mesa, Gold Meadows), Pahute Mesa. CLARK CO.: Nw. Spring Mtns (Clark Cyn, but common and characteristic of S face of mtn range). NYE CO.: E. Kawich Valley and s. Penoyer Valley (below cent. and n. Belted Range), e. Cactus Flat (below s. Kawich Range). 5600—7500 ft. Perennial. July—Sept.

B. trifida Thurb. Occasional populations, crevices in limestone mountain ranges, hills, or outcrops; *Larrea—Ambrosia*, *Larrea—Atriplex*, *Atriplex*, *Coleogyne*. NTS: Striped Hills, w. Spotted Range (Red Mtn), Ranger Mtns, Buried Hills and limestone butte to W. CLARK CO.: Limestone hill below base of N face of nw. Spring Mtns; N-S axis of Spotted Range. 3500—5200 ft. Perennial. Apr—May, some yrs Aug—Sept.

Bromus

B. anomalus Rupr. ex Fourn. Local populations at cliff bases or on talus slopes, *Artemisia*—Pinyon—Juniper and Yellow Pine—Fir. NTS: Frequent in Silent Canyon drainage of n. Pahute Mesa. CLARK CO.: Common, Clark Cyn of nw. Spring Mtns. 6200—9000 ft. Perennial. June—July.

B. carinatus Hook. & Arn. [incl. plants referable to *B. marginatus* Nees]. Frequent local populations around rock outcrops or bases of cliffs, *Artemisia*—Pinyon—Juniper, Yellow Pine—Pinyon. NTS: W. Shoshone Mtn (below cliffs along Forty-Mile Cyn wash), Eleana Range (saddle below S rim of Rainier Mesa). CLARK CO.: Nw. Spring Mtns (near Trough Spg). NYE CO.: Nw. Spring Mtns [around volcanic outcrops near Gold Spg (N slope below Mt. Stirling)], cent. Belted Range (Johnnies Water cyn). (4000—) 5900—8200 ft. Perennial. May—July.

B. ciliatus L. CLARK CO.: Moist or wet soils in Yellow Pine—Fir, nw. Spring Mtns (common near spg, upper Clark Cyn). 9000 ft. Perennial. July—Aug.

B. inermis Leyss. CLARK CO.: Moist soils in Yellow Pine—Fir, nw. Spring Mtns (near spg, upper Clark Cyn). 9000 ft. Introduced perennial. July—Aug.

B. japonicus Thunb. NTS: Local, Mercury townsite. 3700 ft. Introduced annual. June.

B. rigidus Roth [*B. diandrus* Roth] (RIPGUT GRASS). Uncommon and local, washes near springs or disturbed sites. NTS: Cane Spg (w. Frenchman Flat), Topopah Spg (n. Topopah Valley), Forty-Mile Cyn wash (nw. Jackass Flats). CLARK CO.: Indian Springs townsite. NYE CO.: Beatty townsite. 3100—5700 ft. Introduced annual. Apr—May.

B. rubens L. (FOXTAIL CHESS or RED BROMEGRASS). Species of middle elevations, sporadically common to abundant throughout region in undisturbed or disturbed upland *Atriplex*, *Coleogyne*, *Larrea/Coleogyne*, less frequent in upper *Larrea*, occasional or absent in most *Larrea* of lower elevations and *Grayia—Lycium*, and replaced by *B. tectorum* in disturbed *Artemisia* and *Artemisia*—Pinyon—Juniper of higher elevations; in dense stands on burned *Coleogyne* sites. NTS: Mercury Valley, Rock Valley, Jackass Flats, Topopah Valley, Frenchman Flat, Mid Valley, Yucca Flat, Forty-Mile Cyn. NYE CO.: Pahrump Valley, n. Amargosa Valley (Beatty Wash, Bullfrog Hills). 2600—5000 (—6000) ft. Introduced winter annual. Apr—June.

B. tectorum L. (CHEAT GRASS). Common to abundant on disturbed sites in *Artemisia* and *Artemisia*—Pinyon—Juniper, at lower elevations along major washes; occurring in dense stands on burned *Artemisia* sites. NTS: Shoshone Mtn, Buckboard Mesa, Eleana Range, s. Belted Range (Rainier Mesa, Gold Meadows), Pahute Mesa; Mercury townsite, Cane Spg (w. Frenchman Flat), Forty-Mile Cyn wash (nw. Jackass Flats). CLARK CO.: Indian Springs townsite. (3100—) 5000—6500 ft. Introduced winter annual. Apr—June.

B. trinii Desv. in Gay. Locally common on disturbed or undisturbed sites, *Larrea*, *Coleogyne*, *Atriplex canescens*, *A. hymenelytra*, and *Mesquite*. NTS: N. and w. Mercury Valley, e. Rock Valley, nw. Jackass Flats (Forty-Mile Cyn wash), n. Topopah Valley, n. Frenchman Flat, Mid Valley. CLARK CO.: Bases of limestone cliffs, E-W axis of Spotted Range. NYE CO.: W. Pahrump Valley; se. Bullfrog Hills and SW face of Bare Mtn. 2600—5000 ft. Introduced winter annual. Apr—May.

B. willdenovii Kunth [*B. cartharticus* Vahl] (RESCUE GRASS). Local as weed of usually disturbed moist soils, in *Larrea*, *Atriplex*, Pinyon—Juniper. NTS: Mercury townsite. CLARK CO.: Indian Springs and Cactus Springs townsites. ESMERALDA CO.: Roadside S of Tonopah townsite. NYE CO.: Pahrump Valley; nw. Spring Mtns (Crystal Spg cyn); Beatty townsite. 2600—5000 ft. Introduced annual. Apr—Nov.

Chloris

C. virgata Swartz. NTS: Irrigated soils, Mercury townsite. 3700 ft. Introduced annual. July—Oct.

Cynodon

C. dactylon (L.) Pers. (BERMUDA GRASS). Disturbed irrigated sites of lower elevations. NTS: Mercury townsite; Well 5-B of

s. Frenchman Flat. NYE CO.: Spg and irrigation areas of Ash Meadows and Pahrump Valley; s. Hot Creek Valley (Warm Spgs). 2200—5400 ft. Introduced perennial. Apr—Sept.

Dactylis

D. glomerata L. (ORCHARD GRASS). CLARK CO.: Local in Willow thickets along Willow Creek, nw. Spring Mtns. NYE CO.: Disturbed irrigated site of Oasis Valley. 3400—6000 ft. Introduced perennial. May—June.

Deschampsia

D. caespitosa (L.) Beauv. CLARK CO.: Common along stream bank and in water below spg, upper Clark Cyn of nw. Spring Mtns; Yellow Pine—Fir; known also from a wet meadow in the s. Hot Creek Range. 8700—9000 ft. Perennial. June—July.

D. danthonioides (Trin.) Munro ex Benth. NTS: Locally common, seepage sites in *Coleogyne* at Whiterock Spg and *Artemisia* at Tippipah reservoir (se. Forty-Mile Cyn). 5000—5200 ft. Annual. May—June.

Digitaria

D. sanguinalis (L.) Scop. (CRABGRASS). NTS: Occasional, disturbed soils near Mercury townsite. 3700 ft. Introduced annual. July—Sept.

Distichlis

D. spicata (L.) Greene var. **stricta** (Torr.) Scribn. Sod-forming, dominant grass, in areas of seasonally wet alkaline soils. NTS: Local on slope near Cane Spg (w. Frenchman Flat) in *Atriplex canescens*; s. edge of playa, Emigrant Valley. NYE CO.: Widely distributed and abundant in Ash Meadows; Oasis Valley; s. Hot Creek Valley (Warm Spgs); cent. Kawich Range (Stinking Spg). 2200—6600 ft. Perennial. May—Sept.

Echinochloa

E. crusgalli (L.) Beauv. (BARNYARD GRASS). Local, irrigated disturbed soils. NTS: Mercury townsite, and pond in cent. Jackass Flats. NYE CO.: Ash Meadows, Pahrump Valley. 2200—3700 ft. Introduced annual. June—Nov.

Elymus

E. cinereus Scribn. & Merr. Common, usually as scattered small populations, esp. along washes, in depressions, or near springs, in usually volcanic mountain ranges; *Artemisia tridentata*, *Artemisia*—Pinyon—Juniper, and Yellow Pine—Pinyon. NTS: Skull Mtn (Cane Spg), Shoshone Mtn, Timber Mtn, Eleana Range, s. Belted Range (Rainier Mesa, Oak Spring Butte), Pahute Mesa. CLARK CO.: Nw. Spring Mtns (Willow Creek area; upper Clark Cyn). NYE CO.: Tolicha Peak area, Stonewall Mtn (cyn with "Ruins"), n. and s. Kawich Range (Eden Creek cyn and Cedar Pass), n. and cent. Belted Range, White Blotch Spg W of n. Groom Range. In Pahrump Valley (type locality for the species) and Ash Meadows, at 2200—2600 ft, all known populations consist of large clumps of plants with densely velvet—puberulent leaves and culms which, except for plants at Cane Spg, at 4000 ft at the base of Skull Mtn where plants are less densely puberulent, and those of the nearby nw. Spring Mtns (upper Clark Cyn where plants may have either puberulent or glabrous culms), are in distinct contrast with plants elsewhere throughout the region (with glabrous culms, except at the nodes); Pahrump Valley and Ash Meadows populations are associated with Mesquite. (2200—) 5200—7800 ft. Perennial. Late May—July (Ash Meadows and Pahrump Valley, Apr—May).

Putative hybrids, in which *E. cinereus* is an inferred parental species, are known from *Artemisia*—Pinyon—Juniper, at 5600—5800 ft. NTS: One plant in a large and extensive population of *E. cinereus* along Cat Cyn bottoms, E face of Timber Mtn. NYE CO.: One plant, Stonewall Mtn Spg.

E. triticoides Buckl. Small populations at springs or moist canyon bottoms, in usually volcanic mountains, often occurring with *E. cinereus;* in *Coleogyne*, *Atriplex*, or usually *Artemisia* or *Artemisia*—Pinyon—Juniper. NTS: Cane Spg (Skull Mtn), Oak Spg (Oak Spring Butte of s. Belted Range). CLARK CO.: Seepage site between Willow Creek and Cold Creek (nw. Spring Mtns). NYE CO.: Eden Creek cyn bottoms (n. Kawich Range). 4000—7600 ft. Perennial. June—July.

Eragrostis

E. barrelieri Daveau. NTS: Disturbed irrigated soils, Mercury townsite. 3800 ft. Introduced annual. June—July.

E. cilianensis (All.) Mosher. NYE CO.: Locally abundant along irrigation channels, n. Pahrump Valley. 2600 ft. Introduced annual. July—Sept.

E. pectinacea (Michx.) Nees [*E. diffusa* Buckl.] NYE CO.: Common, with *E. cilianensis* along irrigation channel, n. Pahrump Valley. 2600 ft. Annual, probably introduced in this area. July—Sept.

Erioneuron

E. pilosum (Buckl.) Nash [*Tridens p.* (HBK.) A. S. Hitchc.]. NTS: *Atriplex*, s. Buried Hills; *Artemisia nova*, dolomite hill of Eleana Range. 4300–5800 ft. Perennial. May.

E. pulchellum (HBK.) Tateoka [*Tridens p.* (HBK.) A. S. Hitchc.]. Lower canyons of mountain ranges and bajadas below, predictable and esp. abundant below limestone ranges; *Atriplex*, *Larrea—Atriplex*, *Larrea—Ambrosia*, *Coleogyne*, *Grayia—Lycium*, uncommon in *Artemisia*. NTS: In or below Specter Range, w. Spotted Range (incl. Red Mtn, Mercury Ridge), Ranger Mtns, Halfpint Range, hills N of Cane Spg, CP Hills, Yucca Mtn (nw. Jackass Flats); n. Shoshone Mtn, Eleana Range, s. Belted Range (n. Yucca Flat), uplands of Thirsty Cyn. NYE CO.: Foothills of W end of Spring Mtns (e. Amargosa Valley), below Kawich Range (nw. Reveille Valley), Reveille Range (Railroad Valley), W slope of cent. Belted Range (e. Kawich Valley). 3200—5000 (—6200) ft. Perennial. Apr—May, often again in Aug—Sept.

Festuca

F. arundinacea Schreb. Common escape from cultivation in irrigation areas, or in other moist disturbed soils. NTS: Seepage area at Whiterock Spg (nw. Yucca Flat). NYE CO.: Common, Ash Meadows, n. Pahrump Valley, Beatty townsite. 2200—5000 ft. Introduced perennial. May—June.

F. pratensis Huds. [*F. elatior* L.]. NYE CO.: Escape from cultivation, Ash Meadows; disturbed site at Willow Spg, e. Goldfield Hills. 2300–5800 ft. Introduced perennial. June—July.

Hilaria

H. jamesii (Torr.) Benth. Widely distributed, bajadas, hills, and mesas, in *Atriplex*, *A. canescens*, *Atriplex—Ceratoides*, *Coleogyne*, *Grayia—Lycium*, *Artemisia*, *Artemisia*—Pinyon—Juniper. NTS: Jackass Flats (s. Shoshone Mtn, summit of Skull Mtn); Frenchman Flat (hills N of Cane Spg, N slope of w. Spotted Range), n. Mid Valley, Yucca Flat (s. Belted Range, CP Hills), Forty-Mile Cyn (below Timber Mtn, Shoshone Mtn, Eleana Range), Thirsty Cyn, s. Groom Lake (Halfpint Range, nw. Papoose Range); Pahute Mesa. NYE CO.:

Common to abundant on basin floors of Gold Flat, Stonewall Flat (incl. Goldfield Hills), Kawich Valley, Groom Lake, s. Penoyer Valley, w. Railroad Valley (Reveille Range). 3600—6000 ft. Perennial. Apr—June, some yrs July—Sept.

H. rigida (Thurb.) Benth. ex Scribn. Restricted to areas of limestone mountain ranges, often along washes in *Atriplex*, *Larrea—Ambrosia*, *Larrea—Atriplex*, *Coleogyne*. NTS: Specter Range, w. Spotted Range (incl. Red Mtn), esp. below Ranger Mtns, Buried Hills, Halfpint Range (Scarp Cyn), CP Hills. NYE CO.: Foothills W end of Spring Mtns. 2600—4400 ft. Perennial. Apr—May, some yrs Aug—Oct.

Hordeum

H. brachyantherum Nevskii. NYE CO.: Local populations, wet soils of spg areas in *Artemisia*—Pinyon—Juniper, n. Kawich Range (Eden Creek and Longstreet cyns). 7200—8100 ft. Perennial. July—Sept.

H. glaucum Steud. Weed of disturbed moist or irrigated sites. NTS: Near Cane Spg pond (w. Frenchman Flat), Mercury townsite. CLARK CO.: Indian Springs townsite. NYE CO.: Pahrump Valley. 2600—3700 ft. Introduced annual. Apr—May.

H. jubatum L. Local populations, moist soils near springs or swamp areas, in *Atriplex*, *Coleogyne*, *Artemisia*—Pinyon—Juniper. NTS: Whiterock Spg (nw. Yucca Flat). CLARK CO.: Willow Creek (nw. Spring Mtns). LINCOLN CO.: Bald Mtn Spg (cent. Groom Range), Oasis Valley, Willow Spg (e. Goldfield Hills), Johnnies Water (cent. Belted Range). 2600—6500 ft. Perennial. May—Sept.

H. vulgare L. (BARLEY). NYE CO.: Common escape from cultivation, Pahrump Valley; occasional, Ash Meadows and n. Amargosa Valley. 2200—3000 ft. Introduced annual. June.

Koeleria

K. cristata Pers. *Artemisia*—Pinyon—Juniper, usually around boulder-cliffs or rock outcrops, probably more or less common throughout the volcanic mountain ranges of the region. NTS: S. Belted Range (Rainier Mesa), Pahute Mesa. NYE CO.: N. Kawich Range (Eden Creek cyn). 6500—7500 ft. Perennial. Late May—early July.

Leptochloa

L. fascicularis (Lam.) A. Gray. NYE CO.: Disturbed *Atriplex* at Warm Spgs (base of s. Hot Creek Range). 5400 ft. Annual. Sept—Oct.

L. uninervia (Presl) Hitchc. & Chase. Local, moist or irrigated soils, in disturbed *Atriplex* or *Larrea*. NTS: Near Mercury townsite. NYE CO.: Ash Meadows. 2300—3600 ft. Annual, probably introduced in this area. June—Oct.

Lolium

L. perenne L. ssp. **multiflorum** (Lam.) Husnot. (RYEGRASS). Locally common on disturbed irrigated soils, as an escape from cultivation. NTS: Cent. Jackass Flats. NYE CO.: Beatty townsite. 3300—3600 ft. Introduced annual. Aug—Nov.

ssp. **perenne** (PERENNIAL RYEGRASS). Common locally on irrigated sites. NTS: Mercury townsite, cent. Jackass Flats. CLARK CO.: Cold Creek Spg (nw. Spring Mtns). NYE CO.: Beatty townsite. 3300—6200 ft. Introduced perennial. May—Nov.

Melica

M. stricta Bol. NYE CO.: Local in mountain washes, *Artemisia*—Pinyon—Juniper; nw. Spring Mtns (wash below Gold Spg of N slope), Stonewall Mtn (common on bottom of cyn with "Ruins"), n. Kawich Range (upper Eden Creek cyn). 6400—7800 ft. Perennial. Late May—early July.

Muhlenbergia

M. asperifolia (Nees & Meyen) Parodi. NYE CO.: Locally the ground cover in *Atriplex—Haplopappus*, n. Ash Meadows. 2300 ft. Perennial. Sept—Oct.

M. porteri Scribn. Common, esp. locally, bajadas below or lower canyons of limestone mountain ranges, ledges or crevices along washes in *Atriplex*, *Larrea—Atriplex*, *Coleogyne—Atriplex*, *Grayia—Lycium*, *Grayia—Tetradymia*. NTS: W. Spotted Range (incl. Red Mtn), Ranger Mtns, Buried Hills, CP Hills, Eleana Range (nw. Yucca Flat), uplands of Thirsty Cyn. CLARK CO.: E-W axis of Spotted Range. NYE CO.: Cent. Reveille Range (locally common below Fang Ridge). 3500—5800 ft. Perennial. July—Oct.

M. richardsonis (Trin.) Rydb. Uncommon except locally, *Artemisia*—Pinyon—Juniper; moist, sandy soils. NTS: Sandy washes or base of talus slopes, n. Pahute Mesa. CLARK CO.: Nw. Spring Mtns (head of Clark Cyn). NYE CO.: N. and s. Kawich Range (Eden Creek cyn and Cedar Pass). 6400—9200 ft. Perennial. June—Sept.

M. utilis (Torr.) A. S. Hitchc. NYE CO.: Ash Meadows (common at edge of pool, Longstreet Spg). 2300 ft. Perennial. Sept—Oct.

Munroa

M. squarrosa (Nutt.) Torr. Occasional and local, disturbed sites in *Atriplex*, *Artemisia*, *Artemisia*—Pinyon—Juniper. NTS: S. Belted Range (below W face of Rainier Mesa), Pahute Mesa. NYE CO.: Ralston Valley (foothills of s. Monitor Range); common to abundant along roadsides of Stone Cabin Valley, s. Hot Creek Valley, and nw. Reveille Valley (below n. and cent. Kawich Range); White Blotch Spg W of n. Groom Range. 5400—6700 ft. Annual. June—Sept.

Oryzopsis

O. hymenoides (Roem. & Schult.) Ricker (INDIAN RICE-GRASS). Regionally the most widely distributed and common perennial grass, occurring in almost all vegetation types: *Larrea* (esp. *Larrea—Ambrosia), Atriplex*, *A. canescens*, *Atriplex—Ceratoides*, *Lycium pallidum—Grayia*, *Grayia—Lycium*, *Coleogyne*, *Artemisia*, *Artemisia*—Pinyon—Juniper. NTS: Mercury Valley, Rock Valley, Jackass Flats, Frenchman Flat, Mid Valley, Yucca Flat, Forty-Mile Cyn, s. Gold Flat, s. Groom Lake; Rainier Mesa (s. Belted Range), Pahute Mesa. CLARK CO.: N-S axis of Spotted Range. NYE CO.: N. Amargosa Valley [Big Dune, Bullfrog Hills (summit of Sawtooth Mtn)]. 2400—7500 ft. Perennial. Late Apr—June. Two colls., referred to var. *contracta* B. L. Johnson (S rim of Pahute Mesa, and White Blotch Spg W of n. Groom Range), may be hybrids. The following hybrids with *Stipa* species (collectively referred to the genus *Stiporyzopsis*) are known from NTS:

O. hymenoides x **S. pinetorum**. Occasional on Rainier Mesa (s. Belted Range) where *S. pinetorum* is common. 7500 ft. June—July.

O. hymenoides x **S. speciosa**. Plants frequent in most areas where *S. speciosa* is common; n. Mercury Valley, n. Jackass Flats, sw. and se. Frenchman Flat, esp. cent. and s. Mid Valley, n. and w. Yucca Flat, cent. Forty-Mile Cyn. 3600—5500 ft. Apr—June.

O. hymenoides x **S. thurberiana**. Below W face of Rainier Mesa (s. Belted Range), and e. Pahute Mesa. 6600—7000 ft. July.

O. micrantha (Trin. & Rupr.) Thurb. At bases of cliffs or around rock outcrops in *Artemisia*—Pinyon—Juniper. NTS: Frequent in cyns of esp. n. Pahute Mesa. CLARK CO.: Nw. Spring Mtns (lower Clark Cyn). NYE CO.: W slope of cent. Belted Range. 6000—7000 ft. Perennial. Late May—early July.

Panicum

P. virgatum L. NYE CO.: With *Juncus* near pool at spg (destroyed since 1968 coll.), n. Ash Meadows. 2200 ft. Perennial. June—July.

Paspalum

P. distichum L. CLARK CO.: Dense stand along irrigation channel, Indian Springs townsite. 3100 ft. Introduced perennial. June—July.

Phragmites

P. australis (Cav.) Trin. ex Steud. [*P. communis* Trin.] NYE CO.: Scattered colonies in spg areas or other wet sites in *Atriplex*; Ash Meadows, Oasis Valley (in *Atriplex torreyi*), and Bullfrog Hills. 2200—4000 ft. Perennial. Sept—Oct.

Poa

P. annua L. Disturbed wet soils. NTS: Mercury townsite. NYE CO.: Spg in foothills of Grapevine Mtns. 3700—4800 ft. Introduced annual. Apr—June.

P. fendleriana (Steud.) Vasey. The widely distributed and common Bluegrass of *Artemisia*—Pinyon—Juniper and Fir—Pinyon of volcanic and limestone mountain ranges. NTS: Shoshone Mtn, s. Belted Range (Rainier Mesa, Gold Meadows), Pahute Mesa. CLARK CO.: Nw. Spring Mtns (Wheeler Well—Wheeler Pass area, the abundant grass of S slope). LINCOLN CO.: Cent. Groom Range (Bald Mtn). NYE CO.: Nw. Spring Mtns (cyns of both slopes below Mt. Stirling), Bullfrog Hills (summit of Sawtooth Mtn), cent. Grapevine Mtns, Stonewall Mtn (grotto at Spg), n. Kawich Range (cyns of both slopes), n. and cent. Reveille Range, cent. Belted Range (cyns of both slopes and Wheelbarrow Peak). 5000—8500 ft. Perennial. Late Apr—early June.

P. nevadensis Vasey ex Scribn. *Artemisia*—Pinyon—Juniper, occasional and local at bases of cliffs or near springs in volcanic mountain ranges. NTS: Shoshone Mtn (Topopah Spg, Tippipah Reservoir), Eleana Range (Capt. Jack Spg), s. Belted Range (Oak Spg, Whiterock Spg); cyns of n. Pahute Mesa and S rim. NYE CO.: White Blotch Spg W of n. Groom Range. 5000—7100 ft. Perennial. May—July.

P. pratensis L. (KENTUCKY BLUEGRASS). Occurring as bunch- or sodgrass in wet disturbed soils near streams or springs, in *Artemisia* or *Artemisia*—Pinyon—Juniper. NTS: Irrigated sites, Mercury townsite. CLARK CO.: Nw. Spring Mtns (abundant along Willow Creek). LINCOLN CO.: Cent. Groom Range (*Juncus* swamp at Bald Mtn Spg). NYE CO.: Stonewall Mtn Spg; n. and cent. Kawich Range (Eden Creek cyn and Stinking Spg). (3700—) 5600—7100 ft. Perennial, introduced at least at lower elevs. June.

P. sandbergii Vasey. Common, *Artemisia*, *Artemisia*—Pinyon—Juniper, less frequent in *Coleogyne* and *Larrea—Ambrosia*; cliffs and ledges of canyons, and esp. flatrock areas, most frequent in volcanic mountains and mesas. NTS: W. Spotted Range (Mercury Ridge), Shoshone Mtn, Eleana Range, s. Belted Range (Rainier Mesa, Gold Meadows), Pahute Mesa. CLARK CO.: N-S axis of Spotted Range. NYE CO.: Bare Mtn, below nw. Yucca Mtn, Grapevine Mtns. 3800—7400 ft. Perennial. Late Apr—June.

P. scabrella (Thurb.) Benth. ex Vasey. The common and widely distributed Bluegrass of esp. limestone mountains, foothills and upper bajadas at lower elevations; *Larrea—Ambrosia*, *Atriplex*, *Coleogyne*, *Artemisia nova*, and rarely *Artemisia*—Pinyon—Juniper. Nearly all populations characterized by lemmas minutely or sparingly puberulent near base, from the lateral veins to margin, otherwise glabrous or essentially so. NTS: Striped Hills, Little Skull Mtn, rim of Skull Mtn, w. Spotted Range (incl. Red Mtn, Mercury Ridge), Ranger Mtns, Buried Hills, Halfpint Range (incl. French Peak mtn, Banded Mtn, and Cockeyed Ridge), CP Hills, Mine Mtn, Eleana Range (dolomite hills), s. Belted Range (below W face of Rainier Mesa), Shoshone Mtn. NYE CO.: N. and cent. Ash Meadows, nw. Spring Mtns (N slope below Mt. Stirling), Bullfrog Hills, nw. Yucca Mtn (Tates Wash), Stonewall Mtn. 2200—5000 (—6300) ft. Perennial. Apr—early May (—early June).

Polypogon

P. interruptus HBK. NTS: With *P. monspeliensis* on disturbed wet soils, edge of Cane Spg pond (w. Frenchman Flat). 4000 ft. Introduced perennial. May—July.

P. monspeliensis (L.) Desf. Widely distributed and locally common, wet soils near springs or water impoundments in *Atriplex*, *A. canescens*, *Larrea*, *Coleogyne*, *Artemisia*, *Artemisia*—Pinyon—Juniper. NTS: Ponds of Jackass Flats, Cane Spg, Tippipah Reservoir, Whiterock Spg. LINCOLN CO.: Groom Range (Bald Mtn Spg, Cattle Spg). NYE CO.: Ash Meadows, Oasis Valley; nw. Spring Mtns (Crystal Spg), Bullfrog Hills, Cactus Spg (Cactus Range), s. Kawich Range (Cedar Pass), s. Hot Creek Range (Warm Spgs). 2200—7000 ft. Introduced annual. Apr—Nov.

Puccinellia

P. nuttalliana (Schult.) A. S. Hitchc. NYE CO.: Common locally in disturbed *Atriplex* at Antelope Spg, s. Cactus Range. 6200 ft. Perennial. Aug—Sept.

Schismus

S. arabicus Nees. Usually on disturbed sites in *Larrea—Ambrosia*, but some years in mixed populations with *Vulpia octoflora* in undisturbed communities. NTS: N. Amargosa Valley (below W end of Spring Mtns, Specter Range, and Striped Hills); Mercury townsite, and elsewhere in Mercury Valley; w. Jackass Flats. NYE CO.: Ash Meadows. 2300—3800 ft. Introduced winter annual. Mar—June.

Scleropogon

S. brevifolius Phil. NYE CO.: Common to abundant in many large patches in *Atriplex* of flatrock area, upper bajada below N-cent. Belted Range (e. Kawich Valley). 6000—6100 ft. Perennial. July—Aug.

Setaria

S. glauca (L.) Beauv. Uncommon weed of disturbed irrigated sites. NTS: Near ponds, cent. Jackass Flats. CLARK CO.: Indian Springs townsite. 3100—3600 ft. Introduced annual. Sept—Nov.

Sitanion

S. hystrix (Nutt.) J. G. Sm. Widely distributed and common in *Coleogyne*, *Grayia—Lycium*, *Artemisia* (esp. *A. nova*), and *Artemisia*—Pinyon—Juniper, uncommon in *Larrea*, *Atriplex*, and *Atriplex—Ceratoides*; volcanic and limestone mountain ranges, and esp. washes on bajadas below. NTS: Occasional in or below e. Specter Range, w. Spotted Range (Red Mtn), Skull Mtn, Buried Hills, Halfpint Range; common in or below Shoshone Mtn, Timber Mtn, s. Belted Range (Rainier Mesa, Gold Meadows, Oak Spring Butte; s. Groom Lake), nw. Papoose Range (se. Groom Lake); Pahute Mesa. CLARK CO.: Nw. Spring Mtns (upper Clark Cyn). NYE CO.: Nw. Spring Mtns (N slope below Mt. Stirling), n. Pahrump Valley (in Mesquite), Bullfrog Hills, n. Kawich Range, n. and cent. Belted Range, White Blotch Spg W of n. Groom Range. (2600—) 4500—9200 ft. Perennial. Apr—June, some yrs Sept—Oct. One coll., from cent. Mid Valley, is referred to the hybrid *S. hansenii* (Scribn.) J. G. Sm., where parental species would be *Elymus cinereus* and either *S. hystrix* or *S. jubatum*, all of which occur in the area.

S. jubatum J. G. Sm. In *S. hystrix* populations, as scattered plants, or sometimes frequent locally. NTS: Sw. Ranger Mtns, in or below all slopes of Shoshone Mtn, Eleana Range (e. Forty-Mile Cyn). NYE CO.: N. Bare Mtn (Fluorspar Cyn). 3900—6200 ft. Perennial. May—June.

Sorghum

S. bicolor Pers. (L.) Moench. [*S. vulgare* Pers.] (SORGHUM). Common escape from cultivation. CLARK CO.: Indian Springs townsite. NYE CO.: Ash Meadows and Pahrump Valley. 2200—3100 ft. Introduced annual. July—Oct.

S. halepense (L.) Pers. (JOHNSON GRASS). Local on disturbed sites. NTS: Nw. Yucca Flat in disturbed *Coleogyne*. NYE CO.: Common along irrigation channels, some areas of Pahrump Valley. 2800—4800 ft. Introduced perennial. June—Sept.

Spartina

S. gracilis Trin. NYE CO.: Collected by Coville and Funston on the Death Valley Expedition (Coville, F. V., *Contr. U. S. Nat. Herb.*, 4. 1893), in Ash—Screwbean—*Baccharis*, 0.6 mi. S of Devils Hole, n. Ash Meadows. 2300 ft. Perennial. Summer.

Sporobolus

S. airoides (Torr.) Torr. CLARK CO.: Low dunes in Mesquite, Cactus Spgs. NYE CO.: The common perennial grass in *Atriplex* communities of Ash Meadows, Pahrump Valley, Oasis Valley; s. Hot Creek Valley. 2200—5000 ft. Perennial. May—Oct.

S. cryptandrus (Torr.) A. Gray. Widely distributed as small local populations, in or below either limestone or more frequently volcanic mountain ranges, esp. at bases of cliffs, around rock outcrops on level terrain, washes, or disturbed moist sites; *Larrea—Atriplex*, *Atriplex*, *A. canescens*, *Atriplex—Ceratoides*, *Coleogyne*, *Artemisia*, *Artemisia*—Pinyon—Juniper. NTS: In or below w. Spotted Range (Mercury Ridge), Ranger Mtns, Shoshone Mtn, Timber Mtn, Eleana Range, Thirsty Cyn, Pahute Mesa. NYE CO.: Below s. Monitor Range (Ralston Valley), Kawich Range (Stone Cabin Valley, ne. Kawich Valley, s. Penoyer Valley, nw. Reveille Valley), n. Reveille Range (Twin Spgs area, in *Atriplex parryi—Sarcobatus*), W slope of cent. Belted Range (e. Kawich Valley). (3100—) 5000—6500 (—7300) ft. Perennial. June—Oct.

S. flexuosus (Thurb.) Rydb. NTS: Restricted to certain areas of sandy soils; common in *Larrea—Ambrosia* of deep sands W of Buried Hills (ne. Frenchman Flat), and local in sands of Sugar Loaves area (e. Forty-Mile Cyn). 3300—5500 ft. Perennial. June—Oct.

Stipa

S. arida M. E. Jones. Locally common, steep talus slopes and ledges of limestone mountain ranges and calcareous outcrops elsewhere; *Atriplex*, *Coleogyne*, *Artemisia*, *Artemisia*—Pinyon—Juniper. NTS: Specter Range, w. Spotted Range (incl. Red Mtn), Ranger Mtns, Buried Hills and limestone butte to W, CP Hills, Mine Mtn, Eleana Range (nw. Yucca Flat), sw. Shoshone Mtn, s. Belted Range (Rainier Mesa, Oak Spring Butte), Pahute Mesa. CLARK CO.: N-S axis of Spotted Range. NYE CO.: N. Bare Mtn (cliffs near head of Fluorspar Cyn). 4400—6500 ft. Perennial. May—June.

S. **comata** Trin. & Rupr. The common to locally abundant *Stipa* of sandy soils and *Artemisia tridentata*, infrequent in *Artemisia*—Pinyon—Juniper and Yellow Pine—Fir. NTS: In or below Shoshone Mtn, Timber Mtn, Eleana Range, s. Belted Range (Rainier Mesa, Gold Meadows), Pahute Mesa. CLARK CO.: Nw. Spring Mtns (upper Clark Cyn). 5000—9200 ft. Perennial. May—June.

S. **coronata** Thurb. in S. Wats. var. **depauperata** (Jones) A. S. Hitchc. NTS: Local in *Atriplex*, w. Spotted Range. NYE CO.: Common around volcanic outcrops near Gold Spg (N slope of nw. Spring Mtns) in *Artemisia*—Pinyon—Juniper. 4800—5900 ft. Perennial. May—June.

S. **lettermannii** Vasey. CLARK CO.: Near spg in upper Clark Cyn, nw. Spring Mtns, in Yellow Pine—Fir. NYE CO.: Steep cyn wall, cent. Hot Creek Range; to be expected in n. Kawich Range. 8700—9000 ft. Perennial. June—July.

S. **pinetorum** M. E. Jones. Locally common in shallow soils or along washes in *Artemisia*—Pinyon—Juniper of certain areas of volcanic mountain ranges and mesas. NTS: S. Belted Range (Rainier Mesa), s. and e. Pahute Mesa. NYE CO.: N. Kawich Range (Eden Creek cyn). 7200—7800 ft. Perennial. June—July.

S. **speciosa** Trin. & Rupr. The most widely distributed and common *Stipa* of middle elevations of the region, characteristic of *Coleogyne*, less common in *Grayia*—*Lycium* and *Artemisia*; occasional in washes in *Larrea*. NTS: W. Mercury Valley, e. Rock Valley, Jackass Flats, Topopah Valley, w. Frenchman Flat, Mid Valley, Yucca Flat, Forty-Mile Cyn; Pahute Mesa. CLARK CO.: Nw. Spring Mtns (Wheeler Wash, and limestone hill below N slope); N-S axis of Spotted Range. NYE CO.: Bullfrog Hills (summit of Sawtooth Mtn). (3500—) 4000—6000 ft. Perennial. Late Apr—May.

S. **thurberiana** Piper. NTS: The most common and widely distributed *Stipa* of *Artemisia*—Pinyon—Juniper, rarely in washes in *Coleogyne*. NTS: S. Shoshone Mtn, Timber Mtn, s. Belted Range (Rainier Mesa), s. and e. Pahute Mesa and below S rim. (4500—) 6000—7500 ft. Perennial. May—June.

Tridens

T. muticus (Torr.) Nash. Common, crevices, ledges, and esp. along canyon washes, in *Atriplex*, *Larrea—Atriplex*, *Larrea—Ambrosia*, *Coleogyne*, of limestone mountain ranges. NTS: Specter Range, w. Spotted Range (incl. Red Mtn), Ranger Mtns, Buried Hills and limestone butte to W. CLARK CO.: Foothills, W end of Spring Mtns (Johnnie Mine area); N-S axis of Spotted Range. NYE CO.: Bare Mtn (Tungsten Cyn wash). 3300—5000 ft. Perennial. Apr—May; Aug—Sept.

Vulpia

V. microstachys (Nutt.) Benth. var. **pauciflora** (Scribn. ex Beal) Lonard & Gould [*Festuca reflexa* Buckl.]. NTS: Uncommon and local, cliffs or around rock outcrops, in *Coleogyne* and *Artemisia*—Pinyon—Juniper; n. Topopah Valley, nw. Jackass Flats (Yucca Mtn). 4500—5900 ft. Probably winter annual. Apr—May.

V. myuros (L.) K. C. Gmelin var. **hirsuta** Hack. [*Festuca megalura* Nutt.]. NTS: Abundant some yrs on seepage site in *Coleogyne*, Whiterock Spg (nw. Yucca Flat). 5000 ft. Annual. May—June.

V. octoflora (Walt.) Rydb. [incl. var. *octoflora*, the more common var. *hirtella* (Piper) Henr., and intermediates, which occur without geographic definition] [*Festuca o.* Walt.]. Widely distributed, in dense stands along washes or in depressions on bajadas below limestone mountain ranges, local populations throughout much of region; *Larrea*, *Coleogyne*, *Grayia—Lycium*, *Atriplex*, *Artemisia*, uncommonly *Artemisia*—Pinyon—Juniper. NTS: N. Amargosa Valley, Mercury Valley, Rock Valley, Jackass Flats, Topopah Valley, Frenchman Flat, Mid Valley, Yucca Flat, e. Forty-Mile Cyn. NYE CO.: N. Crater Flat, w. Groom Lake. 3100—6600 ft. Winter annual. Apr—June.

POTAMOGETONACEAE. Pondweed Family

Potamogeton

P. pectinatus L. Ponds and reservoirs, probably widely distributed in the region. NTS: Under cat-tails, Cane Spg pond (w. Frenchman Flat); Well C reservoir (s. Yucca Flat); pond near Ammonia Tanks and Well 8 (e. Forty-Mile Cyn). 4000—5800 ft. Perennial. June—July.

Ruppia

R. maritima L. NYE CO.: Common in lake, w. Ash Meadows. 2200 ft. Perennial. July.

TYPHACEAE. Cat-Tail Family

Typha

T. domingensis Pers. Local populations in wet or saturated soils at lower elevations. NTS: Ponds in w. Frenchman Flat (Cane Spg) and e. Forty-Mile Cyn. NYE CO.: Ash Meadows and Oasis Valley. 2200—3900 ft. Perennial. June—July.

T. latifolia L. NTS: Seepage site at Whiterock Spg and margin of Well C reservoir (n. and s. Yucca Flat). 3900—5000 ft. Perennial. June—July.

ZANNICHELLIACEAE. Horned Pondweed Family

Zannichellia

Z. palustris L. NTS: Well 5-B (s. Frenchman Flat). CLARK CO.: Water impoundment on Willow Creek (nw. Spring Mtns). NYE CO.: Pond and water trough, Stinking Spg (cent. Kawich Range). 3100—6600 ft. Perennial. June.

Index to Families, Genera, and Species